Advances in Meat Research

Volume 1
Electrical Stimulation

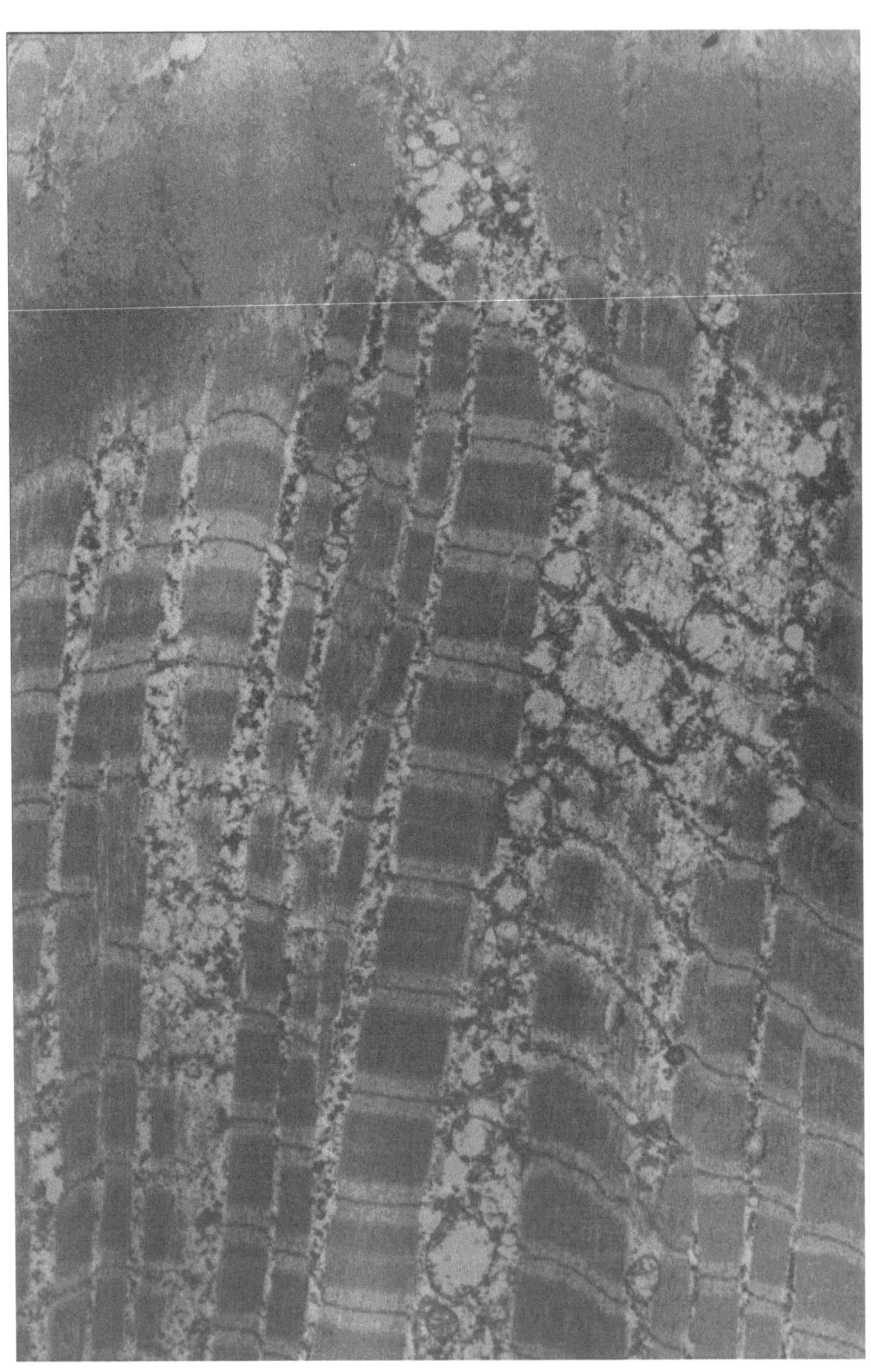

Advances in Meat Research

Volume 1
Electrical Stimulation

A. M. Pearson
T. R. Dutson

Department of Food Science and
Human Nutrition
Michigan State University
East Lansing, Michigan

AVI PUBLISHING COMPANY, INC.
Westport, Connecticut

© Copyright 1985 by
THE AVI PUBLISHING COMPANY, INC.
Softcover reprint of the hardcover 1st edition 1985
250 Post Road East
P.O. Box 831
Westport, Connecticut   06881

All rights reserved. No part of this work
covered by the copyright hereon may be
reproduced or used in any form or by any
means—graphic, electronic, or mechanical,
including photocopying, recording, taping,
or information storage and retrieval systems—
without written permission of the publisher.

ISBN-13: 978-94-011-5941-8   e-ISBN-13: 978-94-011-5939-5
DOI: 10.1007/978-94-011-5939-5

A B C D E   3 2 1 0 9 8 7 6 5 4

Contents

Contributors	ix
Preface	xi

1 Cold-induced Toughness of Meat — 1
R. H. Locker

A Sad Story: New Technology Meant Poorer Quality	2
The Relationship Between Contraction and Toughness	3
Toughening Lamb Carcasses in Blast Freezers	6
Adding Insult to Injury: Thaw Shortening	8
The Cold-Shortening Response	10
The Thaw-Shortening Response	13
The Biochemical Basis for Cold and Thaw Shortening	14
Prevention of Cold or Thaw Shortening	19
Some Doubts on the Relationship Between Shortening and Toughness	26
A Structural Basis for Cold Toughening and a New Theory of Meat Tenderness	29
Conclusion	38
References	40

2 Postmortem Conditioning of Meat — 45
T. R. Dutson and A. M. Pearson

Cold Toughening	47
High Temperature Conditioning	59
Cooler Conditioning	63
Electrical Stimulation	64
Summary	65
References	66

3 Electrical Stimulation: Its Early Development in New Zealand 73
B. B. Chrystall and C. E. Devine

Rigor Mortis	76
Toughness	77
Electrical Stimulation	82
Summary	114
Acknowledgments	115
References	115

4 Effects of Electrical Stimulation on Meat Quality, Color, Grade, Heat Ring, and Palatability 121
G. C. Smith

Tenderness	128
Flavor	140
Lean Color	141
Heat Ring Prevention	142
Marbling and Quality Grade	143
Retail Caselife	146
Processing Properties	147
Low- vs High-voltage Electrical Stimulation	148
Implementation by the Packing Industry	149
References	153

5 Use of Electrical Stimulation for Hot Boning of Meat 159
H. R. Cross and S. C. Seideman

History of Hot Boning Research	161
Electrical Stimulation and Hot Boning	164
Water Holding Capacity	168
Appearance Properties	172
Microbiology	173
Cooking Prerigor Muscle	175
Utilization of Hot Boning Meat for the Production of Ground Beef	176
Why Hasn't Industry Accepted Hot Boning?	177
Economic Implications of Hot Boning	178
Conclusions	180
References	180

6 Scientific Basis for Electrical Stimulation 185
A. M. Pearson and T. R. Dutson

Structure of Muscle	186
Energy Changes in Prerigor Muscle	196
Basic Causes of Cold Shortening	198
Other Gains from Electrical Stimulation	206
Summary	215
References	215

7 Industrial Applications of Electrical Stimulation 219
J. W. Savell

Industry Interest in Electrical Stimulation	220
Electrical Stimulation Equipment	222
Installations of Electrical Stimulators Within the Slaughter–Dressing Sequence	231
Electrical Parameters	232
Safety, Installation, and Sanitation	233
Trade Names and Promotion of Electrically Stimulated Beef	234
Industry Adoption of Electrical Stimulation	234
Summary	235
Acknowledgments	235
References	236

8 Cold Storage Energy Aspects of Electrically Stimulated Hot-boned Meat 237
R. L. Henrickson and A. Asghar

Technologies of Hot Carcass Processing	238
Meat Yield and Losses	239
Quality Characteristics of Meat	240
Tenderness of Hot- and Cold-boned Meat	241
High Temperature Conditioning and Delay in Chilling	242
Electrical Stimulation	243
Innovation in Carcass Cutting	244
Functional Properties of Hot-boned Meat	245
Meat Curing	248
Microbiology of Hot- and Cold-boned Meat	248
Energy Conservation	250
Comparison of Energy Consumption During Chilling of Beef by Conventional and Hot Boning Methods	254
Effect of Plant Location on Transportation Energy	260
Industry Impediments	266
Summary	267
References	268

9 Electrical-Stimulation Research: Present Concepts and Future Directions 277
B. Bruce Marsh

General Considerations	279
Specific Aspects	285
Conclusions	303
References	303

Index 307

Contributors

A. ASGHAR. Oklahoma Agricultural Experiment Station, Oklahoma State University, Stillwater, OK 74078

B. B. CHRYSTALL. Meat Industry Research Institute of New Zealand, Inc., P.O. Box 617, Hamilton, New Zealand

H. R. CROSS. Professor, Meats and Muscle Biology Section, Department of Animal Science, Texas A&M University, College Station, TX 77843-2471

C. E. DEVINE. Meat Industry Research Institute of New Zealand, Inc., P.O. Box 617, Hamilton, New Zealand

T. R. DUTSON. Chairperson, Department of Food Science and Human Nutrition, Michigan State University, East Lansing, MI 48824

R. L. HENRICKSON. Professor, Oklahoma Agricultural Experiment Station, 104 Animal Science Building, Oklahoma State University, Stillwater, OK 74078

R. H. LOCKER. Meat Industry Research Institute of New Zealand, Inc., P.O. Box 617, Hamilton, New Zealand

B. BRUCE MARSH. Professor, Muscle Biology Laboratory, Department of Meat and Animal Science, 1805 Linden Drive, University of Wisconsin, Madison, WI 53706

A. M. PEARSON. Professor, Department of Food Science and Human Nutrition, Michigan State University, East Lansing, MI 48824

J. W. SAVELL. Associate Professor, Meats and Muscle Biology Section, Department of Animal Science, Texas Agricultural

Experiment Station, Texas A&M University, College Station, TX 77843

S. C. SEIDEMAN. Meats Research Unit, Agricultural Research Service, United States Department of Agriculture, Roman L. Hruska U.S. Meat Animal Research Center, P.O. Box 166, Clay Center, NE 68933

G. C. SMITH. Professor and Head, Department of Animal Science, Texas A&M University, College Station, TX 77843

Preface

The *Advances in Meat Research* series has arisen from a perceived need for a comprehensive coverage of certain topics that are pertinent to meat and meat products. We, the editors, have made the decision to concentrate on a series of related topics that are deemed to be important to an understanding of meat, both fresh and processed. It is our sincere hope that by focusing upon areas related to meat science that researchers who contribute to this volume can not only update those involved in academia and industry but also promulgate facts that may lead to solutions of meat industry problems and aid in improving the efficiency of various associated industrial processes.

We have chosen to devote *Volume 1* to electrical stimulation in view of the widespread interest in its meat industry applications. Although the classical study by A. Harsham and Fred Deatherage was published in 1951, it was not accepted by the meat industry owing to a number of factors that are discussed in the text. These investigators did, however, lay the groundwork for modern electrical stimulation of carcasses by their detailed studies on the effects of varying current, voltage, frequency, wave forms, and time. The basic information provided by these workers saved a great amount of experimentation by those who subsequently "rediscovered" electrical stimulation.

Another important study that played a key role in the development of electrical stimulation was the simple observation of R. H. Locker, published in 1960, showing that muscle shortening was related to meat tenderness. This study proved to be the basis for the discovery of cold shortening by Locker and Hagyard in 1963, which led to recognition of the need for speeding up glycolysis to prevent cold shortening in lamb carcasses that were frozen in the prerigor state for export to North American markets from New Zealand.

W. A. Carse, also, at the Meat Industry Research Institute of New Zealand (MIRINZ) published the third landmark paper in 1973 that led others to study electrical stimulation. This was the first study to demonstrate that electrical stimulation accelerated glycolysis and could be used to prevent cold shortening in prerigor meat. This report proved to be the stimulus for all subsequent investigations on electrical stimulation as a means of improving meat quality, and its importance is attested to by the frequency with which it is cited in the voluminous literature on this important subject.

Finally, to put the historical aspects of this subject into perspective, we wish to acknowledge the contribution of B. B. Marsh, formerly of the Meat Industry Research Institute of New Zealand and now of the University of Wisconsin, who, without any prior knowledge of the early work of Harsham and Deatherage, theorized that electrical stimulation might be useful in speeding up glycolysis and thus in preventing cold shortening. He was responsible for stimulating W. A. Carse to initiate his most important investigation. As a final note, one of the editors (A. M. Pearson) publicly acknowledges having delayed Mr. Carse in this most important study while he worked on less promising methods for speeding up glycolysis during the co-editors' stay at MIRINZ in 1971–1972.

A. M. Pearson
T. R. Dutson

REFERENCES

CARSE, W. A. 1973. Meat quality and the acceleration of postmortem glycolysis by electrical stimulation. J. Food Technol. *8*, 163.

HARSHAM, A. and DEATHERAGE, F. 1951. Tenderization of meat. U.S. Pat. 2,544,681. March 13.

LOCKER, R. H. 1960. Degree of muscular contraction as a factor in the tenderness of beef. Food Res. *25*, 304.

LOCKER, R. H. and HAGYARD, C. J. 1963. A cold shortening effect in beef muscles. J. Sci. Food Agric. *14*, 787.

Related AVI Books

BASIC FOOD CHEMISTRY, 2nd Edition
  Lee
COLD AND FREEZER STORAGE MANUAL, 2nd Edition
  Hallowell
COMMERCIAL CHICKEN PRODUCTION MANUAL, 3rd Edition
  North
DIGESTIVE PHYSIOLOGY & METABOLISM IN RUMINANTS
  Ruckebusch and Thivend
ENCYCLOPEDIA OF FOOD SCIENCE
  Peterson and Johnson
ENCYCLOPEDIA OF FOOD TECHNOLOGY
  Johnson and Peterson
FOOD ANALYSIS THEORY & PRACTICE, Revised Edition
  Pomeranz and Meloan
FOOD PRODUCTS FORMULARY SERIES, Vol. 1–4
FUNDAMENTALS OF FOOD FREEZING
  Desrosier and Tressler
LIVESTOCK AND MEAT MARKETING, 2nd Edition
  McCoy
MEAT HANDBOOK, 4th Edition
  Levie
PHYSICAL PROPERTIES OF FOODS
  Peleg and Bagley
PROCESSED MEATS, 2nd Edition
  Pearson and Tauber
PROTEIN QUALITY IN HUMANS: Assessment and In Vitro Estimation
  Bodwell, Adkins, Hopkins
SOURCE BOOK OF FOOD ENZYMOLOGY
  Schwimmer
STATISTICAL METHODS FOR FOOD AND AGRICULTURE
  Bender, Douglass, Kramer
SWINE PRODUCTION AND NUTRITION
  Pond and Maner
THE PSYCHOBIOLOGY OF HUMAN FOOD SELECTION
  Barker

Cold-induced Toughness of Meat

R. H. Locker[1]

A Sad Story: New Technology Meant Poorer Quality
The Relationship Between Contraction and Toughness
Toughening Lamb Carcasses in Blast Freezers
Adding Insult to Injury: Thaw Shortening
The Cold-Shortening Response
The Thaw-Shortening Response
The Biochemical Basis for Cold and Thaw Shortening
Prevention of Cold or Thaw Shortening
Some Doubts on the Relationship Between Shortening and Toughness
A Structural Basis for Cold Toughening and a New Theory of Meat Tenderness
Conclusion
References

Before 1960, ideas about meat tenderness tended to be dominated by the role of connective tissue, of which the quantity, quality, and arrangement were held to be important (Ramsbottom *et al.* 1945). There is still no reason to doubt this judgment. The improvement in tenderness on aging, without much apparent change in the connective tissue, might perhaps have been expected to direct more attention to the myofibrillar component. However, in the late 1950s, the time was only just ripe for serious study of the myofibril. Acceptance of the sliding filament theory of muscular contraction was a recent event in muscle biology. The power of electron microscopy was making an impact in that field but had barely touched meat science. However, it was not so much these exciting developments that led to a new interest in the myofibrillar contribution to toughness, but a much more mundane observation.

In 1958, I was trying to demonstrate proteolysis during the aging process and having little success. In the course of the work, I examined, by phase contrast microscopy, homogenates from a wide range of beef muscles, which had gone into rigor on the carcass under normal abattoir condi-

---

[1]Meat Industry Research Institute of New Zealand, P.O. Box 617, Hamilton, New Zealand.

tions. I was struck by the wide range of sarcomere lengths within and between muscles (Locker 1959). Could this be a factor in meat tenderness, and if so, why had not someone considered it before? Some simple experiments on psoas muscle (in which the contribution of connective tissue is minimal) showed that shortening induced by excision or partial excision after dressing led to pronounced toughening (Locker 1960). Not long afterward, again as a chance observation in work directed to another purpose, I found that excised beef muscles shortened markedly when allowed to go into rigor at 2°C (Locker and Hagyard 1963). These latter two papers, based on quite primitive experiments, were published with hesitation and with the nagging doubt that surely someone must have published such easily observable facts before. Curiously, it seems that it was not so. The two papers have in the long run far exceeded their writer's original estimate of their worth. Taken together, they have opened up a new field in meat science: the relationship of tenderness, contraction, and cold, and have led to a new emphasis on the role of the myofibril.

For this laboratory, then in its formative years, these findings could not have come at a more opportune time. They proved immediately relevant to the first technical challenge offered by industry: the solution of a toughness problem of national significance to our frozen lamb trade. Solving this problem set the course of research here for the next decade and even to the present. Exploring the effects of rigor temperature on contraction state and tenderness was undertaken with vigor by other workers at this institute simultaneously with the search for a practical solution. This has only now reached a satisfactory conclusion in the technique of accelerated conditioning by electrical stimulation. In that period, cold shortening has become a matter of interest to research centers and meat industries in many other places.

A SAD STORY: NEW TECHNOLOGY MEANT POORER QUALITY

Mechanical refrigeration made its first impact on international meat trading in the transatlantic run and soon extended to Australia. New Zealand was not far behind; in fact, 1982 marks the centenary of the first shipment of frozen mutton to London in the sailing ship *Dunedin*. For a young colony with limited resources, it was a godsend. Where previously only the durable products of animal husbandry such as wool, hides, and tallow had been exportable, the carcass had suddenly acquired a value in the receptive markets of industrial Britain. Over a century, the frozen meat trade increased in volume until it became, as it remains, our largest earner of overseas funds. In this trade, lamb still dominates as the meat animal ideally suited to a grassland economy. Conveniently, it harvests the annual flush of pasture without the need for wintering over. Although not the largest producer, New Zealand is the largest exporter of lamb, and by reason of its distance from the market, the carcasses travel almost exclusively in frozen form.

The end of the second world war saw a rapid increase in production, due largely to the development of aerial topdressing of hill country with superphosphate. In the freezing works (packinghouses), the old freezers soon became inadequate for the swollen throughput (usually more than 10,000 head per day and in some works more than 20,000). The old freezers were steadily replaced by new blast freezers. The old technology involved hanging lambs overnight on a ventilated cooling floor to reduce the heat and moisture load on the freezers, to which they were consigned next morning. Carcasses were then frozen slowly beneath grids of brine pipes. The new technology meant that lamb carcasses, sometimes straight from the scales, and never more than a few hours postmortem, were subjected to rapid freezing. The new situation was therefore an extreme one, where small carcasses were subjected to powerful blasts of very cold air, resulting in abnormally high rates of chilling and freezing. Since it is generally true that quick freezing is best for foods, it might have been expected that quality would have improved. Quite the reverse occurred.

About 1960, complaints from Britain of toughness in New Zealand lamb became sufficiently persistent to cause the New Zealand Meat Producers' Board to call for a scientific assessment. It was expected that this would rebut such claims and reestablish our good reputation. However, the very first experiment showed the complaints to be justified and implicated the new freezing technology. Lambs processed in the pre-war manner were found to be invariably tender, while blast-frozen lambs could be tender or very tough or anywhere in between. It became obvious at this point that the observations described in the introduction to this chapter were likely to provide an explanation, and intense study began along these lines. Not long afterward, our small but growing exports to the United States met with resistance for the same reasons, and a solution became urgent. At the time, the only solution seemed to be to hold the carcasses before entry to the freezer. Here freezing works found themselves in a dilemma because most of the cooling floors had vanished. Deemed no longer necessary, they had provided the most obvious space to build the new blast freezers. So the search for alternative solutions began, and that story will be part of the substance of this chapter.

THE RELATIONSHIP BETWEEN CONTRACTION AND TOUGHNESS

The original paper on this relationship (Locker 1960) was based on rather primitive experiments and failed to show its true subtlety. This was revealed by the work of Marsh and Leet (1966A,B) using the cold effect to set beef neck muscles at different lengths, followed by shearing of the cooked material. The curve relating shear force to shortening (Fig. 1.1) showed distinct phases. Toughness scarcely increased in the contraction range of 0–20%, but rose steeply from 20 to 40% and then declined as sharply from 40 to 60%. This relationship has been confirmed on a number of occasions since (Davey *et al.* 1967, 1976).

The dramatic decrease in toughness with extreme shortening is at first sight surprising, but appears to have a perfectly adequate explanation. Micrographs show that such severe contractions are not uniform. Intense nodes of contraction alternate with zones which are stretched and torn (Marsh *et al.* 1974). The muscle has in fact tenderized itself by tearing up its own structure. The tissue may be regarded as a mixture of toughened and self-tenderized fibers, the latter becoming progressively more important as shortening increases. This region of the curve is not normally

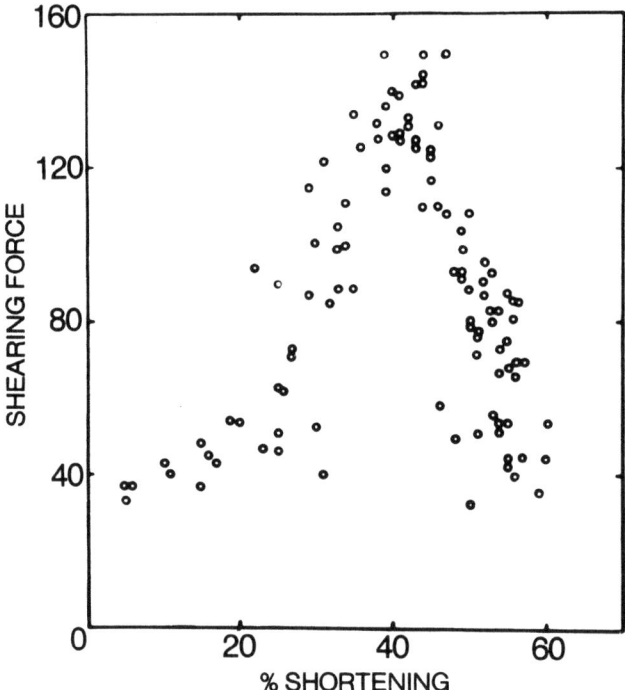

FIG. 1.1. Relative tenderness in relation to the shortening induced by transfer of samples from room temperature to 2°C at intervals during rigor onset. Cold shortening as percentage initial excised length.
Reprinted from *J. Food Sci.*, Marsh and Leet (1966B) 31:455. Copyright © by Institute of Food Technologists.

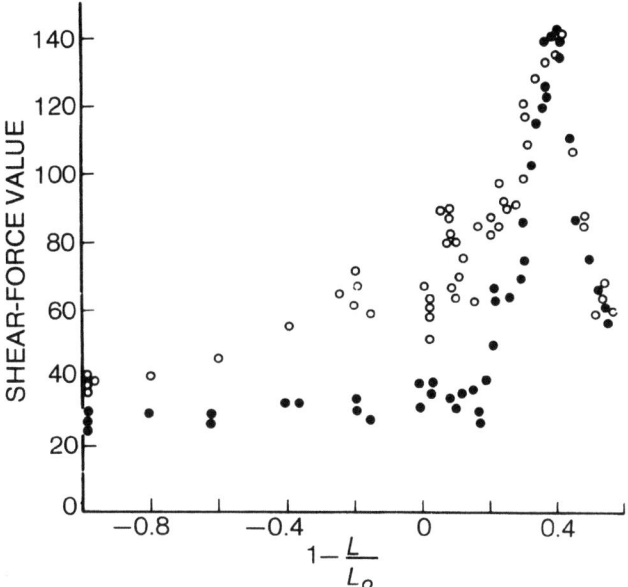

FIG. 1.2. The effect of degree of shortening on shear force values of bovine sternomandibularis muscles. ○ Mean of five determinations, unaged samples. ● Mean of five determinations, maximum aging (3 days, 15°C).
From Davey et al. (1967).

encountered in commercial practice, although it is not hard to achieve by severe chilling or thaw shortening of excised muscles.

The region of most practical interest is the toughening phase (0–40% shortening), which, on the basis of a sarcomere length of 2.1 μm for the equilibrium excised condition (Marsh and Carse 1974), corresponds to sarcomeres averaging down to 1.3 μm. The reasons for this phase have been discussed by Marsh and Carse (1974) and by Voyle (1969). These will be considered in detail later in this chapter along with newer ideas. For the moment, it should be emphasized that the curve is based on gross length changes, which disguise the fact that sarcomere lengths are highly disperse. Voyle (1969) has shown a wide variety of responses to cold shortening in individual fibers, ranging from slightly contracted crimpled fibers to extreme contractions with tearing. This is my own experience, and it also holds true for heat shortened muscle. I have tried unsuccessfully to relate the response to "red" or "white" muscle fibers in beef sternomandibularis. The cold shortening response seems to apply in all beef muscles, no matter what the fiber mix.

Marsh and Carse (1974) extended the length-toughness relationship to include stretch (up to 100%). A small peak amounting to only a 20%

increase in shear force was seen at 25% stretch. Values fell slowly away with further stretch until, at 100%, shear force was slightly lower than in unstretched muscle.

An interesting and highly significant corollary to the shortening-toughening curve is the finding of Davey *et al.* (1967) that the same beef neck muscle at the peak of toughening does not age at all (Fig. 1.2). The tenderizing achieved in muscle at normal lengths declines quickly over the range of 20–40% shortening. Herring *et al.* (1967) did not obtain such a clearcut result with beef semitendinosus, finding that at all points between 48% stretch and 48% contraction, aging reduced shear force significantly. However, at the higher levels of contraction, aging was less effective and failed to produce an acceptable level of tenderness.

TOUGHENING LAMB CARCASSES IN BLAST FREEZERS

The very first experiments on toughness in frozen lamb clearly implicated blast freezing (Marsh *et al.* 1968). The toughness of longissimus muscle was assessed both mechanically and by panel after cooking from the thawed state. This muscle was a happy choice because it has considerable freedom to shorten. Only one insertion is directly onto the skeleton (vertebrae) while the other is not, over most of the back. The marked

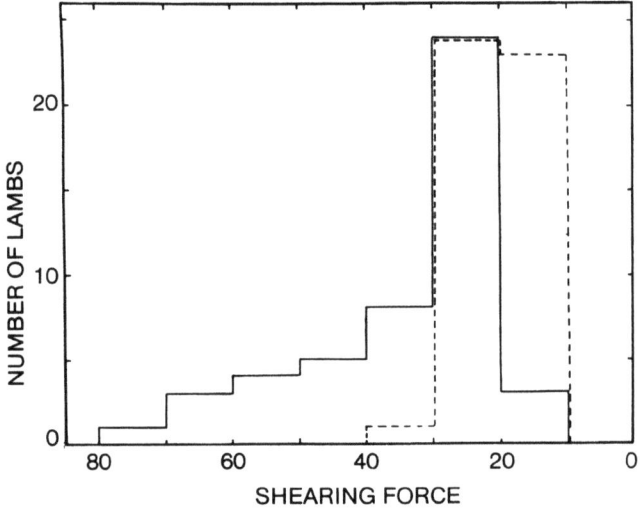

FIG. 1.3. The effect of postmortem treatment on shearing force requirement. ——— Early rapid freezing. - - - - - Delayed slow freezing.
Reprinted from J. Food Sci., Marsh et al. (1968) 33:12. Copyright © by Institute of Food Technologists.

difference in the distribution of shear force values derived from modern blast freezing and those from the old delayed slow freezing is shown in Fig. 1.3. All of the latter samples lie within the acceptable range of 10–40 shear force units (MIRINZ tenderometer; Macfarlane and Marer 1966). While most of the blast frozen samples also lie within this limit, the bias is to the higher end, and other values extend up to twice the limit of acceptable shear force.

A similar result from subjective assessment is seen in Fig. 1.4, where a steep rise from unacceptable to acceptable scores is evident as delay before blast freezing is increased from 6 to 16 hr.

A more detailed study involving assessment of a range of muscles confirmed these findings (McCrae et al. 1971). In this case, muscles were cooked from the frozen state. Results presented in Fig. 1.5 show that for four major muscles, a 16 hr delay was necessary before blast freezing to obtain an acceptable shear force. On the other hand, four other muscles were largely unaffected by delay. In all of the latter group, the skeleton imposes restraint. The importance of this was demonstrated by cutting one end of the semitendinosus and triceps muscles, after which the shear values trebled.

These carcass experiments demonstrate clearly that freezing in itself is not deleterious to tenderness, provided that the carcass is in or near rigor mortis. They do not, however, answer the question of whether freezing is necessary to damage prerigor carcasses. A few experiments were done on

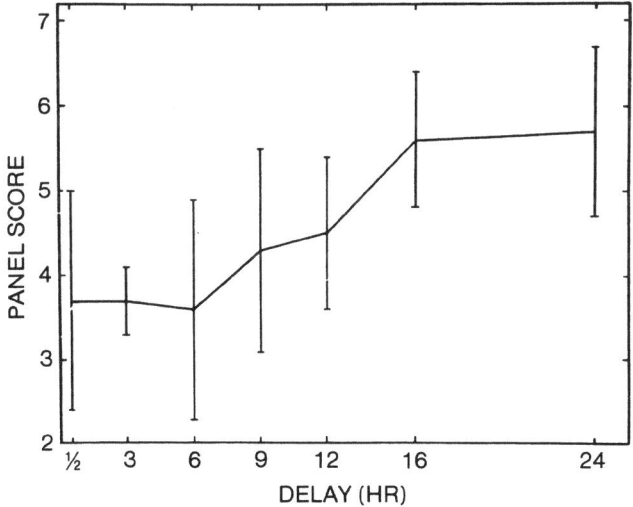

FIG. 1.4. The effect on panel tenderness assessment of delaying carcasses at 18°C–24°C.
Reprinted from J. Food Sci., Marsh et al. (1968) 33:12. Copyright © by Institute of Food Technologists.

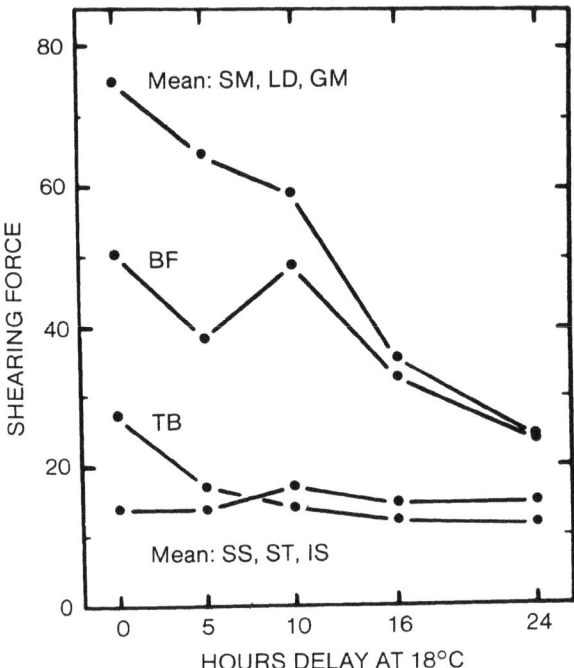

FIG. 1.5. Effect of delay before freezing on shear force. Twenty animals in each muscle and delay group. Loin: LD—Longissimus dorsi. Leg: SM—Semimembranosus; GM—Gluteus medius; BF—Biceps femoris; ST—Semitendinosus. Shoulder: TB—Triceps brachii; SS—Supraspinatus; IS—Infraspinatus. Reprinted from J. Food Sci., McCrae et al. (1971) 36:566. Copyright © by Institute of Food Technologists.

hot lamb carcasses subjected to very rapid chilling (Locker 1963). Sides were held in a blast freezer until the loin was just above the freezing point, or else immersed in a slurry of ice and water. In both cases, rigor was completed at 2°C. Both treatments produced loins much tougher than the control sides held overnight at 15°C. This demonstration that blast cooling alone could toughen established the link with cold shortening, but this was soon shown to be only part of the sad story.

ADDING INSULT TO INJURY: THAW SHORTENING

Marsh, who had enjoyed the unusual experience of a season on the whaling factory ship *Balaena*, was familiar with the drastic effects of thaw shortening in whale meat, frozen immediately after flensing. These were manifest in extreme shortening and drip loss. He surmised correctly that at least some of the musculature of a lamb would freeze prerigor in a blast freezer, and was concerned at the possibility of similar deleterious effects.

A paper on thaw rigor in lamb, published two years before the toughening problem emerged (Marsh and Thompson 1958), showed that lamb could also undergo substantial, although less extreme, shortening and drip loss thaw rigor. Thaw shortening did not occur if the carcass was thawed intact, due to the restraint of internal ice and the skeleton. These authors anticipated a future problem:

Thus major problems associated with thaw rigor are unlikely to be encountered, even if freezing rates increase considerably, so long as the present day commercial thawing procedure—a slow thaw of the whole carcass—is maintained, but possible variations in current commercial thawing practice might be attended by difficulties if thaw rigor is disregarded. A carcass frozen before rigor set onset might be broken down to prepackaged cuts while still frozen, for instance, or faster thawing methods might be used. . . .

Although tenderness was not considered, this warning proved to be prophetic.

Subsequent work has shown clearly that in blast-frozen lamb the effects of cold shortening and thaw shortening can be cumulative and devastating to tenderness. The accumulated results in Table 1.1 show this effect clearly. The difference between blast frozen and "conditioned" lambs when cooked from the thawed state is a measure of the effect of cold shortening alone. The difference in shear force between blast-frozen and "conditioned and aged" lambs when cooked from the frozen state is a measure of the two effects combined. At that time, New Zealand lamb loins were reaching British supermarket cabinets as chops cut from the frozen carcass, so the warning of Marsh and Thompson (1958) seemed highly relevant.

However, it has subsequently been found that the breakdown of ATP progresses steadily at $-12°C$, a store temperature common in the industry (Davey and Gilbert 1976). Within 2–3 weeks at $-12°C$, the pH had fallen below 6.0 and toughening due to thaw shortening had almost disappeared after roasting from the frozen state. Some toughening persisted in grilled leg steaks but had become negligible by 3 months. Thus, it became clear that for New Zealand's lamb trade, involving considerable periods in freez-

TABLE 1.1. Lamb Tenderness Results from Different Prefreezing Treatments

Pretreatment	Number of animals	% acceptable (SF < 40)	Mean shear force
Loins roasted from frozen state			
blast-frozen at 0–6 hr	218	30	53
conditioned 16–24 hr	150	65	36
conditioned and aged	178	94	22
Loins roasted from thawed state			
blast-frozen at 0–1 hr	48	73	33
conditioned 24 hr, slow-frozen	48	100	21

Source: Reprinted from Locker et al. (1975).

ing stores or at sea, thaw shortening was unlikely to be a significant factor. However, this history remains a warning to anyone with a fast turnover in consumer packs who may be tempted to market a frozen chop, cut from a carcass, recently blast frozen without the benefits of electrical stimulation. In the absence of this or other prophylactic measures, the problem of cold shortening remains. It is not helped by storage or by slow thawing of intact carcasses, nor much by subsequent aging. It remains as permanent damage to quality.

THE COLD-SHORTENING RESPONSE

Cold shortening has already been the subject of a number of reviews. Two excellent concise reviews are by Marsh (1972) and by Newbold and Harris (1972). A much more extensive review is by Locker et al. (1975) and the most recent by Davey and Winger (1980). It is not proposed to cover again in detail all the ground covered by the 1975 review, but rather to summarize the major features of cold shortening. For more detailed and complete references, readers should refer to Locker et al. (1975).

The effect might have been discovered earlier had it not been that Bendall (1951) used the wrong animal (the rabbit) and the wrong temperatures (not low enough) in his rigor studies. When Marsh (1954) studied rigor in beef, he used Bendall's techniques and temperatures, so likewise did not observe the effect. Cold shortening was discovered accidentally in beef sternomandibularis muscle as the result of an attempt to use Bendall's observation that a higher rigor temperature induces greater shortening. It seemed a convenient method of achieving different degrees of shortening in rigor. I anticipated that a sample at 2°C would show a minimal response but was surprised to find maximal shortening at that temperature. Repetition showed it was no accident, but a reliably reproducible effect. This led to more detailed study (Locker and Hagyard 1963) and the finding that shortening in rigor showed a minimum of about 10% in the region of 15°C–20°C (Fig. 1.6). The maximum shortening of near 50% at 0°C was well above that at the highest temperature used (43°C). Analysis of the time course (Fig. 1.7) showed that the two parts of the curve in Fig. 1.6 represented quite different effects: a cold shortening, which was an immediate response to chilling, and a delayed shortening at higher temperatures, corresponding to the onset of rigor mortis. It was this latter arm of the curve that had been previously observed.

Cold shortening begins with cooling and is most rapid in the first 1–2 hr. Shortening then becomes progressively slower but continues throughout the course of rigor. It should be pointed out that what is often called "cold shortening," the ultimate shortening achieved between entry to the cold and completion of rigor (which takes about 48 hr at 2°C in beef sternomandibularis) is actually the sum of the cold shortening and rigor shortening, and the latter contribution is substantial.

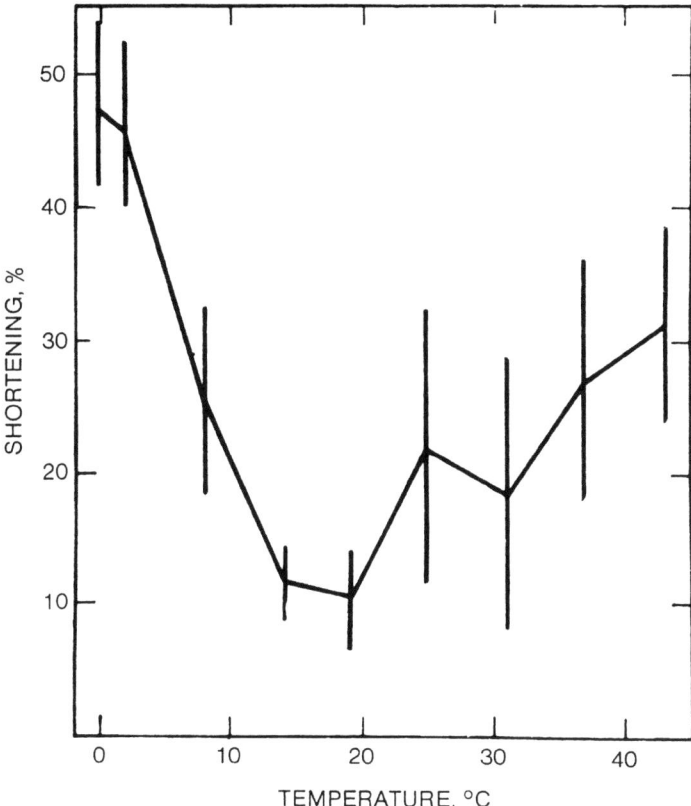

FIG. 1.6. Shortening and temperature.
From Locker and Hagyard (1963).

Cold shortening is not unlike a slow physiological response to nervous stimulation, but nerves do not appear to be involved, as twitch response to electrical shock fades out in the cold just as the rapid phase of contraction is beginning. In beef sternomandibularis, the ability to cold shorten persists for about 10 hr at 25°C, declining rapidly toward the end of this period. The contraction is at first fully reversible, but becomes progressively less so during this 10 hr period. It appears that the response disappears when the pH of the muscle has fallen to about 6.3. It is at about this pH that rigor shortening begins.

The tension generated by cold shortening can be substantial. It is higher than the tension generated during rigor shortening at 37°C, but lower than that due to thaw shortening, which in turn is lower than the capacity of the muscle in the live animal (Locker et al. 1975). In beef, the tension generated depends on the maturity of the muscle. It may be as low as 1 g/cm^2 in young calf sternomandibularis muscles and as high as 200 g/cm^2

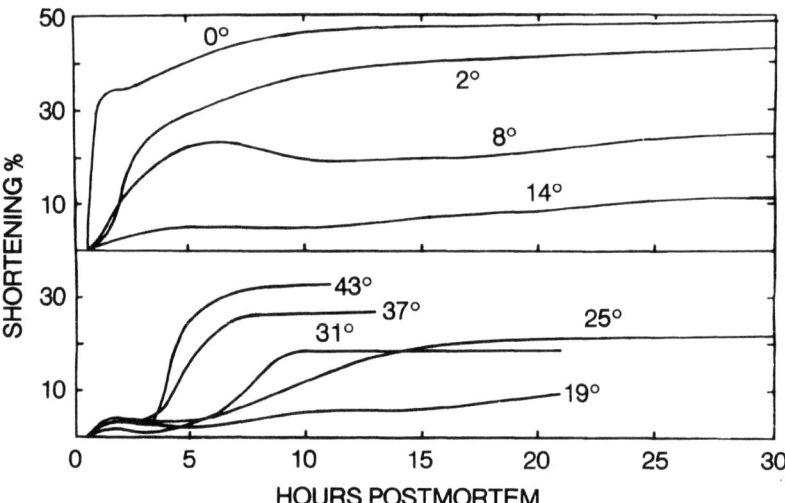

FIG. 1.7. Shortening after postmortem.
From Locker and Hagyard (1963).

in the same muscle from mature bulls according to Davey and Gilbert (1975). They also reported a similar gradation in degree of shortening, yet the greatest cold shortening I have ever observed was in the same muscle from fetal calves (up to 70%).

The emphasis in this review on beef (in the laboratory) and on lambs (in industry) arises from the historical development of the research. Cold shortening and its effect on tenderness have since been studied in many other laboratories, where the earlier results have been confirmed for the ovine and bovine species and shown to be relevant to other meat animals such as chickens, turkeys, and pigs (marginally), but not to rabbits (Locker et al. 1975). Cold shortening appears to be a general hazard in the meat industry.

Cold shortening also has relevance to research in muscle biology and medicine. Unfortunately, this is largely unrecognized because the results are embedded in the literature of meat science and little known elsewhere. Cold shortening is a hazard in the preparation of myofibrils from prerigor muscle, when blending it in cold solutions, as indeed is the act of blending, itself (Locker et al. 1976). A major risk is in histological studies, where fixatives are used in the cold. It is not difficult to find micrographs in the literature of muscle (or of meat science) where cold shortening has gone unrecognized as a major modification of the material. Nor is it difficult to find in medical textbooks pictures of "pathological" muscle samples which are really acute cases of cold shortening, arising from dropping human biopsies into cold fixatives, with the best of intentions. Thaw shortening can also be a serious problem in sections of prerigor muscle sectioned on a cryostat. Unless there are good reasons against it, it is far better to use rigor muscle for histological or histochemical studies.

It is common in biochemistry to "freeze" metabolic states in muscle by freezing at extreme rates. A number of workers have reported that immersion in liquid air, liquid nitrogen, or Dry Ice-alcohol results in shortening during the freezing process (Locker and Hagyard 1963). The muscle is no longer in its original state.

THE THAW-SHORTENING RESPONSE

Perry (1950), using frog sartorius muscle, was the first to seriously study the drastic shortening which occurs on thawing muscle, frozen before the onset of rigor. The effect was first studied in meat on the whaling factory ship *Balaena* during the seasons 1946–1949 by scientists from the Low Temperature Research Station at Cambridge. One of many peculiarities of whale meat is its ability to maintain a physiological pH for long periods postmortem. Marsh (1952) records that in whale meat cooling from 35°C to 22°C, the pH remained at 7.2 for 18 hr before beginning to fall. Since at death the musculature of a whale becomes perfused with gut bacteria via the blood vessels, early freezing is necessary, and in spite of the large size of the blocks, this pH "dwell" ensures that prerigor freezing is common. On thawing, up to 47% of the total weight may be exuded as drip (Sharp and Marsh 1953) during the severe contraction. I have observed 80% shortening on thawing whale meat frozen at a shore station here. A good account of thaw shortening, which includes pig and chicken muscles, is given by Luyet (1966).

Back in New Zealand, Marsh found lamb to be in a position potentially almost as serious as that of whale meat. A study on excised lamb longissimus muscle (Marsh and Thompson 1958) showed that in strips frozen without delay and thawed at 16°–20°C, there was an average shortening of 72% with a 27% drip. In contrast, muscle frozen in rigor, when thawed, shortened by only 5% and lost but 3% drip. However, if muscle frozen prerigor was thawed at -3°C for 4 days, it did not shorten or drip. The amount of drip rose steeply as the ambient temperature during thawing was increased in the range of 5°–10°C.

Although neither carcass meat nor tenderness was studied, it was clear from the rate of pH fall that blast freezers were quite capable of freezing these small carcasses before rigor was complete. However, lamb differs from the whale meat in having its skeletal attachments intact during processing, and during the slow thawing of carcasses, shortening and drip are largely prevented by the restraint of the skeleton and by internal ice.

It should be emphasized that a degree of rigor which is adequate to remove any danger of cold shortening may still be inadequate to ensure no possibility of thaw shortening. For example, pH 6.3 was found to be the minimum for cold shortening in excised beef muscles (Locker and Hagyard 1963). Experience here with electrically stimulated lamb carcasses has shown that pH 6.0 is a safe industrial standard for eliminating any possibility of toughness induced by cold shortening. Yet Brunton and Gilbert (1972) found a thaw toughening effect when lambs were conditioned for 16 hr at 7°C (mean pH 5.74) and blast frozen, but not when conditioned for 24 hr (mean pH 5.58). The difference of a mere 0.16 pH units meant that when the loins were roasted without prior thawing, there was a doubling of shear force in the 16 hr samples, but negligible change in the 24 hr samples. Davey and Gilbert (1976) found that lamb longissimus muscles that had reached a mean pH of 5.90, when newly frozen had a mean ATP content of 0.4 μmoles/g and shortened by 55% on thawing. When the ATP had fallen to 0.1 μmoles/g (at the same pH) after storage for a month at $-12°C$, no significant shortening occurred on thawing. It appears that a quite small residue of ATP is capable of producing thaw shortening with consequent toughening.

THE BIOCHEMICAL BASIS FOR COLD AND THAW SHORTENING

The biochemical events during onset of rigor mortis at 1°–37°C have been reviewed by Newbold (1966). His results (Fig. 1.8) show clearly that at 1°C, cold shortening in beef sternomandibularis is a rapid response, preceding glycolytic rundown of the muscle. In the most rapid phase (the

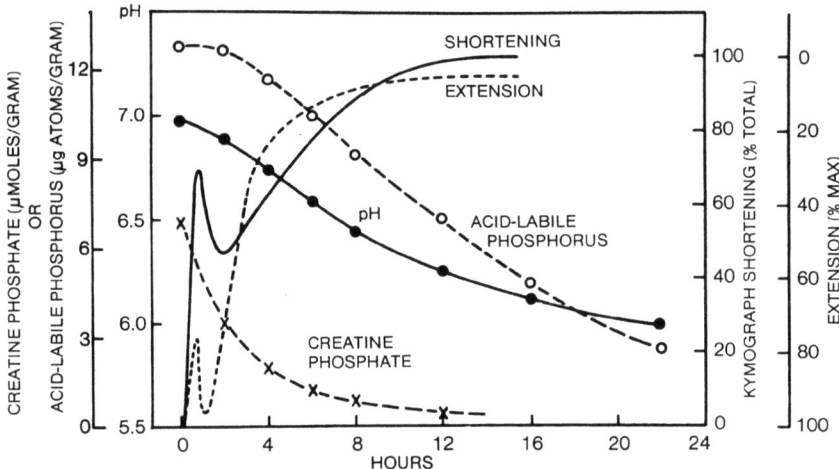

FIG. 1.8. Chemical and physical changes in beef sternomandibularis muscle held at 1°C.
From Newbold (1966).

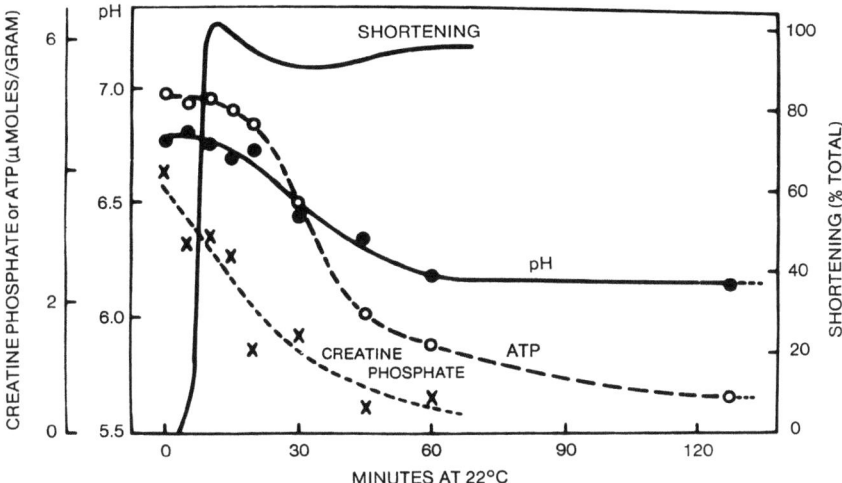

FIG. 1.9. Chemical and physical changes in beef sternomandibularis frozen 2 hr postmortem.
From Newbold (1966).

first hour), pH and acid-labile phosphorus were unchanged and even the greater part of creatine phosphate remained. The kymograph extensibility remained high in this phase, a fact compatible with the reversibility of shortening (Locker and Hagyard 1963). The relaxation in the second hour on the kymograph (involving periodic loading with 60 g/cm^2, 8 min on, 8 min off) is interesting, and was attributed by Newbold to an increase in inorganic phosphate's enhancing Ca^{2+}-binding by the sarcoplasmic reticulum. It is reflected only as a step in the 2°C curve of Locker and Hagyard (1963) and more distinctly in their 8°C curve. It is also seen in the thaw shortening curve of Newbold (Fig. 1.9). In fact, the relationship between thaw shortening and its biochemical events, although compressed in time, so closely resembles the 1°C data (Fig. 1.8) that Newbold suggests "that the underlying reasons for the effects observed with beef muscle at 1°C and for thaw rigor are the same."

The rate of glycolytic rundown, as measured by pH or by ATP disappearance, shows that in muscles which do not cold shorten, the rate declines steadily with temperature. For example, it is minimal in rabbit psoas muscle at 0°C (Bendall 1951). In chicken breast, which does cold shorten, a minimum in rate of ATP breakdown occurs at 10°–30°C (de Fremery and Pool 1960). Cassens and Newbold (1967) found the rate of pH fall in beef sternomandibularis at 1°C to be similar to that at 15°C, but faster than at 5°C, an effect which I also have observed. Bendall (1972) found the rate of ATP turnover in this muscle was minimal at 5°C, whereas at 1°C it was equal to that at 15°C. He attributed this upturn to the superimposing of cold-shortening changes on the resting rate. Davey and Gilbert (1974A) have also shown the rate of ATP breakdown in this muscle to be similar at 0°C to that at 15°C if the muscle was unloaded, but on loading to only 40 g/cm^2 at 0°C, the rate of breakdown was more than doubled, while the same loading at 15°C (where there is no cold shortening) does not affect the rate.

There seems to be little doubt that the trigger for cold or thaw shortening is the release of Ca^{2+}, as for other contractions. Davey and Gilbert (1974A) showed that the minimal level of organization needed for cold response is the fiber piece. Myofibrils do not cold shorten, nor do fibers in the presence of EGTA (ethylene glycol tetraacetate). A pCa^{2+} value of 6.5 was needed to make fibers shorten at 0°C, and 6.3 for myofibrils.

The major question is, where does the Ca^{2+} come from? The two candidates are the sarcoplasmic reticulum (SR) and the mitochondria. Both have considerable capacity for accumulating Ca^{2+}, and both can be induced to disgorge it. The SR was earlier favored, particularly in view of its decline in binding ability with temperature (Martinosi and Feretos 1964). The potential of mitochondria as the effective Ca^{2+} source was first emphasized by some simple but elegant experiments on the original material of cold shortening, beef sternomandibularis. Using thin strips (2–3 mm in diameter), Buege and Marsh (1975) showed that cold shortening was normal in a nitrogen atmosphere at 2°C, but much reduced in oxygen. This inhibition was entirely removed by dipping the strips in inhibitors or uncouplers of oxidative phosphorylation. The authors concluded that in this muscle an anaerobic condition in the cold causes mitochondria to lose Ca^{2+} at a rate faster than the SR can mop it up. Their conclusions are in line with the general tendency of red muscles to cold shorten more effectively than white ones.

There have been three major contributions to the debate on the relative importance of sarcoplasmic reticulum and mitochondria. In a review paper, Newbold (1979) quotes earlier work showing that SR loaded with Ca^{2+} loses about half its load on cooling to 0°C. Inorganic phosphate has a fairly strong inhibitory effect in rabbit white muscle but not in ox muscle. He suggests that failure of rabbit muscle to cold shorten may be due to a faster metabolism postmortem, leading to a phosphate concentration sufficient to prevent cold-induced release of Ca^{2+}. He found that ox mitochondria, preloaded with Ca^{2+} to 10–30% of their capacity in the presence of ATP, remained loaded at 23°C and lost only small amounts of Ca^{2+} over 3 hr at 0°C. Under anaerobic conditions, however, 75% of the Ca^{2+} was lost in 5 hr in the absence of ATP, but in its presence there was no loss in 5 hr at 23° or 2°C. The addition of azide or dinitrophenol (DNP) to loaded mitochondria caused total release at 23°C in less than an hour or more slowly at 0°C. The failure of anaerobic mitochondria to release Ca^{2+} if ATP is present led Newbold (1979) to conclude that cold shortening was due to cold-induced release of Ca^{2+} from the SR.

Whiting (1980) was more concerned with survival of Ca^{2+}-binding ability under various conditions than with actual performance under early postmortem conditions. However, his data show the binding ability in air of beef SR and mitochondria at various temperatures for pH values and ATP concentrations near the physiological.

These data are rearranged in Table 1.2. He took the trouble to type his beef biceps femoris muscle (34% β-red, 60% α-white, and 6% α-red fibers, see Chapter 6) and to give yields of the organelles, which, being based on simple centrifuge cuts, are probably a fair indication of total amounts in the muscle. These yields have been used in Table 1.2 to give the total capacity within the muscle of each organelle. It can be seen that SR dominates and also loses its binding ability more slowly as temperature falls. Mitochondria had no Ca^{2+}-binding ability at 1°C and little at 15°C. On the basis of these figures, and making the questionable assumption that mitochondria were fully charged at 37°C, the mitochondria should, on cooling to 2°C, dump more Ca^{2+} than the SR could deal with, even if the SR were initially empty. These aerobic data would predict a cold shortening in the aerobic state, which, according to Buege and Marsh (1975), does not occur, at least in sternomandibularis. However, the degree of Ca^{2+} charge in the organelles is too uncertain to speculate. Whiting (1980) favored mitochondria as the more probable initial agents of Ca^{2+} release.

Cornforth et al. (1980) found much lower Ca^{2+}-binding values overall for sternomandibularis, with comparable values for both organelles at

TABLE 1.2. Comparison of Ca^{2+}-binding Ability of SR and Mitochondria of Beef Muscle at Various Temperatures (μmoles Ca^{2+}/mg Protein). Collected Results from Three Papers. Binding Values All Assessed Essentially by the Millipore Filter Method of Martinosi and Feretos (1964)

	Muscle of beef			23°C	
	Sternomandibularis[a]				
	SR			0.08*	
	Mitochondria			2.0	
		0°C	15°C		37°C
	Sternomandibularis[b]				
	SR	0.06	0.09		0.55
	Mitochondria	0.07	0.32		0.38
		1°C	15°C	25°C	37°C
	Biceps femoris[c]				
	SR	0.5	3.8	5.7	5.6
	Mitochondria	0.00	0.09	0.57	1.23
(μmol Ca^{2+}/g muscle)**					
	SR	1.1	8.4	12.5	12.3
	Mitochondria	0.00	0.11	0.68	1.48

[a] Newbold (1979); Table 1.
[b] Cornforth et al. (1980); Fig. 8.
[c] Whiting (1980); Fig. 3.
*>3.0 in the presence of 5–10 mM phosphate; mitochondrial uptake unchanged.
**Using reported yields of 2.2 mg SR, 1.2 mg mitochondria/g muscle.

37°C. Again, a rapid decline was seen with temperature, but in these experiments, binding by SR faded first, to a low level by 15°C (Table 1.2). Yet they found that most of the Ca^{2+} was retained where Ca^{2+}-loaded SR from rabbit longissimus or beef sternomandibularis muscle was cooled from 37° to 0°C. This was also true for both sorts of mitochondria in air. Replacing air with nitrogen caused only a 28% discharge of Ca^{2+} from beef mitochondria and only 6% from rabbit mitochondria. Although their results are not very conclusive in this respect, the authors favor the theory of Buege and Marsh (1975), adding the points that discharge of Ca^{2+} from mitochondria occurs at the time of anoxia and long before the muscle is chilled, but at a temperature where the SR can cope. They also attribute the reversibility of cold shortening to a regain of efficiency in the Ca^{2+} pump of the SR on rewarming.

The disparity of the results shown in Table 1.2 makes a conclusion difficult. My own inclination is toward the hypothesis of Buege and Marsh (1975), whose experiments were simple and clearcut and used intact tissue. Experiments on organelles involve uncertainties due to isolation and measurement under artificial conditions and usually lack the quantitative data on yield needed for an overall assessment of the interactions between SR and mitochondria.

However, recent results from Marsh's laboratory on isolated mitochondria of beef sternomandibularis do not on the whole support his first hypothesis. The mitochondria liberated Ca^{2+} on becoming anaerobic, to an extent dependent on initial load (Mickelson and Marsh 1980). Inorganic phosphate, which stimulates Ca^{2+} release in heart and liver mitochondria, in this case halved the release. More recently, Mickelson (1982) found, in agreement with Newbold (1979), that the release of Ca^{2+} caused by anaerobiosis (or by electron transport blocking agents) was inhibited by addition of ATP. Only when the mitochondrial ATPase was inhibited, or the entry of ATP to the matrix space blocked, did the mitochondria release Ca^{2+} in the presence of ATP. He also measured the concentration in sternomandibularis of palmitoyl CoA, which inhibits ATP transfer and is known to accumulate in ischemic heart muscle, but concluded that this agent could not be responsible for cold shortening.

There is a large literature on Ca^{2+} transport in mitochondria which cannot be covered here but is accessible through the references already given. Two further aspects should be noted.

The experiments of Patriarca and Carafoli (1969) emphasize the dominant role of mitochondria in the intracellular movements of Ca^{2+} in red muscle. The specific activity of mitochondria in muscle, labelled 15 min before death with $^{45}Ca^{2+}$, was 60 times higher in the mitochondria than in the SR of rabbit red muscle, but only 1.5 times higher in white muscle. They suggested a regulatory role for mitochondria in the contraction-relaxation cycle of red muscle.

Finally, Mickelson and Marsh (1980) point out that it requires release of only 1 n mole Ca^{2+}/mg of mitochondrial protein (2 n moles by my calculation from their mitochondrial yield from sternomandibularis) to produce $10^{-5}M$ Ca^{2+} in muscle, the level needed to fully activate contraction. This is a tiny amount compared with loading capacities in the μmole/mg range for both mitochondria and SR (see Table 1.2). Whether or not contraction is activated is determined by a very fine balance between ability of the organelles to release or mop up Ca^{2+}. This may account for the lack of a decisive answer from the attempts described to explain cold shortening in terms of calcium transport.

The effects of freezing and thawing on Ca^{2+}-binding require some final comment. Newbold (1979) found heavily loaded SR or mitochondria lost one-third to one-half of the Ca^{2+} on freezing and thawing and this was not subsequently taken up again. If lightly loaded, both organelles release Ca^{2+}, which is rapidly and completely resorbed by SR, but only slowly and incompletely by mitochondria. Whiting (1980) found the Ca^{2+}-uptake abilities for both organelles to be badly impaired by freezing and thawing in plain buffers, particularly for SR. Mitochondria were completely protected by 1% bovine serum albumin in the buffer while only 40% of the SR activity survived. It appears that the combined effects of damage to both organelles suffice to explain the drastic response of thaw-rigor.

The shortening during onset of rigor seems to be adequately explained by the loss of ATP, leading to failure of the Ca^{2+}-pump of the SR and a reduced Ca^{2+}-binding capacity in the mitochondria.

PREVENTION OF COLD OR THAW SHORTENING

There are several possible approaches to preventing damage to tenderness by shortening. The most obvious is to ensure that the meat goes into rigor before entry to a chiller or freezer. Another is to reduce chilling rates or temperatures to a safe level. The third is to take advantage of the restraints of the skeleton, possibly with some rearrangement of posture.

For a number of years, it seemed to workers in this laboratory that the only practical approach to this problem was to ensure that rigor was near completion before freezing. The old cooling floors, where they survived, were no longer adequate in terms of control or hygiene for this task. New air-conditioned holding rooms had to be constructed, although this proved too formidable a proposition for the whole national kill of about 27 million lambs. A "New Zealand Specification for the Conditioning and Ageing of Lamb" was developed by Marsh (see Table III in Locker *et al.* 1975). This was adhered to only for the limited North American trade. Essentially, it allowed completion of rigor ("conditioning"), plus a period of aging to ensure greater uniformity in tenderness. Several temperature regimes were allowed as alternatives, the one most commonly used being a one-temperature process (40–48 hr at 13°C). Other regimes used include temperatures of 16° or 18°C for 16–27 hr, followed by chilling to 3°C for 38–96 hr. This specification has remained satisfactorily in use until recently superseded by electrical stimulation. For 178 lamb loins, treated to specification and roasted from the frozen state, the mean shear value on the MIRINZ tenderometer was 22 (acceptable limit 40).

The threat of European Economic Community (EEC) regulations requiring early chilling to 7°C prompted an investigation of conditioning at that temperature. It was found that holding for 24 hr at 7°C did in fact produce a satisfactory product (Brunton and Gilbert 1972). Although this temperature seemed dangerously low from our studies on beef neck muscle, it appears that there is more latitude for muscle on the skeleton.

A bid to speed up conditioning was made by Davey and Gilbert (1973), who accelerated rigor by retaining the animal heat, in fact by raising the temperature to 45°C. The method had been used previously on beef (Roschen et al. 1950; Wilson et al. 1960) to produce rapid aging in beef. The temperature coefficient of the pH fall rises steeply in this region, but 45°C sets a limit above which discoloration occurs. The treatment effectively produced rigor in 3 hr, but both the leg and the loin were moderately tough, presumably due to heat shortening. If, however, the carcass was maintained in a standing posture, very tender meat was obtained. The experiment might have been more successful with rigor at 37°C, where experience with beef has shown no toughening in spite of shortening (Locker and Daines 1975). These high temperatures do not present the microbiological problems that might be imagined since the lag phase of growth is longer than the treatment. It might, however, be difficult to convince the makers of regulations on this score.

The problem of cold shortening is not confined to small carcasses. Using 50 prime steers, Davey (1970) showed clearly that chillers of average efficiency were able to toughen beef sides, in spite of cooling rates far below those for lamb in blast freezers (Fig. 1.10). A slow chill (deep leg to 17°C in 18 hr) produced about 36% of cube rolls with an acceptable shear value of 35 (MIRINZ tenderometer), but all values lay below 70. A fast chill which is close to current practice (deep leg to 8°C in 18 hr), produced only 25% within the acceptable range, with values up to 120. After aging for 24 hr at 20°C, nearly all of the slowly chilled sides were acceptable (none were above 40), whereas about half of the fast-chilled sides were still unacceptable, with values ranging up to 110.

Buchter (1972) confirmed these results by showing that in young bulls, cows, and calves there was progressive toughening in the loins as chilling rate increased. Extended aging barely brought rapidly chilled beef back to the initial tenderness of slowly chilled meat.

Since no discriminating market could be expected to accept such variability in prime cuts, Davey developed a "New Zealand Specification for Conditioned and Aged Beef" (see Table IV in Locker et al. 1975). Essentially, this involved chilling sides for 18–24 hr at 10°C, then boning out and aging in Cryovac packs for a further 66–72 hr at 10°C. Beef so processed has proved extremely acceptable in the quality trade.

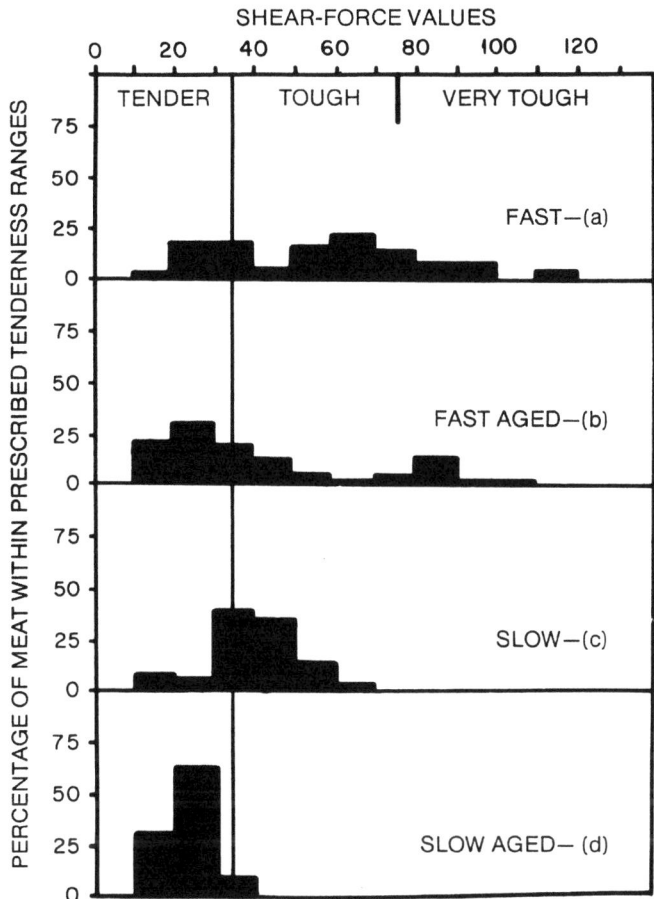

FIG. 1.10. The distribution of tenderness and toughness in cube rolls from beef sides.
From Davey (1970).

The protection afforded by reduced chilling rates applies in practice not only by technical design but also as a natural consequence of animal variation. There is a long established preference in the American market for grain-finished beef with good fat cover and marbling. This is based on tenderness as well as other palatability factors. This choice now appears to be soundly based, but largely as the result of an indirect effect. The superior fat cover in well finished carcasses is effective as an insulator, resulting in substantially lower chilling rates. A higher total mass also helps. A convincing laboratory demonstration of the effect of insulation on cold response was provided by Marsh and Leet (1966B) who insulated the two ends of a restrained beef neck muscle before transferring it to 2°C. The difference between the final state of contraction in the unprotected central part and the insulated ends was dramatic.

From their carcass studies, Merkel and Pearson (1975) concluded that "the major differences in tenderness between fat higher grading and thin lower grading beef are due to the effects that fat, including marbling, has on slowing down the rate of heat dissipation during carcass chilling." Bowling et al. (1977) showed that chilling at higher temperatures brought the tenderness of forage finished beef nearer to that of grain finished beef, chilled normally, but a margin still remained. The most detailed study has been done on lambs with different degrees of finish (Smith et al. 1976), chilled at 1°C. Tenderness in the loin and two major leg muscles was substantially modified by fat cover. In some sides the effect was confirmed by stripping the fat cover. In spite of the tenderness changes, it must be noted that mean sarcomere lengths varied only over quite a small range (1.72–1.86 μm). This was also the experience of Lochner et al. (1980) who confirmed tenderness effects in beef resulting from variations in fat cover and chilling regimes. I shall return to this point later in this chapter.

The fat cover of the pig, together with its compact conformation, provides the extreme example of natural protection from fast chilling. This factor, combined with the rather limited cold-shortening response of pork muscles, even when excised (Locker et al. 1975), appears to eliminate cold shortening as a problem when pigs are handled in carcass form.

A different approach to preventing cold shortening is to utilize the natural restraint of the skeleton. In the limbs of the living animal, there is a balance between opposing adductor and extensor muscles. It is possible to make use of this balance or to modify it in determining the final set of the muscles in rigor. This is particularly important for the valuable hind limb, in which highly unnatural strains are imposed when it is used as a point of suspension. Some muscles are greatly stretched whereas others are allowed scope for considerable shortening.

The modification of skeletal restraint was first hinted at in the original paper relating contraction and toughness (Locker 1960) which concluded that "it should be possible to improve the quality of the longissimus for example, by hanging the carcass in such a way that this muscle is stretched and prevented from shortening." The writer held little hope that the trade would consider such innovation. The idea was first taken up by Herring et al. (1965) who showed a marked improvement in this muscle when a side of beef was laid horizontally with the spine straight. These results have been confirmed in several laboratories for a variety of beef muscles (Stouffer et al. 1971; Hostetler et al. 1972; Bouton et al. 1973). The detailed study of the Texas workers (who coined the name "Tenderstretch Process") is summarized in Table 1.3. There was little to choose between the horizontal position and hanging from the pelvis, the latter being the obvious choice for practical reasons. The greatest improvement in the unconventional positions was in longissimus, but most of the major mus-

TABLE 1.3. Tenderness of 9 Beef Muscles from Sides Suspended in 5 Ways

Muscle	Mean shear force[a] (kg)				
	Vertical	Horizontal	Neck-tied[b]	Hip-free[c]	Hip-tied[d]
Longissimus	6.0[e]	4.7[f]	4.8[f]	4.9[f]	4.9[f]
Semimembranosus	5.0[e]	4.4[f]	4.7[e,f]	4.3[f]	4.5[e,f]
Semitendinosus	5.3[e]	5.0[e,f]	4.9[e,f]	4.9[e,f]	4.4[f]
Biceps femoris	4.2[f]	4.3[f]	4.3[f]	5.0[e]	4.6[e,f]
Rectus femoris	4.7[e]	4.4[e]	4.9[e]	3.8[f]	3.8[f]
Adductor	4.7[e]	3.8[f]	4.4[e,f]	4.0[f]	4.3[e,f]
Gluteus medius	4.9[e]	3.9[f,g]	4.3[f]	3.7[g]	3.6[g]
Psoas major	3.7	3.8	4.3	4.2	4.4
Triceps brachii	4.7[f]	4.9[f]	5.4[e]	4.8[f]	5.5[e]
Mean of means	4.8[e]	4.4[f]	4.7[e]	4.4[f]	4.4[f]

Source: Reprinted from J. Food Sci., Hostetler et al. (1972) 37:132–135. Copyright © by Institute of Food Technologists.
[a] Warner-Bratzler, mean of 16 sides.
[b] Hung from neck, fore, and hind limbs tied together.
[c] Hung from aitch bone, limbs free.
[d] Hung from aitch bone, limbs tied together.
[e,f,g] Means on horizontal lines with different superscripts differ significantly ($P < 0.05$).

cles of the leg were also improved. Similar work by Bouton et al. (1973) produced comparable results. These workers used the normal suspension as control and three variations of pelvic suspension (one being the rather sophisticated set of metal splints patented by Stouffer et al. 1971). The simple pelvic suspension was clearly the method of choice. Their data have been rearranged as Table 1.4 to compare only this method of hanging with

TABLE 1.4. Mean Warner-Bratzler Shear Values and Sarcomere Lengths Obtained for Muscles Taken at 2–3 Days Postmortem from Sides Subjected to Different Hanging Treatments. Cooked at 80°C for 1.5 hr

Muscle	Shear force (kg)		Sarcomere length (μm)	
	Achilles tendon	Aitch bone	Achilles tendon	Aitch bone
Semimembranosus	8.35 (5.49)	5.13 (4.04)	1.70	2.76
Gluteus medius	7.99 (5.22)	4.00 (3.54)	1.68	1.39
Longissimus	11.12 (6.04)	5.72 (3.90)	1.80	2.00
Vastus lateralis	8.81	5.36	1.74	2.80
Biceps femoris	6.54	6.67	1.78	2.93
Semitendinosus	6.13	5.95	2.30	2.28
Infraspinatus	6.40	6.04	2.10	1.96
Psoas major	3.63	4.99	3.20	2.29

Source: Reprinted from Bouton et al. (1973). J. Food Technol. 8:39. Copyright © by Inst. of Food Sci. and Technol. (U.K.).
Figures in brackets represent shear values obtained for meat samples aged 21 days at 0°–1°C.
Least significant shear differences (5%) (1) for the comparison of hanging treatments on: longissimus dorsi, 3.0; all other muscles, 1.6; (2) for comparison of aging treatments on: longissimus dorsi, 3.6; semimembranosus and gluteus medius, 1.1.
Least significant difference in sarcomere length (5%) for the comparison of hanging treatments is 0.05.

the normal. Large increases in sarcomere lengths resulted, in parallel with substantial reductions in shear force. Only the psoas muscle did not benefit, although it did not suffer badly. Comparing these values with aging results from the earlier experiments, they concluded that "aitch bone suspension gave muscles which, when cooked at 2–3 days, gave shear values which were equivalent to those obtained after 2–3 weeks ageing of similar samples from muscles hung conventionally."

The alteration of posture has also proved effective for lamb. The first experiments in this laboratory supported lambs in a natural standing posture on a broomstick during conditioning with improvement to loin and leg muscles (Davey and Gilbert 1974B), but this was soon abandoned for hanging from the pelvis with similar results (Davey and Curson 1971). A leg at right angles to the spine gave the best results.

American and Australian workers have both used pelvic suspension of lamb with good results. Quarrier et al. (1972) found improvements in tenderness associated with longer sarcomere lengths in the leg and loin of lambs suspended from the pelvis. Baxter et al. (1972) found a similar situation in hoggets, the margins persisting after aging for 2 weeks at 0°C. In both these cases, the carcasses were immediately chilled at 0°–2°C, which induces cold toughening in the normally hung controls, thus accentuating the benefits of the altered posture. Experience in our laboratory shows that in lambs, if proper chilling precautions are adopted, the improvements due to altered posture alone are real but small. In the worst situation, immediate blast freezing, the benefits are great, in fact the "altered posture" approach offered a quite satisfactory solution to the problems of New Zealand frozen lamb (Locker et al. 1975). It was the industrial solution favored here at the time when the potential of electrical stimulation first became evident. There was some resistance in the trade to the shape of the altered posture leg, which appeared to arise from conservatism rather than any intrinsic failings. The joint is in fact more compact, and has advantages for roasting and carving. The awkward configuration of a carcass with the leg perpendicular to the spine was easily overcome by pinning the shanks back with an extra gambrel, passing behind the pelvic hook. This "crouching-posture" carcass is actually shorter and more compact for stacking.

A curious recent result which appears to contradict previous experience is the finding by Davey and Garnett (1980) that prerigor lambs, normally hung, can be frozen immediately without toughening the loin if the freezing rate is extreme (less than 4 hr). The method relies on a period of frozen storage to eliminate thaw toughening effects. It appears that the shell freezing of the carcass supplies the necessary restraint (exterior fat becomes very rigid under such conditions). It seems less likely that results would have been as favorable in this posture for leg muscles, for which no results were recorded, although the authors hint at some success in this direction.

A quite different approach to skeletal restraint is that of hot cutting of lambs to primal joints. This was explored in this laboratory after successful early experiments in hot boning of beef, followed by holding at 15°C (Schmidt and Gilbert 1970). Sides of lamb were cut to leg, loin, and shoulder immediately after dressing, and the joints were then shrink-wrapped and conditioned at 10°C for 24 hr before freezing. When joints were roasted without thawing, muscles from the loin and shoulder were as tender as those from the control sides, conditioned in the normal hanging position, while the hot-cut legs were significantly more tender than the controls (McLeod et al. 1973). If the joints were conditioned in a carton before freezing (24 hr in air at 1°C), similar results were obtained, but immediate freezing in the carton toughened the loins (McLeod et al. 1974). The shrinkwrap is an essential part of the process, converting the initially untidy cuts into rounded shapes by molding warm fat and flesh. The conformation of the leg is particularly attractive, resembling a ham (and particularly suitable for curing). The method was also tested on old ewes, with great success in the leg. Shear values comparable to those from lamb legs were obtained from roasted ewe legs, derived from animals up to 9 years old (Wenham et al. 1973).

The essential factor in hot cutting to joints is that the muscles of the isolated leg take up the same configuration as in life or in "altered-posture" carcasses, with the same protection for tenderness. This is less true for the loin. It should be pointed out that in the shoulder of lambs, the muscle configuration is always good, no matter what the posture. The shoulder is therefore relatively immune to damage by bad refrigeration practice. This probably also applies to beef, although bovine shoulder muscles have been ignored in past studies. It appears that the shoulder, although the least favored joint of lamb, is the safest buy!

The last word in this catalogue of stratagems for protecting carcasses against too early exposure to cold is due to Carse (1973), a technician at the Meat Industry Research Institute of New Zealand. He chanced to observe that excised beef muscle subjected to electrical shock went more rapidly into rigor. Grasping the practical potential of this observation, he was soon subjecting newly dressed lamb carcasses to AC voltage, direct from the power mains. At the time, he was unaware of previous laboratory experiments on rabbit and pig muscle or of an unused patent for the rapid conditioning of beef by the same means (Carse 1973). His was, in fact, a rediscovery, but one which led for the first time to industrial application. The work is still in progress here and around the world, now essentially fine tuning a process that shows every sign of becoming universal in meat processing. The process, "electrical stimulation," is the title of this volume and the substance of its other chapters. It benefits extend beyond mere protection against cold or thaw toughening. One of the greatest is that hot cutting, which seemed just possible in early beef experiments, using temperature as the only control (Schmidt and Gilbert 1970), can now be done with confidence that the product will be at least the equal of beef boned in rigor (Seideman et al. 1979).

While the simplicity and cheapness of electrical stimulation make it a clear choice, it would be a pity if it resulted in the neglect of some of the other work here described, which was directed toward the same end. The obvious benefits to beef from pelvic hanging as seen in Tables 1.3 and 1.4 suggest that any operator in quality trade might do well to use both techniques. After all, pelvic hanging is also a simple and inexpensive process.

SOME DOUBTS ON THE RELATIONSHIP BETWEEN SHORTENING AND TOUGHNESS

Definition of a precise relationship between degree of shortening and shear force led to the conclusion that ultimate sarcomere length is the important factor in the myofibrillar contribution, and not the path by which this length is reached. This was understandable since in the original work of Marsh and Leet (1966B) a variety of strategems were used to produce gradations of contraction, and the resulting curves were nearly identical. There is now reason to doubt a straightforward relationship.

All the treatments used by Marsh and Leet (1966B) involved exposure to cold, such as early or delayed chilling, or thaw shortening. Since muscles may also be shortened by elevated temperatures during rigor, this too might be expected to produce toughness. The question is of real technical importance, and there is some evidence for a positive effect. The muscles of some beef animals reach low pH values (5.8–6.2) by the end of dressing and these produce tougher meat (Khan and Lentz 1973; Khan and Ballantyne 1973). Lamb carcasses subjected to accelerated rigor at a high temperature (2 hr at 43°C), toughen markedly in a normal hanging posture, but not if hung from the pelvis (Davey and Gilbert 1973). Excised poultry muscle held at 30°–32°C also toughens (de Fremery and Poole 1960; Khan 1971). However, in other cases heat toughening has not been detected, e.g., in beef (Busch et al. 1967) or in electrically stimulated lamb (Chrystall and Hagyard 1975).

In light of the potential importance of this matter to the emerging technique of electrical stimulation, which causes carcasses to go into rigor mortis at temperatures near the physiological, we studied rigor mortis in beef sternomandibularis and rectus abdominis muscles at temperatures up to 37°C (Locker and Daines 1975). In sternomandibularis, 25% shortening during rigor at 34°C did not increase shear force over a 15°C control, while at 37°C, a 32% shortening caused a significant decline in shear force. If the muscle was restrained at 37°C, it was a little tougher than if unrestrained, although still more tender than a 15°C control. Rectus muscle, after shortening at 37°C, was tougher than 15°C controls, but less so than would have been expected from the degree of shortening. Some effort was made to decide whether the results at 37°C were due to aging, and it was concluded that they were not.

A more surprising result was obtained with sternomandibularis, which was allowed to shorten for 24 hr at 2°C and was then transferred to 37°C for 3 hr for completion of rigor (Locker and Daines 1976). In other pieces left at 2°C for 48 hr for completion of rigor, the shear force was twice that of 15°C rigor controls, but in the cold-shortened muscle in which rigor was completed at 37°C, shear force reverted to the values of the 15°C controls. Thus, the temperature at the very last stages of rigor (the final 0.3 units of pH fall) was quite critical. This astonishing result implies that at exactly the same final shortening, shear force differed by a factor of two, according to whether rigor was completed at 2° or at 37°C. It was concluded that conditions during the last stages of rigor mortis are more important to tenderness than the rest of the postmortem history, and overshadow the effects of sarcomere length. Again, the effect did not appear to be due to aging.

We have made subsequent attempts to explain this disconcerting experiment, but with little success. Muscle that has gone into rigor at 37°C is often, but not invariably, paler in color and softer in texture, but the two effects need not go together (Locker and Daines 1975). The effect is somewhat reminiscent of PSE (pale, soft, exudative) pork. In a search of an explanation, we have subjected muscle to various rigor temperature regimes and examined the tensile and histological properties in the raw state (Locker and Wild 1982A) and after cooking (Locker et al. 1982). The results have not been very rewarding in explaining the phenomenon. We found (Locker and Wild 1982A) that yield point was distinctly lower after rigor at 37°C, in agreement with earlier observations (Locker and Daines 1975). The fact that a dramatic decline in yield values is associated with aging (Locker and Wild 1982A,B) has led us to wonder whether, in spite of our earlier assertions to the contrary, a degree of aging may be involved in producing heat-shortened muscle which is tender.

There are a number of reported cases where tenderness and sarcomere length do not relate well. Culler et al. (1978) found no significant difference in mean sarcomere length among three diverse groups of beef loin rectus muscles, (tough, intermediate, and tender), chilled and aged at 2°C. Again, the same laboratory found large variations in tenderness of loin steaks with degree of fatness at identical sarcomere lengths. The leaner (and faster cooling) carcasses were the tougher (Parrish et al. 1979). The effect of fat cover and cooling rate was examined in more detail by Lochner et al. (1980), who concluded that the tenderness of loin steaks was highly dependent on the muscle temperature attained at 2 hr postmortem, while cold shortening was not a significant factor, except in the leanest, most rapidly chilled carcasses. As these authors point out, their result recalls the claim of Roschen et al. (1950) for tenderizing beef by maintaining 37°C for the first 4–5 hr postmortem before chilling. The Wisconsin group extended their work by holding loins at 37°C for the first 3 hr after slaughter, with benefit to tenderness (Marsh et al. 1981). A slow rate of glycolysis during this period promoted tenderness; in fact, a premature plunge toward rigor pH, induced by electrical stimulation, eliminated the beneficial effect. They came to the revolutionary conclusion that "it is wrong to suppose that ageing commences only after the full achievement of rigor mortis . . ." and that "our observations are entirely compatible with the view that tenderising is due primarily to an enzyme or enzyme system that is highly (perhaps optimally) active at a near neutral pH and a temperature of about 37°C." This throws out an exciting new challenge to meat research on an issue that has long remained unresolved: when does aging begin? The answer, as far as shear force is concerned, appears to demand an impossible experiment. Tenderness relates to cooked meat, but cooking in a prerigor state involves such drastic modification of the myofibrils that the resultant material is sheer artifact. However, this does not exclude the possibility of demonstrating proteolysis in the raw state.

Marsh et al. (1981) may well be right since there is abundant evidence of neutral proteases in muscle. In the first few hours postmortem, beef muscles, apart from their anaerobic state, are likely to be in a near physiological condition in regard to temperature, pH, ATP, and Ca^{2+} concentration. Under these conditions, cathepsins will have low activity and indeed are likely still to be bound within the lysosomes.

The role of the neutral third protease "calcium-activated factor" (CAF) has become complicated by the discovery of a second enzyme with a much lower calcium requirement (50 μM for maximum activation) as well as an inhibitor, which is effective against both enzymes at the Ca^{2+} concentrations that activate them (Szpacenko et al. 1981). This is only a special case of a situation which now seems to be universal in the protein catabolism of cells, where two "calpains" and "calpastatin" are found, and show a close parallel to the behavior of calmodulin and its inhibitor (Murachi 1982). Free Ca^{2+} in the hot muscles seems likely to be at the the "relaxed" level ($10^{-7}M$) where neither form of CAF could be active.

Other reports of neutral proteases in muscle are too numerous to review here, but two of the better researched ones may be noted. Murakami and Uchida (1980) prepared from rat heart a serine protease which attacks a variety of structural proteins (myosin most rapidly, also desmin, but actin not at all). At pH 7 in EDTA, Z- and M-lines (see Chapter 6) were removed from myofibrils with fragmentation. It should be noted, however, that neutral proteolytic activity is far higher in the cardiac and skeletal muscle of the rat than in that of ox or of rabbit (Drabikowski et al. 1977). Another, apparently distinct, serine protease has been crystallized from both rat intestinal and rabbit skeletal muscle. It attacks myosin most rapidly and actin only slowly. At pH 8 in EDTA, it causes early loss of the Z-line in myofibrils without fragmentation, followed by disappearance of the A-band (Sanada et al. 1979).

Since aging must involve a higher level of proteolytic breakdown than in life, the question arises: What could be a trigger for this enhanced activity? The only obvious departure from *in vivo* conditions is the anaerobic state and it is not clear how this could affect a neutral protease.

In our experiment on cold-shortened meat, in which rigor was finished at 37°C, the effect could more accurately be claimed as a lack of toughening than as a tenderization, although it could be tenderization of a toughened material. The significant effect occurred in the pH range 5.8–5.5, more favorable to cathepsins than to neutral proteases. It has been claimed that rigor onset at higher temperatures favors release of lysosomal enzymes (Dutson et al. 1977), although the evidence is not very convincing.

There is further discussion of aging, for which a new mechanism is proposed, in the next section.

This section reveals a current situation, which is quite unsatisfactory, but points to important new avenues in research on tenderness. Clearly, more attention must be given to the effects of elevated temperature both in the early postmortem period and in the last stages of rigor. Cold shortening remains an important factor in tenderness, but it is now clear that sarcomere length cannot be taken in isolation as a measure of expected tenderness.

A STRUCTURAL BASIS FOR COLD TOUGHENING AND A NEW THEORY OF MEAT TENDERNESS

Having discussed the reasons why cold shortening occurs, it remains to offer some explanation for the resultant toughening, in terms of the protein filaments of the myofibril. An attempt to do this in terms of the sliding filament theory of contraction has already been made by Marsh and Carse (1974), whereas a rather different picture arises from the work of Voyle (1969). From this work emerges the notion that the sarcomere in contracture pattern is the culprit in toughening. Although there seems no reason to question this idea, a reassessment of the structural basis of toughness is due in light of recent evidence about filaments of muscle and their survival on cooking.

This new situation arises from advancing knowledge of the "gap filaments" of muscles, known for 21 years, but neglected as irrelevant to the sliding filament theory of contraction, now universally accepted. Recent advances in the morphology of gap filaments have come from observations on highly stretched beef fibers in this laboratory. A scheme for the connections of these gap filaments (hereafter referred to as "G-filaments") has been proposed (Fig. 1.11). A fortuitous but welcome by-product of this work was a growing realization that these filaments must be important to the tensile properties of cooked meat. A first step toward a new theory of meat tenderness in such terms was published seven years ago (Locker et al. 1977) but has not yet made any significant impact on the literature. More recent results have supported and developed these ideas to the point where it is now possible to claim that in the cooked myofibril the G-filaments are the only survivors, and hence the only contributors to structural continuity and tensile strength. All the known modifiers of tenderness can be explained in terms of these filaments. It is not proposed to discuss the evidence for the connections of G-filaments or to give a full account of their significance to tenderness, but rather to summarize the new ideas with emphasis on their relevance to the 0–40% shortening phase of the shortening-toughening diagram (Fig. 1.1) and to the earlier ideas on cold toughening. A comprehensive account of the new theory of tenderness in terms of G-filaments has recently been published (Locker 1982A) as well as a somewhat more concise version (Locker 1982B). Reference should be made to these for fuller detail and sources.

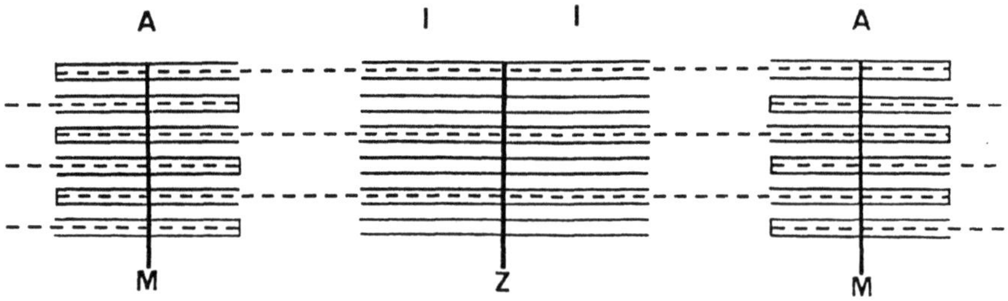

FIG. 1.11. A scheme for the connections of "G-filaments."
From Locker and Leet (1976).

The model (Fig. 1.11) for the connections of G-filaments (Locker and Leet 1976A), proposes that each G-filament forms a core to an A-filament, but emerges at one end only, passing between the I-filaments, through the Z-line, and between the I-filaments of the next sarcomere, to terminate as a second core to a thick filament. In this unusual symmetry, centered on the Z-line, G-filaments link half the A-filaments in an A-band to those in one neighboring sarcomere and half to those in the other. G-filaments are continuous through the Z-line, but not through the sarcomere, and therefore depend on adequate anchoring of their ends to maintain structural continuity under stress. They are resistant to a variety of powerful protein solvents, but dissolve in some in the presence of thiol reducers. They are vulnerable to indigenous or exogenous proteases. It seems that the protein of G-filaments is an unusual one, of very high molecular weight, and can be identified with the "connectin" of Maruyama (1980) or the "titin" of Wang et al. (1979). The biological functions of G-filaments are unclear.

Since prerigor muscle offers such slight resistance to stretch, G-filaments must be readily extensible (as is the crimped collagen net). The stiffness of rigor muscle is purely a function of myosin-actin bonds. This resistance fails at the "yield point" of muscle when the I-filaments tear out of the Z-line, but G-filaments simply stretch. If the muscle has been previously aged, both sets of filaments snap together when the muscle yields.

In considering tenderness, we are usually dealing with cooked meat.

The term "cooked" covers many degrees of denaturation, with large changes in the properties and relative importance of the different filaments. These changes have been fully discussed (Locker 1982A) but may be summarized as follows.

Within an hour at 60°C, "cooked character" has emerged to the extent that the myofibril may be regarded as cooked, while, by a narrow margin, the collagen net remains undenatured. The A-filaments have dissociated and coagulate, the I-filaments appear intact, although their strength is doubtful, while G-filaments have denatured, but remain strong and elastic (although less readily extensible than in the native state).

After an hour at 70°C, cooked character is largely complete. The steepest changes in hydration and length have occurred, although there are further substantial changes up to 80°C. The collagen net has undergone thermal shrinkage to a taut "crossed diagonal" configuration and the previously inelastic collagen fibers have become elasticized. Tensile strains now involve the stretching in unison of elastic gap filaments and pretensioned elastic connective tissue.

Curiously, no one seems to have asked how the I-band of cooked meat maintains its integrity, when I-filaments have disintegrated by the time 70°C is reached. In fact, a sparser array of gap filaments, often carrying the coagulated remains of I-filaments, survives in the I-band. The G-filaments, unlike collagen fibers, are highly resistant to further cooking.

In earlier ideas on the structural basis for the shortening-toughening relationship, the A-band has been given a major role. I shall now consider these ideas in relation to new evidence on the fate of structure within the A-band. The more important region is the 0–40% contraction range, which covers the degrees of shortening most likely to be encountered in commercial practice during the handling (or mishandling) of carcass or boned-out meats. Marsh and Carse (1974) interpreted their curve in terms of the relative positions of A- and I-filaments at specific degrees of shortening, to which a specific sarcomere length was assigned, based on a length at excision of 2.1 μm. It is implicit in their argument that muscle remains homogeneous in sarcomere length throughout the shortening range. While this is near the truth for the excised muscle, prerigor or after rigor at 15°C, it becomes less and less true as cold shortening progresses. Voyle (1969) observed that cold-shortened muscle contained a wide variety of contraction states. Some fibers were extremely contracted, showing contraction nodes with stretched or broken areas. Others were more uniformly contracted along the fiber, but the degree varied greatly from fiber to fiber, the least contracted being thrown into a crimp.

At a sarcomere length of 1.6 μm, the A-filaments touch the Z-line and at 1.5 μm or less, they crumple against it or penetrate it, to overlap A-filaments in the next sarcomere. The peak toughening actually occurs at a mean sarcomere length of 1.3 μm. Marsh and Carse (1974) appear to conclude that the overlap of A-filaments from adjoining sarcomeres confers the extreme of toughness. Although they barely discuss the changes induced by cooking, they seem to believe that the degree of actin-myosin interaction determines strength in rigor, and that this persists, determining strength in the cooked state. Voyle (1969) does not mention changes on cooking, but likewise appears to associate peak toughening with "compression of the myosin filaments by the Z-lines." Although he does not say so, one must assume from his demonstration of heterogeneity in contraction, that if the presence of sarcomeres showing contracture bands is the reason for the toughening, the rising phase of the toughening curve must be interpreted statistically as an increasing incidence of sarcomeres in this state.

It should be pointed out at this stage that cold toughening is seen only in cooked meat. Rhodes and Dransfield (1974) found that for beef sternomandibularis or semitendinosus in the raw rigor state, the strength per unit cross section as measured by shearing between blunt teeth declined steadily from 50% stretch to 50% contraction. It seems from their data that the degree of decline may be attributed simply to the decline in number of collagen and muscle fibers per cm^2. Davey and Gilbert (1975) found for ox sternomandibularis that samples that showed a marked peak in the shortening shear force curve in the cooked state showed only a decline in shear force with shortening when raw (Fig. 1.12). A more selective measure of myofibrillar strength is yield point, which also decreases with cold shortening, although when corrected for the packing density of fibers, the decrease is eliminated (Locker and Wild 1982A).

To summarize, there is a very limited literature on the structural basis of toughening, and this has paid rather scant attention to the modifications induced by cooking. The most satisfactory conclusion that can be drawn from it is that the toughening associated with cold shortening relates to a statistical decrease in sarcomeres containing I-bands as weak links, in favor of an increasing incidence of sarcomeres without I-bands, which cook to a strong continuum of A-bands. This interpretation of cold toughening is basically sound, but demands a new concept on the nature of the continuum.

This new view is based on recent experiments in which muscle, set in rigor under various conditions, including cold shortening, was cooked under restraint for 40 min at 80°C and stretched to its limit. It recovered and was then restretched to 50–70% extension for fixation. Whether the muscle had been shortened or not, the A-filaments had disappeared, leaving behind an array of fine parallel filaments embedded in fragments of coagulum (Locker and Wild 1982C). My interpretation of this result is that A-filaments are dissociated by heat into myosin molecules which then form an actomyosin gel with the actin from disintegrated I-filaments. Only the G-filament cores of the original A-filaments remain. These stretch elastically, whereas the gel fragments.

It is not surprising that A-filaments "explode." Studies on the fragments of myosin produced by proteolysis show that "heads" (S-1) may coagulate at a temperature as low as 35°C, whereas the "tails" (myosin rod) gel at 60°C in 0.6 M KCl, pH 6.0, with major loss of α-helical structure (Samejima *et al.* 1981). The fact that a mixed solution of actin and myosin forms a stronger gel on heating than a solution of myosin alone (Yasui *et al.* 1980) strongly suggests the survival of actomyosin links during coagulation. Thus, the matrix is likely to consist largely of coagulated actomyosin, with myosin-to-myosin links also playing a part. It is possible that links may be established between G-filaments and the other coagulating proteins.

The idea of the I-band as a weak link is compatible with the G-filament model (Fig. 1.11), since this requires twice as many G-filaments in the A-band as in the I-band. The sparser array in the I-band is confirmed in the actual photographs of unshortened, cooked muscle (Locker and Wild 1982C). The G-filament array of the A-band is, therefore, superior in strength because of a double ration of filaments and is further reinforced by a matrix of actomyosin gel.

Thus, the notion that strength in cold-shortened, cooked muscle is due to

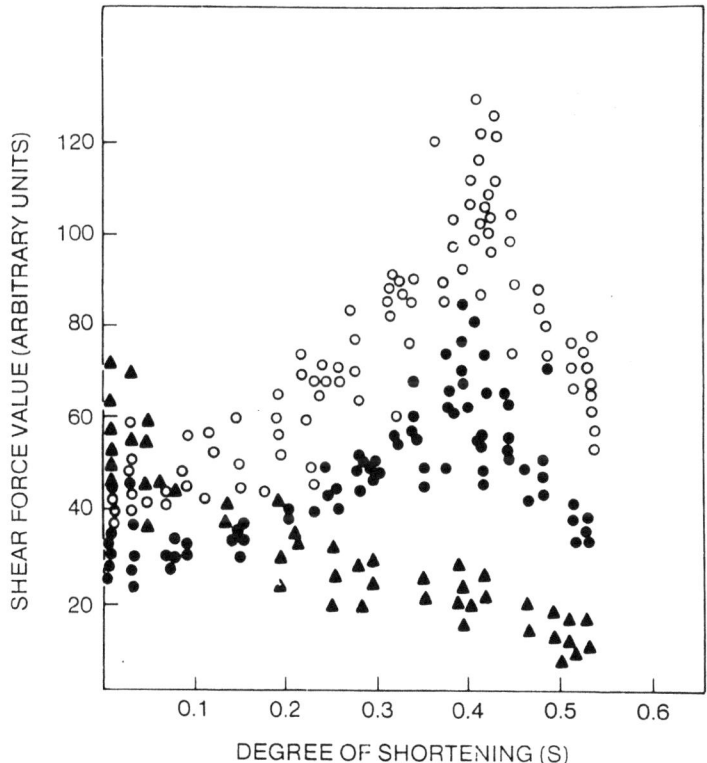

FIG. 1.12. Tenderness in relation to the cold-induced shortening in neck muscles from young ox. ○ Meat cooked to 80°C. ● Meat cooked to 60°C. ▲ Raw meat. *From Davey and Gilbert (1975).*

a continuum of denatured A-filaments must be replaced by the concept of a strong continuum of denatured G-filaments. In the earlier idea, strength depended on lateral adhesion of A-filaments from adjoining sarcomeres, either by preservation of actin-myosin links as Marsh and Carse (1974) believed, or by direct cross-linking of myosin heads in adjacent filaments during coagulation. The G-filaments should form a stronger continuum on denaturation than might have been expected of A-filaments, since although half depend only on overlap across the Z-line, the other half, according to the model, are truly continuous through the Z-line. Additional evidence for this was provided by Locker and Leet (1976B). However, since G-filaments each pass through only one Z-line, they cannot form a true long-range continuum but must depend on lateral linkages for continuity within the A-band. In the raw state, some such restraint certainly exists near the M-lines, since in grossly overstretched, prerigor fibers, the A-filaments become dislocated but never separate, retaining an overlap of 0.6 μm, even in an 11 μm sarcomere (Locker and Leet 1975).

The shortening range of 0–20% deserves closer scrutiny. The original version of the curve (Marsh and Leet 1966A) had few points in this region, but it was concluded that it represents a threshold region beyond which toughening increased rapidly. The diagrams in a more detailed paper (Marsh and Leet 1966B) where shortening was obtained by a variety of methods have plenty of points in this region and lead to the same conclusion (Fig. 1.1). However, the later curve of Marsh and Carse (1974) shows a marked rise in shear force from the beginning of shortening, although the gradient is steeper in the 20–30% region. This latter situation seems to be the case in subsequent curves presented by Davey and co-workers (Davey *et al.* 1967, 1976; Davey and Gilbert 1975), (see Fig. 1.12 and 1.13). All these curves are based on the same muscle, beef sternomandibularis, but there was a difference in cooking technique, the Marsh curves being derived from meat cooked slowly in a bath rising over 1 hr to 80°C, whereas the Davey curves were derived from meat cooked in a water bath at 80°C for 40 min. If the toughening is purely a function of an increasing incidence of sarcomeres of 1.5 μm or less, it might be expected that there would be little effect at first, since such sarcomeres would only begin to appear after appreciable shortening but rapidly become abundant in the later stages (20% shortening is equivalent to a mean sarcomere length of 1.7 μm and 29% shortening to 1.5 μm). Thus, a threshold effect might be expected.

It is interesting in the respect that in two diagrams that include aged meat (Fig. 1.2 and 1.13), there is no toughening at all in the shortening range of 0–20% in aged meat, but a steep rise begins with increasing contraction. In the same meat, unaged, shear force values increase from the beginning of contraction. The flatness of the 0–20% region of the original Marsh and Leet (1966A) curves may also have been due to accelerated aging during the process of slow cooking (simulated roasting). The elimination of the 0–20% rise by aging may be interpreted in the following

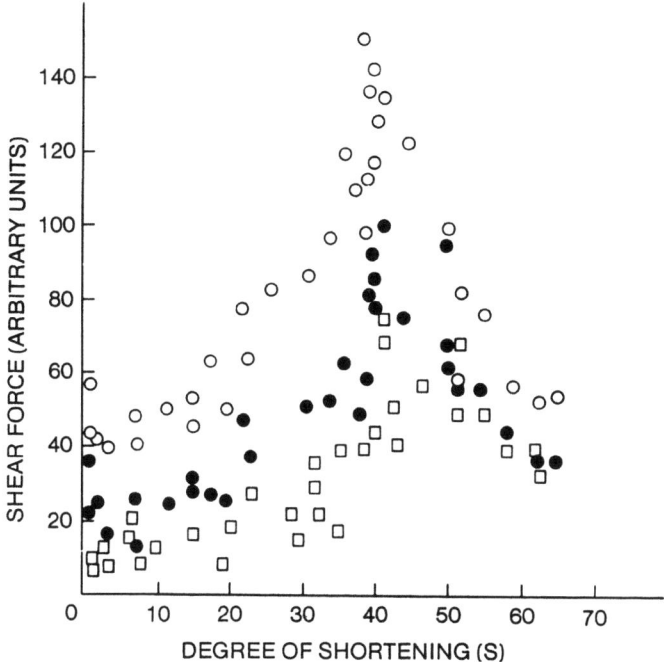

FIG. 1.13. The relationship between degree of shortening (S) and SF values for beef neck muscle. ○ Unaged meat cooked 40 min, 80°C. ● Unaged meat cooked 24 hr, 80°C. □ Meat aged 3 days at 15°C, then cooked 24 hr, 80°C.
From Davey and Gilbert (1975).

way. The extreme aging used (3 days at 15°C) would completely destroy G-filaments where I-bands exist and hence demolish the myofibrillar contribution. The sudden rise in shear force beyond 20% marks the emergence of sarcomeres in contracture pattern, which resist aging and hence contribute to shear force. This degree of contraction, corresponding to a mean sarcomere length of 1.7 μm, seems a reasonable stage for the contracture pattern to become significant. It follows that the increase in shear force in unaged meat prior to this stage, that is, in the 0–20% region, must be a function of myofibrils with distinct but progressively shrinking I-bands. Marsh and Carse (1974), by their mathematical treatment of their curve in this region, also imply that a short I-band leads to greater toughness than a long one. A possible new explanation for this arises from the G-filament theory. The tensile strength of the heat-denatured G-filament, which alone provides continuity in a cooked I-band, may well vary with degree of stretch at the time of denaturation. Stretch must reduce diameter and possibly strength at the same time.

Further consistencies with the G-filament theory of tenderness may be gleaned from the Davey curves. I have earlier claimed that G-filaments denature steadily within an hour at 60°C, before collagen undergoes thermal shrinkage. Conveniently, Davey and Gilbert (1975) present diagrams that compare meat cooked 40 min at 60° and at 80°C (Fig. 1.12). The 60°C curve follows exactly the shape of the 80°C curve, but at lower shear values. Peak toughness remains at the same degree of shortening. The curves shown are for young ox neck muscle, but a similar result was obtained for old bull neck muscle. If the G-filaments determine the properties of cooked myofibrils, then it seems reasonable to conclude that the essential changes leading to the characteristic curve have been achieved by a mild cooking, sufficient to denature G-filaments, but not the collagen net, which remains slack and contributes little to the shear force measurement on the MIRINZ tenderometer. This results in an overall lowering of values. Another way of removing the collagen contribution is by long cooking, and it is significant that exactly the same result as just described from cooking at 60°C is obtained from meat cooked 24 hr at 80°C. (Fig. 1.12). Again, the shortening–toughening curve is parallel to, but lower than, that for meat cooked for only 40 min at 80°C (Davey et al. 1976).

This last result emphasizes another property of G-filaments, their extreme resistance to heat. In fact, shear force in this muscle hardly changes between 3 and 5 hr at 100° or 120°C, or between 10 and 24 hr at 80°C (Davey and Niederer 1977). The stability of G-filaments, as actually seen in the "gap," has also been demonstrated for muscle stretched to the maximum and cooked 4 hr at 100°C (Locker et al. 1977).

Any theory of tenderness must be able to account for the effects of aging, and for its lack of effect on cold-shortened meat. The current view is that the Z-lines are particularly vulnerable to aging, and in fact ultimately disappear, leaving a line of weakness (Davey and Gilbert 1967). This effect has been attributed to the action of a "Ca^{2+}-activated factor" (CAF) (Busch et al. 1972).

I believe the disappearance of the Z-line is a largely irrelevant event, for the following reasons. Firstly, most of the tenderization has occurred before this happens, as, for example, in the experiment of Davey and Graafhuis (1976), which is described in the following paragraph. Secondly, Hattori and Takahashi (1982) have recently distinguished two events in the Z-line during autolysis of rabbit myofibrils, using calpastatin, the endogenous inhibitor of calpain or CAF (Murachi 1983). Removal of the Z-line with release of α-actinin occurred optimally at pH 7.0 and in mM Ca^{2+}, as was true for the caseinolytic activity of calpain. However, the weakening at the Z-line, as manifested by easier fragmentation of myofibrils at this site, occurred optimally at pH 6.5 and 0.1 mM Ca^{2+}. Hattori and Takahashi (1982) concluded that the weakening at the Z-line is the predominant effect in postmortem muscle and is not due to the proteolytic activity of calpain. Finally, on my model, the G-filaments are continuous through the Z-line, and if these in fact determine tenderness, the removal of Z-line material should not matter. Severe damage to G-filaments at pH

7 in the presence of EDTA has been shown when glycerinated stretched fibers are incubated (Locker et al. 1977), and this cannot be due to CAF.

The first clues to the involvement of G-filaments in meat tenderness came from the micrographs of Davey and Graafhuis (1976), who showed that, in fully stretched sternomandibularis muscle, gap filaments survived aging in the raw meat but disappeared on cooking, whereas, in unaged muscle, they were intact after cooking. It may be noted that the severe aging (3 days at 15°C) left the Z-lines apparently intact. This experiment was repeated using stretched glycerinated muscle, incubated in a crude muscle protease extract. Similar results were obtained on cooking, whether the incubation occurred in the presence of Ca^{2+} or of EDTA (Locker et al. 1977).

Attack on G-filaments also occurs in muscle at normal sarcomere lengths, as demonstrated by the drastic drop in yield point on aging. Histological study showed this to be due to a clean break across the A-I junction, in contrast to yield in unaged muscle where I-filaments pull out of the Z-line as G-filaments stretch (Locker and Wild 1982A). Thus, it seems that the mechanism of aging is an attack on G-filaments in the I-band where they are readily accessible to proteases. Although G-filaments seem vulnerable to proteases in general (Locker et al. 1977), it seems most likely that cathepsins are the principal agents in the aging of meat postrigor.

The failure of cold-shortened meat to age can also be explained by the disappearance of I-bands, leaving the G-filaments located entirely within the A-filaments as cores, where they are fully protected by a tight sheath of the resistant α-helical tails of myosin. This has been discussed more fully elsewhere (Locker 1982A).

Failure of cold-shortened meat to age in terms of shear force disguises the fact that proteolysis is active. We have found (Locker and Wild 1984B) that the most sensitive criterion of aging, yield point, shows a decline when cold-shortened meat is aged, almost as drastic as in unshortened meat. The aged cold-shortened muscles when cooked showed only a slight fall in shear force. Thus, it appears that I-filaments, the determinants of yield point, are attacked within the A-band, but this is irrelevant in the cooked state, where undamaged G-filaments survive.

The only treatment effective in restoring tenderness to cold-shortened meat appears to be the pressure-heat (P-H) treatment described by Australian workers. High pressure tenderizes prerigor meat in a few minutes at ambient temperature. It also tenderizes rigor meat if longer times and elevated temperatures are used. Only rigor meat will be considered here. The results are described in a series of papers, of which the most recent is by Macfarlane et al. (1981). Pressures of 100–150 MPa at 50°–60°C for periods of 30–60 min are effective. The more rigorous treatments reduce shear force of cold-shortened meat to the level of aged unshortened meat, whereas temperature controls remain unchanged. Prior aging does not enhance the effect, and it was suggested that the same structure is attacked in both treatments.

We have recently shown that a relatively mild P-H treatment (60 MPa, 20 min, 55°C) suffices to cause G-filaments to fracture during yield point testing of raw strips, while the filaments remain quite intact in 55°C controls (Locker 1982A; Locker and Wild 1984A). More rigorous treatment (60 MPa, 1 hr, 60°C) produced some histological changes in both shortened and unshortened meat, whether cooked 40 min at 80°C or not, but in no case was structural continuity lost. However, weakening of essential structure was clearly evident in greater extensibilities and lower breaking loads. After such P-H treatment and cooking, unshortened muscle broke between the N_2-line and the Z-line; treated cold-shortened muscle always broke at the M-line (presumably because of less actomyosin reinforcement there), whereas in controls, breaks were fairly evenly distributed between the M-line and the Z-line.

It seems reasonable to conclude that the treatment works by weakening G-filaments. Connective tissue does not appear to be affected. It is still unclear whether the mechanism is a purely physical effect such as depolymerization, or whether a proteolytic attack accelerated by elevated temperature is involved, for example, in cold-shortened muscle, where G-filaments might be stripped of their protective sheath by disassembly of myosin. The situation is a complex one in which depolymerization by pressure of various filament proteins, their progressive heat denaturation, and proteolytic activity are all likely to be occurring simultaneously. We have found that tenderizing stops in cold-shortened meat with release of pressure at 20 min, but usually continues in unshortened meat if 60°C is maintained. The Australians found pre-incubation at 60°C for an hour or more made the meat immune to P-H treatment, and this I have earlier claimed to be due to denaturation of the G-filaments, which are thus rendered insensitive to pressure. The method is an interesting research tool and merits further study.

It seems fair to claim that the hypotheses based on G-filaments presented in this section offer reasonable mechanisms for the known moderators of meat tenderness. The only exception at the present time is a failure to account for the tenderness of meat that has cold shortened but has completed rigor onset at 37°C.[2]

CONCLUSION

This chapter was not intended to be a comprehensive review of a subject, which now has a rather large literature. It is an attempt to give a histor-

---

[2]Since this was written, gel electrophoresis studies, here and elsewhere, on changes in the very high-molecular-weight proteins of the myofibril during aging and subsequent cooking indicate a need for some revision of the G-filament theory of tenderness. This may have to wait until new notions about the third filaments of the sarcomere have become clearer (Lusby et al. 1983; King 1984; Locker and Wild 1984c; Locker 1984).

ical perspective of the development of a distinct aspect of meat tenderness, to provide a more readable account than an earlier review (Locker *et al.* 1975), and to bring it up-to-date in terms of theory and practice.

Cold-induced toughening has served to transfer attention from connective to contractile tissue, and because its exploration has coincided with tremendous strides in the understanding of fine structure and function in muscle, it has also served as a stimulus in relating the academic advances to meat tenderness. The G-filament theory proposed here is based on tensile properties of the myofibril, which are of course the most relevant to cold toughening. It has, however, been acknowledged elsewhere (Locker 1982A,B) that lateral strength and adhesion in muscles must also be considered in a comprehensive theory of meat tenderness.

The history of cold toughening is a salutary reminder that progress in engineering, which has made fast refrigeration possible, and the zeal of the veterinarian regulators, who insist on its desirability, is positively dangerous when applied without due regard to the responses of a sensitive and still-living tissue. Ironically, it implies that in practice, over the last 20 years or so, the consumer whose meat has come from the most poorly equipped abattoirs has had the best eating, and may yet, until electrical stimulation becomes universal. At present in New Zealand, electrical stimulators can be seen in routine use in a large modern packing house on export lines, while in the same plant the home kill goes unstimulated to efficient chillers, to be converted to second-rate products for the local consumer. Such anachronisms must pass.

The inspiration of W. A. Carse has developed into the most significant practical advance in meat technology in recent times. Few innovations in meat handling have spread around the world so quickly. In the end it may well change the face of the meat industry. An interest in hot cutting, linked to shrink-wrapping, seems to be emerging in parallel with interest in electrical stimulation. The concept of on-line hot cutting as the normal routine process for a packing house no longer seems a pipe dream. The stimulated plunge into rigor not only protects the excised product from cold toughening but is so rapid that aging may begin while the meat is hot and already securely encapsulated in plastic.

Whether or not such exciting possibilities become the future pattern of the industry remains to be seen. What does seem certain is that, to remain competitive, packing houses will take more trouble to ensure that the potential for consumer pleasure in the animal that walks into the yards is fully realized. The investment in time, energy, and money that goes into raising the beast, particularly on a feedlot, demands no less. As a matter of routine, cold-induced toughening can and will be eliminated from meat processing by intelligent use of the elegant methods now available. The outcome of a long tussle with a technology-created problem in remote sheep-farming islands should prove an enduring benefit to both the meat industry and the consumer.

REFERENCES

BAXTER, R.I., BOUTON, P.E., FISHER, A.L. and HARRIS, P.V. 1972. Evaluation of hanging methods for improving the tenderness of beef and mutton. CSIRO Meat Res. Rep. 2/72.

BENDALL, J.R. 1951. The shortening of rabbit muscle during rigor mortis: Its relation to the breakdown of ATP, creatine phosphate and to muscular contraction. J. Physiol. 114, 71.

BENDALL, J.R. 1972. Consumption of oxygen by the muscles of beef animals and related species and its effects on the colour of meat. I. Oxygen consumption in prerigor meat. J. Sci. Food Agric. 23, 61.

BOUTON, P.E., FISHER, A.L., HARRIS, P.V. and BAXTER, R.I. 1973. A comparison of the effects of some post-slaughter treatments on the tenderness of beef. J. Food Technol. 8, 39.

BOWLING, R.A., SMITH, G.C., CARPENTER, Z.L., DUTSON, T.R. and OLIVER, W.M. 1977. Comparison of forage-finished and grain-finished beef carcasses. J. Anim. Sci. 45, 209.

BRUNTON, W.G. and GILBERT, K.V. 1972. Interim studies in lamb tenderness. Evaluation of lamb conditioned and aged at 10°C and 7°C. Meat Ind. Res. Inst. N.Z. Publ. 275.

BUCHTER, L. 1972. The influence of chilling temperature on the toughness of loin muscles from young calves and bulls. In Symp. 2. p. 451. Meat Res. Inst., Langford, England, 1972.

BUEGE, D.R. and MARSH, B.B. 1975. Mitochondrial calcium and post mortem muscle shortening. Biochem. Biophys. Res. Comm. 65, 478.

BUSCH, W.A., PARRISH, F.C. and GOLL, D.E. 1967. Molecular properties of post mortem muscle. 4. Effect of temperature on ATP degradation, isometric tension and shear resistance of bovine muscle. J. Food Sci. 32, 390.

BUSH, W.A., STROMER, M.H., GOLL, D.E. and SUZUKI, A. 1972. Ca^{++} specific removal of Z-lines from rabbit skeletal muscle. J. Cell Biol. 52, 367.

CARSE, W.A. 1973. Meat quality and the acceleration of post mortem glycolysis by electrical stimulation. J. Food Technol. 8, 163.

CASSENS, R.G. and NEWBOLD, R. P. 1967. Temperature dependence of pH changes in ox muscles post-mortem. J. Food Sci. 32, 13.

CHRYSTALL, B.B. and HAGYARD, C.J. 1975. Accelerated conditioning of lamb. N.Z. J. Agric. 130 (6) 7.

CORNFORTH, D.P., PEARSON, A.M. and MERKEL, R.A. 1980. Relationship of mitochondria and sarcoplasmic reticulum to cold shortening. Meat Sci. 4, 103.

CULLER, R.D., PARRISH, F.C., SMITH, G.C. and CROSS, H.R. 1978. Relationship of myofibril fragmentation index to certain chemical, physical and sensory characteristics of bovine muscle. J. Food Sci. 43, 1177.

DAVEY, C.L. 1970. Beef processing and aging. Proc. 12th Meat Ind. Res. Inst. N.Z. Conf. Publ. 199, 73, Hamilton, New Zealand, 1982.

DAVEY, C.L. and CURSON, P. 1971. Interim studies in lamb tenderness. 1. Evaluation of high temperature conditioning and aging of lamb. 2. The effect of carcass posture on tenderness of lamb muscles. Meat Ind. Res. Inst. N.Z. Publ. 215.

DAVEY, C.L. and GARNETT, K.J. 1980. Rapid freezing, frozen storage and the tenderness of lamb. Meat Sci. 4, 319.

DAVEY, C.L. and GILBERT, K.V. 1967. Structural changes in meat during aging. J. Food Technol. 2, 57.

DAVEY, C.L. and GILBERT, K.V. 1973. The effect of carcass posture on cold, heat and thaw shortening in lamb. J. Food Technol. 8, 445.

DAVEY, C.L. and GILBERT, K.V. 1974A. The mechanism of cold-induced shortening in beef muscle. J. Food Technol. 9, 51.

DAVEY, C.L. and GILBERT, K.V. 1974B. Carcass posture and tenderness in frozen lamb. J. Sci. Food Agric. 25, 923.

DAVEY, C.L. and GILBERT, K.V. 1975. The tenderness of cooked and raw meat from young and old animals. J. Sci. Food Agric. 26, 953.
DAVEY, C.L. and GILBERT, K.V. 1976. Thaw contracture and the disappearance of ATP in frozen lamb. J. Sci. Food Agric. 27, 1085.
DAVEY, C.L. and GRAAFHUIS, A.E. 1976. Structural changes in beef muscle during ageing. J. Sci. Food Agric. 27, 301.
DAVEY, C.L. and NIEDERER, A.F. 1977. Cooking tenderising in beef. Meat Sci. 1, 271.
DAVEY, C.L. and WINGER, R.J. 1980. The structure of skeletal muscle and meat toughness. In Fibrous Proteins: Scientific, Industrial and Medical Aspects, Vol. 1. p. 97. D.A.D. Parry and L.K. Creamer (Editors). Academic Press, London.
DAVEY, C.L., KUTTLE, H. and GILBERT, K.V. 1967. Shortening as a factor in meat aging. J. Food Technol. 2, 53.
DAVEY, C.L., NIEDERER, A. F. and GRAAFHUIS, A.E. 1976. Effects of ageing and cooking on tenderness of beef muscle. J. Sci. Food Agric. 27, 251.
DE FREMERY, D. and POOL, M.F. 1960. Biochemistry of chicken muscle as related to rigor mortis and tenderization. Food Res. 25, 73–87.
DRABIKOWSKI, W., GORÉCKA, A. and JAKUBIEC-PUKA, A. 1977. Endogenous proteases in vertebrate skeletal muscle. Int. J. Biochem. 8, 61.
DUTSON, T.R., YATES, L.F., SMITH, G.C., CARPENTER, Z.L. and HOSTETLER, R.L. 1977. Rigor onset before chilling. Proc. Recip. Meat Conf. 30, 79.
GRAINGER, B.L. and LAZARIDES, E. 1978. The existence of an insoluble Z-disc scaffold in chicken skeletal muscle. Cell 15, 1253.
HATTORI, A. and TAKAHASHI, K. 1982. Calcium-induced weakening of the Z-disc. J. Biochem. 92, 381.
HERRING, H.K., CASSENS, R.G. and BRISKEY, E.J. 1965. Further studies on bovine tenderness as influenced by carcass position, sarcomere length, and fiber diameter. J. Food Sci. 30, 1049.
HERRING, H.K., CASSENS, R.G., SUESS, G.G., BRUNGARDT, V.H. and BRISKEY, E.J. 1967. Tenderness and associated characteristics of stretched and contracted bovine muscles. J. Food Sci. 32, 317.
HOSTETLER, R.L., LINK, B.A., LANDMANN, W.A. and FITZHUGH, H.A. 1972. Effect of carcass suspension on sarcomere lengths and shear force of some major bovine muscles. J. Food Sci. 37, 132.
KHAN, A.W. 1971. Effect of temperature during post-mortem glycolysis and dephosphorylation of high energy phosphate on poultry meat tenderness. J. Food Sci. 36, 120.
KHAN, A.W. and BALLANTYNE, W.W. 1973. Post slaughter pH variation in beef. J. Food Sci. 38, 710.
KHAN, A.W. and LENTZ, C.P., 1973. Influence of ante-mortem glycolysis and dephosphorylation of high energy phosphates on beef ageing and tenderness. J. Food Sci. 38, 56.
KING, N.L. 1984. Breakdown of connectin during cooking of meat. Meat Sci. 11, 27.
LOCHNER, J.V., KAUFFMAN, R.G. and MARSH, B.B. 1980. Early post mortem cooling rate and beef tenderness. Meat Sci. 4, 227.
LOCKER, R.H. 1959. Striation patterns of ox muscle in rigor mortis. J. Biophys. Biochem. Cytol. 6, 419.
LOCKER, R.H. 1960. Degree of muscular contraction as a factor in the tenderness of beef. Food Res. 25, 304.
LOCKER, R.H. 1963. Unpublished results. MIRINZ, P.O. Box 617, Hamilton, New Zealand.
LOCKER, R.H. 1982a. A new theory of meat tenderness, based on the gap filaments. Proc. Recip. Meat Conf. 35, 92.
LOCKER, R.H. 1982B. A new basis for meat tenderness in terms of gap filaments. Proc. 28th Congr. Eur. Meat Res. Workers, Madrid, 1982, 117.

LOCKER, R.H. 1984. The role of gap filaments in muscle and meat. Food Microstruct. 3, 17.
LOCKER, R.H. and DAINES, G.J. 1975. Rigor mortis in beef sternomandibularis muscle at 37°C. J. Sci. Food Agric. 26, 1721.
LOCKER, R. H. and DAINES, G.J. 1976. Tenderness in relation to the temperature of rigor onset in cold shortened beef. J. Sci. Food Agric. 27, 193.
LOCKER, R.H. and HAGYARD, C.J. 1963. A cold shortening effect in beef muscles. J. Sci. Food Agric. 14, 787.
LOCKER, R.H. and LEET, N.G. 1975. Histology of highly stretched beef muscle. I. The fine structure of grossly stretched single fibres. J. Ultrastruct. Res. 52, 64.
LOCKER, R.H. and LEET, N.G. 1976A. Histology of highly stretched beef muscle. II. Further evidence on the nature and location of the G-filaments. J. Ultrastruct. Res. 55, 157.
LOCKER, R.H. and LEET, N.G. 1976B. Histology of highly-stretched beef muscle. IV. Evidence for movement of gap filaments through the Z-line, using the N_2-line and M-line as markers. J. Ultrastruct. Res. 56, 31.
LOCKER, R.H. and WILD, D.J.C. 1982A. Yield point in raw beef muscle. The effects of ageing, rigor temperature and stretch. Meat Sci. 7, 45.
LOCKER, R.H. and WILD, D.J.C. 1982B. A machine for measuring yield point in raw meat. J. Texture Stud. 13, 71.
LOCKER, R.H. and WILD, D.J.C. 1982C. Myofibrils of cooked meat are a continuum of gap filaments. Meat Sci. 7, 61.
LOCKER, R.H. and WILD, D.J.C. 1984A. "Ageing" of cold shortened meat depends on the criterion. Meat Sci. 10, 235.
LOCKER, R.H. and WILD, D.J.C. 1984B. Tenderization of meat by pressure-heat treatment involves weakening of the gap filaments in the myofibril. Meat Sci. 10, 207.
LOCKER, R.H. and WILD, D.J.C. 1984C. The fate of the large proteins of the myofibril during tenderising treatments. Meat Sci. 11, 89.
LOCKER, R.H., DAVEY, C.L., NOTTINGHAM, P.M., HAUGHEY, D.P. and LAW, N.H. 1975. New concepts in meat processing. Adv. Food Res. 21, 157.
LOCKER, R.H., DAINES, G.J. and LEET, N.G. 1976. Histology of highly-stretched beef muscle. III. Abnormal contraction patterns in ox muscle, produced by overstretching during pre-rigor blending. J. Ultrastruc. Res. 55, 173.
LOCKER, R.H., DAINES, G.J., CARSE, W.A. and LEET, N.G. 1977. Meat tenderness and the gap filaments. Meat Sci. 1, 87.
LOCKER, R.H., WILD, D.J.C. and DAINES, G.J. 1982. Tensile properties of cooked beef in relation to rigor temperature and tenderness. Meat Sci. 8, 283.
LUSBY, M.L., RIDPATH, J.F., PARRISH, F.C. and ROBSON, R.M. 1983. Effect of post mortem storage on degradation of the myofibrillar protein, titin, in bovine longissimus muscle. J. Food Sci. 48, 1787.
LUYET, B.J. 1966. Behaviour of muscle subjected to freezing and thawing. In Physiology and Biochemistry of Muscle as a Food. p. 353. E.J. Briskey, R.G. Cassens and J.C. Trautman (Editors). Univ. of Wisconsin Press, Madison.
MACFARLANE, J.J., McKENZIE, I.J. and TURNER, R.H. 1981. Pressure treatment of meat. Effects on thermal transitions and shear values. Meat Sci. 5, 307.
MACFARLANE, P.G. and MARER, J.M. 1966. An apparatus for determining the tenderness of meat. Food Technol. 20, 838.
MARSH, B.B. 1952. Observations on rigor mortis in whale muscle. Biochim. Biophys. Acta 9, 127.
MARSH, B.B. 1954. Rigor mortis in beef. J. Sci. Food Agric. 5, 70.
MARSH, B.B. 1972. Post mortem shortening and meat tenderness. Proc. Meat Ind. Res. Conf., Chicago, 109.
MARSH, B.B. and CARSE, W.A. 1974. Meat tenderness and the sliding filament hypothesis. J. Food Technol. 9, 129.

MARSH, B.B. and LEET, N.G. 1966A. Resistance to shearing of heat-denatured muscle in relation to shortening. Nature *211*, 635.
MARSH, B.B. and LEET, N.G. 1966B. Studies in meat tenderness. III. The effects of cold shortening on tenderness. J. Food Sci. *31*, 450.
MARSH, B.B. and THOMPSON, J.F. 1958. Rigor mortis and thaw rigor in lamb. J. Sci. Food Agric. *9*, 417.
MARSH, B.B., WOODHAMS, P.R. and LEET, N.G. 1968. Studies in meat tenderness. 5. The effects on tenderness of carcass cooling and freezing before the completion of rigor mortis. J. Food Sci. *33*, 12.
MARSH, B.B., LEET, N.G. and DICKSON, M.R. 1974. The ultrastructure and tenderness of highly cold shortened muscle. J. Food Technol. *9*, 141.
MARSH, B.B., LOCHNER, J.V., TAKAHASHI, G. and KRAGNESS, D.D. 1981. Effects of early post mortem pH and temperature on beef tenderness. Meat Sci. *5*, 479.
MARTINOSI, A. and FERETOS, R. 1964. Sarcoplasmic reticulum. 1. The uptake of Ca^{++} by sarcoplasmic reticulum fragments. J. Biol. Chem. *243*, 61.
MARUYAMA, K. 1980. Elastic structure of connectin in muscle. In Regulatory Mechanisms of Muscle Contraction. S. Ebashi *et al.* (Editors). Japan Sci. Soc. Press, Tokyo. (Springer-Verlag, Berlin.)
McCRAE, S.E., SECCOMBE, C.G., MARSH, B.B. and CARSE, W.A. 1971. Studies in meat tenderness. 9. The tenderness of various lamb muscles in relation to their skeletal restraint and delay before freezing. J. Food Sci. *36*, 566.
McLEOD, K., GILBERT, K.V., WYBORN, R., WENHAM, L.M., DAVEY, C.L. and LOCKER, R.H. 1973. Hot cutting of lamb and mutton. J. Food Technol. *8*, 71.
McLEOD, K., GILBERT, K.V., FAIRBAIRN, S.J. and LOCKER, R.H. 1974. Further experiments on the hot cutting of lamb. J. Food Technol. *9*, 179.
MERKEL, R.A. and PEARSON, A.M. 1975. Meat Ind. *27*, 62.
MICKELSON, J.R. 1982. Ph.D. Thesis. Univ. of Wisconsin, Madison.
MICKELSON, J.R. and Marsh, B.B. 1980. Calcium uptake and release by skeletal muscle mitochondria. Cell Calcium *1*, 119.
MURACHI, T. 1983. Calpain and calpastatin. Trends Biochem. Sci., May, 167.
MURAKAMI, U. and UCHIDA, K. 1980. Ultrastructural alteration of rat cardiac myofibrils caused by a myosin-cleaving protease. J. Biochem. *88*, 877.
NEWBOLD, R.P. 1966. Changes associated with rigor mortis. In The Physiology and Biochemistry of Muscle as a Food. p. 213. E.J. Briskey, R.G. Cassens and J.C. Trautman (Editors). Univ. of Wisconsin Press, Madison.
NEWBOLD, R.P. 1979. Calcium uptake and release by skeletal muscle mitochondria and sarcoplasmic reticulum. Proc. Recipr. Meat Conf. *32*, 70
NEWBOLD, R.P. and HARRIS, P.V. 1972. Effect of pre-rigor changes on meat tenderness. A review. J. Food Sci. *37*, 337.
PARRISH, F.C., VANDELL, C.J. and CULLER, R.D. 1979. Effect of maturity and marbling on the myofibril fragmentation index of bovine longissimus muscle. J. Food Sci. *44*, 1668.
PATRIARCA, P. and CARAFOLI, E. 1969. A comparative study of the intracellular Ca^{++} movements in red and white muscle. Experientia *25*, 598.
PERRY, S.V. 1950. Studies on the rigor resulting from the thawing of frozen frog sartorius muscle. J. Gen. Physiol. *33*, 563.
QUARRIER, E., CARPENTER, Z.L. and SMITH, G.C. 1972. A physical method to increase tenderness in lamb carcasses. J. Food Sci. *37*, 130.
RAMSBOTTOM, J.M., STRANDINE, E.J. and KOONZ, C.H. 1945. Comparative tenderness of representative beef muscles. Food Res. *10*, 497.
RHODES, D.N. and DRANSFIELD, E. 1974. Mechanical strength of raw beef from cold shortened muscles. J. Sci. Food Agric. *25*, 1163.
ROSCHEN, H.L., ORTSCHEID, B.J. and RAMSBOTTOM, J.M. 1950. Tenderizing meats. U.S. Pat. 2,519,931. Aug. 22.

SAMEJIMA, K., ISHIOROSHI, M. and YASUI, T. 1981. Relative roles of the head and tail portions of the molecule in the heat induced gelation of myosin. J. Food Sci. *46*, 1412.

SANADA, Y., YASOGAWA, N. and KATUNUMA, N. 1979. Effect of a serine protease on isolated myofibrils. J. Biochem. *85*, 481.

SCHMIDT, G.R. and GILBERT, K.V. 1970. The effect of muscle excision before the onset of rigor mortis on the palatability of beef. J. Food Technol. *5*, 331.

SEIDEMANN, S.C., SMITH, G.C., DUTSON, T.R. and CARPENTER, Z.L. 1979. Physical, chemical and palatability traits of electrically stimulated, hot-boned, vacuum-packaged beef. J. Food Protect. *42*, 651.

SHARP, J.G. and MARSH, B.B. 1953. Whalemeat; Production and preservation. Spec. Rep. Food Invest. Bd., London *58*, 33.

SMITH, G.C., DUTSON, T.R., HOSTETLER, R.L. and CARPENTER, Z.L. 1976. Fatness, rate of chilling and tenderness of lamb. J. Food Sci. *41*, 748.

STOUFFER, J.R., BUEGE, D.R. and GILLIS, W.A. 1971. Meat tenderising method. U.S. Pat. 3,579,716. May 25.

SZPACENKO, A., KAY, J., GOLL, D.E. and OTSUKA, Y. 1981. A different form of the Ca^{++} dependent proteinase activated by micromolar levels of Ca^{++}. *In* Proteinases and Their Inhibitors. p. 151. V. Turk and L. Vitale (Editors). Pergamon Press, Oxford.

VOYLE, C.A. 1969. Some observations on the histology of cold shortened muscle. J. Food Technol. *4*, 275.

WANG, K., McCLURE, J. and TU, A. 1979. Titin: Major myofibrillar components of striated muscle. Proc. Natl. Acad. Sci. U.S.A. *76*, 3698.

WENHAM, L.M., FAIRBAIRN, S.J., McLEOD, K., CARSE, W.A., PEARSON, A.M. and LOCKER, R.H. 1973. The eating quality of mutton compared with lamb and its relationship to freezing practice. J. Anim. Sci. *36*, 1081.

WHITING, R.C. 1980. Calcium uptake by bovine mitochondria and sarcoplasmic reticulum. J. Food Sci. *45*, 288.

WILSON, G.D., BROWN, P.D., CHESBRO, W.R., GINGER, B. and WEIR, C.E. 1960. The use of antibiotics and gamma irradiation in the ageing of steaks at high temperatures. Food Technol. *14*, 143.

YASUI, T., ISHIOROSHI, M. and NAKANO, H. 1980. Heat induced gelation of myosin in the presence of actin. J. Food Biochem. *4*, 61.

2

Postmortem Conditioning of Meat[1]

T. R. Dutson[2] and A. M. Pearson[3]

Cold Toughening
High Temperature Conditioning
Cooler Conditioning
Electrical Stimulation
Summary
References

Holding unprocessed meat for various lengths of time at temperatures above freezing has commonly been referred to as "aging" or "conditioning" (Lawrie 1979). Since the term "aging" has often been confused with physiological aging, "conditioning" is a more preferred term (Penny 1980) and will be used in this chapter to denote the holding of carcasses or meat cuts at various temperatures above freezing to cause improvements in meat quality. "Cooler conditioning" should be used when referring to studies where carcasses are held at 0°–5°C, and "high temperature conditioning" should be used when referring to studies where carcasses are held at temperatures above 5°C (usually ranging from 15° to 40°C).

Conditioning has been recognized for many years as a method of increasing the tenderness and flavor of meat (Bouley 1874; Lehmann 1907; Hoagland *et al.* 1917). Although the improvement in meat tenderness following conditioning was observed over a century ago, meat scientists are still studying means by which the effects of conditioning can be enhanced and the mechanisms by which conditioning causes improvements in meat quality. Of the methods devised to enhance conditioning effects, two stand out as being the most studied and most efficient. These two methods are high temperature conditioning and electrical stimulation, which seem to produce enhanced conditioning effects by some common mechanisms (Dutson *et al.* 1977).

---

[1]Technical Article 18300 of the Texas Agricultural Experiment Station. This project was partially supported by grants from King Ranch, Inc., Kingsville, TX, and the Center for Energy and Mineral Resources, College Station, TX.
[2]Chairperson, Department of Food Science and Human Nutrition, Michigan State University, East Lansing, MI.
[3]Professor, Department of Food Science and Human Nutrition, Michigan State University, East Lansing, MI.

Much of the early research on high temperature conditioning was conducted on muscles that had previously entered rigor mortis at normal chill room temperatures of 0° to 40°C (Sleeth et al. 1957, 1958; Wilson et al. 1960; Parrish et al. 1973). However, since the discovery that the degree of muscle shortening is a factor in meat tenderness (Locker 1960) and that muscle shortening is increased as prerigor muscle temperatures approach 0°C (Locker and Hagyard 1963), considerable research has been conducted on the effects of increased prerigor temperature on the tenderness of muscle (Locker and Hagyard 1963; Marsh and Leet 1966A,B; Marsh et al. 1968; McCrae et al. 1971; Busch et al. 1967; Parrish et al. 1969; Smith et al. 1971, 1976; Bouton et al. 1973A,B,C, 1974; Harris 1975; Fields et al. 1976; Dutson et al. 1975, 1977; Locker and Daines 1976; Lochner et al. 1980). A majority of these studies were conducted to evaluate the use of elevated postmortem temperatures in reducing cold shortening. However, Dutson et al. (1975, 1977), Locker and Daines (1976), and Lochner et al. (1980) conducted experiments in which no difference in muscle shortening was observed between controls (0° to 4°C) and high temperature-treated samples. In spite of the fact that no shortening differences were produced in these studies, the tenderness values of high temperature-treated samples were significantly improved from those of the cold-treated controls.

Of special note in these experiments is that temperatures were elevated prior to the onset of rigor mortis; the high temperature incubations of Dutson et al. (1975, 1977) and Lochner et al. (1980) were conducted at very early postmortem times. Most of the researchers conducting experiments on electrical stimulation have also reported that tenderness is improved with no evident difference in muscle shortening (Dutson et al. 1977, 1980B). The data of Dutson et al. (1975, 1977), Locker and Daines (1976), Lochner et al. (1980) and those on electrical stimulation (Dutson et al. 1977, 1980B) indicate that tenderness improvements can be realized by means unrelated to shortening. The most plausible mechanism for this "other tenderizing factor" is muscle autolysis (Dutson et al. 1977; Lochner et al. 1980; Dutson 1983), which appears to be most operative at high temperatures.

The term "cold shortening" was originally used to describe the events that occur in muscle chilled below 10°C, producing shortening (Locker and Hagyard 1963). The association of cold shortening with toughening of meat, described by Marsh and Leet (1966A), has expanded the meaning of cold shortening to include the toughening due to cold temperatures. However, since the term "cold shortening" describes the effect of cold temperatures on muscle shortening and the associated toughening without consideration of the effect of cold on muscle autolysis, another term must be used to describe the total effect of cold temperatures on meat tenderness. A useful term would be "cold toughening," since it encompasses the factor of "cold shortening," the factor of "cold-reduced autolysis," plus any other factor that might be related to the reduction of tenderness due to cold temperatures (Dutson 1983). Therefore, "cold toughening" should be used when discussing the effects of cold temperatures on the tenderness of meat.

This chapter will discuss the effects of carcass chilling on the tenderness of meat (cold toughening) and will attempt to show possible mechanisms by which elevated temperatures (high temperature conditioning) cause reductions in cold toughening. The effects of conditioning at normal chill room temperatures (cooler conditioning) on increasing the tenderness of cold-toughened meat will also be discussed. The interaction of electrical stimulation and high temperature conditioning and the similarity of the mechanisms each utilizes to reduce cold toughening will be presented.

COLD TOUGHENING

As mentioned previously, cold toughening encompasses both cold shortening and cold-reduced autolysis. Although these two factors of cold toughening normally occur simultaneously when muscles are rapidly chilled, they will be discussed separately because they are independent phenomena and, under certain conditions, one may occur without the other.

Cold Shortening

Some of the basic mechanisms of cold shortening are presented in Chapters 1 and 6. The discussion in the present chapter includes a portion of the same material, but this material will be discussed from a slightly different viewpoint. This discussion will emphasize the effects of temperature on shortening, the ensuing effects on meat tenderness, and the mechanism by which shortening causes reductions in tenderness.

Temperature–Shortening Relationship. The temperature dependence of muscle shortening during the onset of rigor mortis at temperatures between 0° and 43°C was described by Locker and Hagyard (1963). Other authors had previously described the relationship between shortening and high (between 17° and 37°C) temperatures (Bendall 1951; Marsh 1965) but had not included the effects of cold temperatures. From the data presented in Fig. 1.6, it is evident that shortening is indeed caused by cold temperatures and that cold is a more effective inductor of shortening than is heat. Locker and Hagyard (1963) also stated that cold (0°C) shortening occurs immediately, whereas high temperature (37°C) shortening is delayed up to 5 hr. These authors also stated that muscles have more force of contraction at 2° than at 37°C, which may allow muscles to shorten more at cold temperatures, although restrained by the skeleton.

Shortening–Tenderness Relationship. The relationship of muscle shortening to meat tenderness was first described by Locker (1960). His research demonstrated that the psoas muscle, normally one of the more tender muscles in the carcass, could become very tough if allowed to shorten during rigor development. Previous to the discovery of the relationship between shortening and tenderness (Locker 1960), other researchers had shown that prerigor excision of muscle resulted in toughening (Lowe 1948; Ramsbottom and Strandine 1949). However, they did not recognize that the shortening of muscle fibers due to the exposure of excised muscles to rapid chilling was causing the observed toughness, since none of these authors suggested that this mechanism was responsible for the reduction in tenderness.

Locker and Hagyard (1963) suggested that cold-induced shortening of muscles, as observed in that experiment, may be the reason for tenderness differences in muscles from different carcasses, particularly if the muscles are exposed to rapid chilling or freezing. Marsh and Leet (1966A) conducted a number of experiments to investigate the relationship of cold shortening to tenderness. These authors demonstrated that, when muscles cold shorten to approximately 20% of the excised resting length, no appreciable reduction in tenderness ensues (Fig. 2.1). Herring et al. (1967) also showed that only small changes in tenderness were produced when sarcomere lengths of muscles from the "A" maturity animals were shortened from a lengthened value of 3.25 to 2.0 μm (Fig. 2.1). The 2.0 μm sarcomere length of Herring et al. (1967) probably corresponds to the 20% muscle shortening value reported by Marsh and Leet (1966A). Above 20% shortening, tenderness was shown to decrease very rapidly to a peak value (maximum toughness) at approximately 40% shortening (Marsh and Leet 1966A, as shown in Fig. 2.1).

Similar increases in toughness were found by Herring et al. (1967) when muscles were shortened from sarcomere lengths of 2.0 μm to less than 1.5 μm (Fig. 2.1). The extreme increase in toughness of muscle between 20 and 40% shortening (sarcomere lengths of 2.0 and 1.5 μm, respectively) and the negligible amount of toughening up to 20% shortening are difficult to explain, although Marsh and Carse (1974) suggest that the amount of overlap of the various myofilaments and the number of rigor bonds formed could be the basis for this interesting phenomenon. In contrast to this explanation, Voyle (1969) stated that the ratio of actively shortened to passively shortened muscle fibers may be an important factor in the tenderness of shortened muscle, and proposed that physical changes at the fiber level (possible connective tissue in the endomysium) and macromolecular changes at the myofilament level (probably the contact of thick filaments with the Z-lines) in the actively shortened fibers are responsible for tenderness changes.

Increases in tenderness occur when muscles are cold shortened above 40% (Marsh and Leet 1966A), with a tenderness value equal to that of unshortened muscle being reached at 60% shortening (Fig. 2.1). Marsh et al. (1974) have stated that tenderness increases at levels of cold shortening above 40% are caused by myofibril disruption. These authors attributed

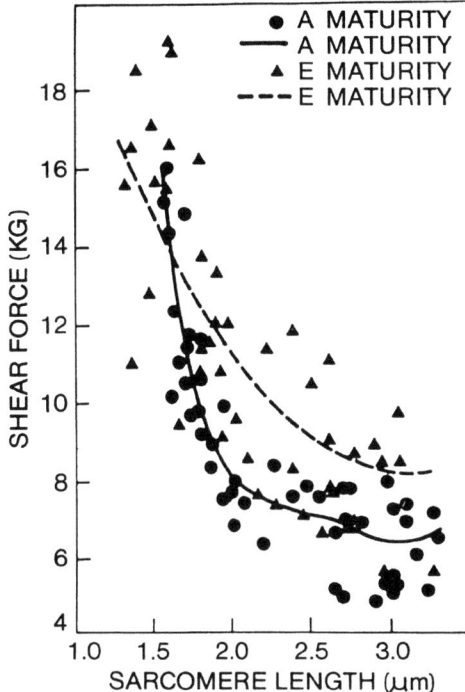

FIG. 2.1. Sarcomere length compared with shear force.
Reprinted from J. Food Sci., Herring et al. (1967) 32, 317. Copyright © by Institute of Food Technologists.

the improvement in tenderness to structural alterations (observed in extreme cold-shortened muscles as contracture bands, similar to those formed in electrically stimulated muscle; Chapter 6) and fiber fracture.

Reduction of Cold Shortening. Since finding that muscle shortening is related to tenderness (Locker 1960) and that cold temperatures cause muscle shortening (Locker and Hagyard 1963), many authors have substantiated the theory that the amount of muscle contraction during rigor development causes muscle toughness and have shown that reducing the extent of contraction improves tenderness. Herring et al. (1965A) evaluated the differences in restrained and unrestrained, excised muscle held at cold temperatures. Sarcomere lengths of unrestrained muscle were shorter than those of restrained muscle, and the restrained muscles were more tender than the unrestrained muscles. Herring et al. (1965B) indicated that carcass position placed tension on various muscles, with those muscles under greater tension having longer sarcomere lengths and increased tenderness. These authors related the differences in shortening and tenderness to the amount of cold contracture produced in the muscle, and concluded that cold shortening, and thus toughness, could be circumvented by increasing the tension on various muscles of the carcass. This led to development of the technique of hanging carcasses by the obturator foramen (Hostetler et al. 1970, 1972; Harris and MacFarlane 1971; Davey and Gilbert 1973; Bouton et al. 1973C, 1974; Quarrier et al. 1972; Stouffer 1977).

The tenderness of muscle has been related to cold shortening in a number of studies (Marsh and Leet 1966A,B; Marsh et al. 1968; Klose et al. 1970; Smith et al. 1971; Newbold and Harris 1972; Bouton 1973B,C, 1974; Purchas and Davies 1974A,B; MacFarlane et al. 1974; Rhodes and Dransfield 1974; Harris 1975; Davey and Gilbert 1975A,B; Fields et al. 1976; Smith et al. 1976; Marsh 1977; Dutson et al. 1977; Asghar and Yeates 1978; Bendall 1978A,B). Marsh and Leet (1966A) and Marsh et al. (1968) demonstrated that cold shortening and toughening can be minimized if chilling is delayed for a period of time sufficient to allow a relatively substantial reduction of ATP, which is reflected in a reduction of the muscle pH. Smith et al. (1971) also found that holding carcasses at higher temperatures (16°C) for 16 hr increased tenderness 40 to 47% and that similar tenderness improvements could be achieved without the use of elevated temperatures by stretching muscles in the carcass during the early chilling process. These authors demonstrated that the muscle stretching treatments increased sarcomere lengths from approximately 1.8 to 2.1 μm, whereas the high temperature-treated carcasses exhibited virtually no change in sarcomere length (the length varied from only 1.8 to 1.9 μm). Therefore, the improvement in tenderness attributed to increased prerigor mortis carcass temperatures is not entirely caused by a reduction in muscle shortening. Prerigor stretching of muscles by hanging from the obturator foramen produces tenderness values at 2 to 3 days postmortem similar to those values for nonstretched muscles which have been conditioned for 21 days at cooler temperatures (Bouton et al. 1973B). Bouton et al. (1973C) have shown that muscle shortening can be reduced either by stretching muscles or by increasing temperatures during the early postmortem period; and that tenderness is related to sarcomere length in the stretched muscles (up to a sarcomere length of 2.0 μm), but that increasing sarcomere length above 2.0 μm has little effect in improving tenderness. These data are similar to those reported by Marsh and Leet (1966A) and Herring et al. (1967).

Purchas and Davies (1974A,B) have shown that increased carcass fatness increases muscle tenderness. These authors stated that the increased fatness probably reduces cold shortening, making the muscles more tender. Smith et al. (1976) indicated that fatness of lambs reduced the rate of chilling and increased the tenderness of muscles. Generally, the increase in tenderness was accompanied by greater sarcomere lengths. In some instances, however, increased tenderness occurred without concomitant changes in sarcomere length. Muscles from older animals have been shown to have a greater capacity to cold shorten than those from younger animals; however, the shortening tendency has been reduced by increasing prerigor carcass temperatures in both young and old animals (Davey and Gilbert 1975A,B). Fields *et al.* (1976) concluded that shortening can be reduced by increasing carcass temperatures, and that this reduced muscle shortening is related to increased tenderness. However, sarcomere length changes in this study were small and, in one comparison, not different, although the tenderness difference between high temperature-conditioned and rapidly chilled carcasses was still significant. Rhodes and Dransfield (1974) found that cold-shortened raw muscles containing considerable connective tissue decreased in shear force as compared with nonshortened controls and concluded that cooking is required to produce the toughness associated with cold-shortened muscle. These authors reasoned that the shortening of muscle dilutes the connective tissue component and that myofibrillar proteins have no effect on the shear force value since these proteins are not coagulated by heat.

Locker (1982 and Chapter 1) has postulated that the effect of cold shortening on tenderness of nonconditioned muscle is due to the effect of stretch on the gap filaments. This author stated that gap filaments are quite likely reduced in diameter on stretching, thus reducing their strength, and that the reduced strength of gap filaments is more related to tenderness of stretched muscle than are the number of rigor bonds or the reduction in the width of the I-bands as postulated by Marsh and Carse (1974). Although the involvement of gap filaments in the shortening–tenderness relationship as proposed by Locker (1982 and Chapter 1) is a reasonable hypothesis, to date no direct experimental evidence has been presented to support this hypothesis.

Cold-reduced Autolysis

Tenderness of Shortened Muscle at Elevated Temperatures. As discussed in the previous section, cold shortening is definitely related to meat tenderness, with increased shortening causing tougher muscles when the shortened and unshortened muscles are held at cold temperatures. However, some of the studies reported in the previous section (Smith *et al.* 1971, 1976; Fields *et al.* 1976; Bouton *et al.* 1973C) presented data that, when analyzed closely, show that muscles with higher prerigor temperatures were more tender than those that had been rapidly chilled, but that sarcomere length in many cases was not different for muscles chilled at different rates. Data from the studies of Smith *et al.* (1971) and Bouton *et al.* (1973C) are presented in Table 2.1. These authors mentioned that a portion of the difference in tenderness between the muscles with different prerigor temperatures might be due to muscle proteolysis. However, they did not discuss the fact that no shortening differences were present in some of their treatments, and that all of the tenderness differences in these nonshortened treatments must be due to reduction of autolysis by cold temperatures.

Locker and Daines (1975) showed that raising the prerigor holding tem-

TABLE 2.1. Comparisons of Sarcomere Length and Tenderness of Muscles Held at Low (0°C to 2°C) and Elevated (16°C) Prerigor Temperatures

Study	Sarcomere length (μm)		Shear force (kg)	
	Cold temp.	Elevated temp.	Cold temp.	Elevated temp.
Smith et al. (1971)[a]	1.96 NS	1.89	6.23 **	3.27
Smith et al. (1971)[a]	1.83 NS	1.93	3.73 *	2.54
Bouton et al. (1973C)[b]	1.78	1.78	9.5	5.5

[a] Data from Table 8 of Smith et al. (1971).
[b] Data from Bouton et al. (1973C) for sarcomere length between 1.65 and 1.90 μm.
NS = Not significant ($P > 0.05$). * = Significant ($P < 0.05$). ** = Significant ($P < 0.01$) differences between low and elevated temperature treatments.

perature of unrestrained muscle above 15°C produced some muscles with greater shortening coupled with greater tenderness as compared with unrestrained muscle held at 5°C. Subsequent to this study, Locker and Daines (1976) demonstrated that, compared with muscles which were held at 2°C for the entire postmortem period, muscles that were incubated for a 3 hr period at 37°C had similar shortening values, but a two-fold reduction in shear force (Table 2.2). Since Z-lines were still intact in muscles that

TABLE 2.2. Effect on Shear Force of Warming Cold-Shortened Muscle to 37°C in the Last Stages of Rigor Mortis (12 Animals)

		Time–temperature treatment			
		24 hr, 15°C + 24 hr, 2°C	48 hr, 2°C	24 hr, 2°C + 3 hr, 37°C + 21 hr, 2°C	24 hr, 2°C + 7 hr, 37°C + 17 hr, 2°C
pH	Mean[a]	5.49 (48 hr)		5.76 (24 hr)	
	S.D.	0.03		0.06	
Shortening (%)					
(1) after 24 hr	Mean			23.3	24.6
	S.D.			5.0	5.2
(2) after 1 hr more at 37°C	Mean			23.5	25.5
	S.D.			7.1	8.3
(3) after 3 hr at 37°C	Mean			29.5	30.6
	S.D.			5.4	6.3
(4) after 48 hr total	Mean	1.1	34.5	32.0	33.5
	S.D.	4.5	5.2	5.3	5.5
Shear force	Mean	53.7	114.0	60.9	54.5
	S.D.	5.1	10.7	15.2	15.4
Cooking shortening (%)	Mean	23.8	13.8	13.7	9.8
	S.D.	2.9	3.2	3.6	3.2
Cooking loss (%)	Mean	23.7	19.4	23.4	23.0
	S.D.	3.2	1.5	2.6	2.7

[a] Mean of 8 animals.
Source: Adapted from Locker and Daines (1976), with permission of the authors and publishers.

had been incubated at 37°C, Locker and Daines (1976) concluded that proteolysis (aging) was not a factor in the tenderness increase; however, they did not consider proteolytic alterations in other muscle proteins. More recently, Locker and Wild (1982A,B) stated that conditioning (proteolysis) may be a factor in producing tender meat from muscles which have shortened, then have been held for a period at higher temperatures (see Chapter 1). Dutson et al. (1977) conducted an experiment using carcasses with one side incubated at elevated prerigor temperatures (22°C for 4 hr and 15°C for 8 hr), while the other side was held at normal chill room temperatures (2°C). Both sides were suspended by the obturator foramen (aitch bone) to prevent shortening of the longissimus muscle. In this study, significant differences were found in overall sensory panel tenderness and shear force values between high temperature- and cold-treated samples; however, no difference was found in sarcomere lengths (Table 2.3). Combining these data with other data presented in their paper, Dutson et al. (1977) concluded that proteolytic changes in the muscle are probably responsible for tenderness differences attributable to elevated temperatures in the absence of shortening. Data supporting this theory are presented in subsequent papers from the same laboratory (Dutson and Yates 1978; Dutson et al. 1980B; Dutson 1983).

Lochner et al. (1980) have shown that increased tenderness is highly related to early-postmortem muscle temperature (Fig. 2.2). These authors stated that, in carcasses held at elevated temperatures at very early postmortem times, there is little or no reduction in shortening, but a marked increase in tenderness occurs. Thus, the data of Lochner et al. (1980), obtained under experimental temperature conditions similar to those used by Dutson et al. (1977) and Moeller et al. (1976, 1977), suggest that reduced autolysis may be a major factor in cold toughening of muscle, and that cold shortening may be a minor factor, particularly among fed cattle in the United States.

Lysosomal Enzymes. The incubation of carcasses at an elevated temperature in the study of Dutson et al. (1977; Table 2.3) caused an increase in the free activity of lysosomal enzymes, allowing them to be available to act on the muscle proteins at a higher temperature and a lower pH (Moeller et al. 1976). These conditions cause increased activity of lysosomal enzymes, which may be a factor in the improved tenderness of muscles exhibiting no differences in shortening (Dutson et al. 1977). Moeller et al. (1977) found that the higher temperature and more rapid reduction in pH associated with the incubation of muscle samples at 37°C also resulted in increased free activity of lysosomal enzymes. This increased activity of lysosomal enzymes (Moeller et al. 1977) was related to increased fragmentation values, which have been shown to be a measure of meat tenderness. Thus, increased activity of lysosomal enzymes may also have been a factor in the increased tenderness of meat in the study of Locker and Daines (1976) in which muscles were incubated at 37°C.

TABLE 2.3. Mean Sarcomere Length, Overall Tenderness, and Shear Force Values for Longissimus Muscles from Control (C) and High Temperature-Conditioned (HT) Sides

N	Animals		Method of carcass suspension	Sarcomere length (μm)		Overall tenderness[b]		Shear force (N)[c]	
	Age	Description		C	HT[a]	C	HT	C	HT
17	2–4 weeks	Veal	Achilles	1.75 ***	1.92	5.1 ***	6.3	63.2 ***	45.8
6	9–12 months	Beef	Achilles	1.77 ***	1.96	5.0 *	6.3	76.1	68.1
20	18–26 months	Beef	Achilles	1.78 ***	1.93	4.8 ***	5.6	84.1 ***	69.4
10	9–12 months	Beef	Obturator foramen	2.22	2.20	6.7 *	7.1	48.9 **	42.3

Source: Dutson et al. (1977).
[a] HT sides were held at 22°C for 4 hr, 16°C for 8 hr, and 1°C for 36 hr, whereas the opposite sides (control) were held at 1°C for the entire 48 hr period.
[b] Determined by an 8-member sensory panel (8 = extremely tender, 1 = extremely tough).
[c] Force expressed as Newtons.
*$P < 0.05$, **$P < 0.01$, ***$P < 0.001$, between C and HT values on the same line.

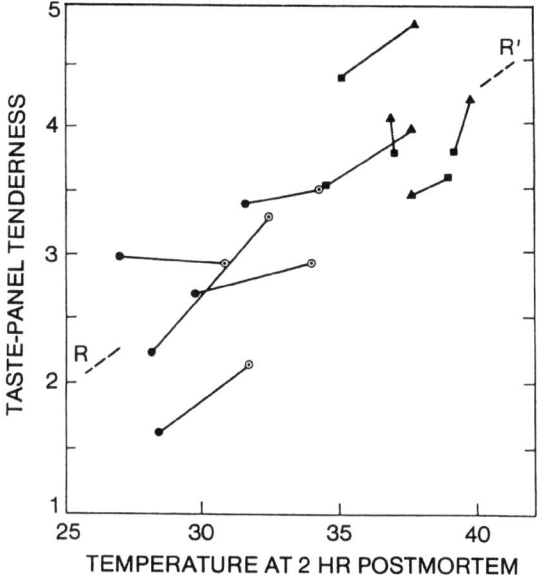

FIG. 2.2. Taste panel tenderness compared with temperature at 2 hr postmortem. R- - - - -R'—Line of regression of tenderness on temperature. ● LR. ○ LS. ▲ FR. ■ FS.
From Lochner et al. (1980) with permission of the authors and publishers.

The mechanism by which improvement of tenderness occurs at elevated holding temperatures during the early-postmortem period has not been elucidated, but proteolysis of muscle proteins appears to be the most logical primary contributor in this process. The data showing increased free activity of lysosomal enzymes, and elevated muscle temperatures early-postmortem, which creates a milieu more conducive for action of these enzymes, give associative evidence that the lysosomal enzymes may be responsible for the observed tenderness improvement (Moeller et al. 1976, 1977; Dutson et al. 1977; Dutson 1982, 1983). Support for the theory that high muscle temperatures cause increased muscle proteolysis has been documented in many articles, as reviewed by Dutson (1983). Some of these data are discussed in the following sections of this chapter.

Troponin-T Degradation. Since the production of a 30,000 dalton protein fragment on SDS gels of myofibrillar proteins has been associated with meat tenderness (MacBride and Parrish 1977; Parrish 1977; Penny and Dransfield 1979), it is logical to assume that this protein fragment would be increased by elevated postmortem temperatures. This protein fragment is thought to be produced by hydrolysis of troponin-T, and a plot of the relationship of troponin-T degradation to tenderness is presented in

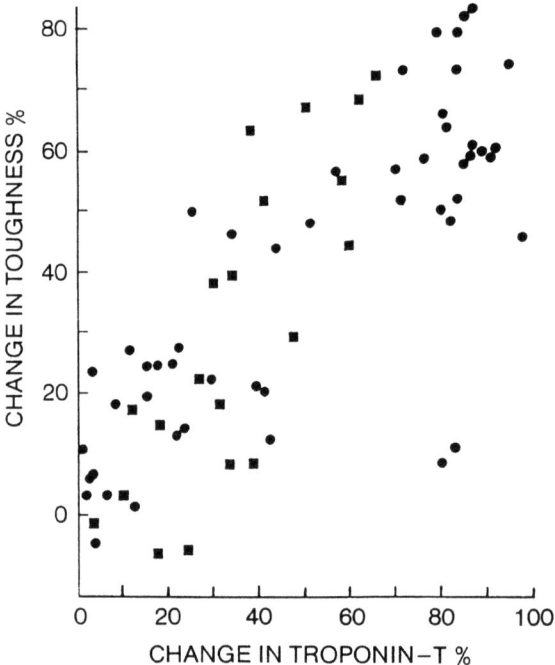

FIG. 2.3. The relationship between the change in toughness and the change in troponin-T during conditioning of beef m. semimembranosus (●) and m. longissimus dorsi (■).
From Penny and Dransfield (1979) with permission of the authors and publishers.

Fig. 2.3. Penny and Dransfield (1979) have shown that the 30,000 dalton protein fragment is increased as muscle temperatures are increased between 5° and 35°C. Samejima and Wolfe (1976) also have shown that the 30,000 dalton fragment is increased when muscle homogenates are incubated at 40°C, particularly when the pH is at 5.4. A study conducted by Yates et al. (1983) demonstrated that when intact muscle is held at 37°C, there is a significant increase in the amount of protein in the 28,000 to 30,000 dalton region as compared with intact muscle incubated at 4°C. Other data presented by Yates et al. (1983) are also of interest and demonstrate that more troponin-T degradation occurs at pH 5.4 (a pH conducive to acid proteolysis) than at pH 7.0 [a pH conducive to neutral or calcium-activated factor (CAF) proteolysis]. These data would lead one to question whether CAF is involved in muscle tenderization at elevated temperatures.

Myosin Degradation. Several studies have shown that elevated temperatures enhance the degradation of myosin in intact muscle (Arakawa *et* al. 1976; Penny and Dransfield 1979; Yates et al. 1983). Other researchers have shown that myosin is also degraded by specific muscle proteases during incubation of myofibrils, with these enzymes at temperatures in the range of 35° to 37°C (Schwartz and Bird 1977; Robbins et al. 1979; Matsukura et al. 1981). These studies show that various molecular fragments in the range of 50,000 to 170,000 daltons are produced when myosin is degraded by the action of endogenous proteases on intact muscle proteins and by the action of added proteases on myofibrils and purified myosin. Table 2.4 shows differences in percentages of myofibrillar proteins in myofibrils isolated from muscles incubated at 4° and at 37°C. The differences in amount of proteins isolated at the different temperatures (Table 2.4) were related to increased proteolysis of myofibrillar proteins, particularly myosin, at the higher temperature (Yates et al. 1983).

TABLE 2.4. Mean Values for the Percentage Protein Composition as Determined by Densitometry of SDS Polyacrylamide Gels of the Myofibrillar Fraction Prepared from the 4° and 37°C Incubated Longissimus Muscle Samples

Region	Percentage of myofibrillar protein	
	4°C	37°C
Myosin	29.5[a]	24.8[b]
M-proteins	8.3[a]	9.0[a]
C-proteins	3.0[a]	3.2[a]
α-Actinin	3.6[a]	6.6[b]
50–100 K daltons	1.6[a]	3.0[b]
Actin	27.8[a]	26.6[a]
Troponin-T and tropomyosin	17.3[a]	16.0[a]
28–23 K daltons	1.1[a]	1.8[b]
Myosin LC-1	2.4[a]	3.2[b]
Troponin-I	2.3[a]	2.4[a]
Troponin-C and myosin LC-2	3.4[a]	3.5[a]

Source: From Yates et al. (1983) with permission of the authors and publishers.
[a,b] Mean values in the same row bearing a common superscript letter are not significantly different ($P > 0.05$).

Connectin and Gap Filament Degradation. Gap filaments are myofibrillar elements which have recently been identified in highly stretched skeletal muscle (Locker and Leet 1976A,B). These filaments are thought to be composed of the proteins connectin or titin and have been implicated as a major factor in determining meat tenderness (see Chapter 1).

King et al. (1981) have demonstrated that connectin is hydrolyzed by incubation at 55°C, and King and Harris (1982) have attributed connectin degradation to a carboxyl protease. However, Locker (1982) has stated that gap filaments are stable when heated to 100°C for as long as 4 hr. Therefore, either the gap filaments are not composed entirely of connectin or else the experiments of King et al. (1981) were unable to measure connectin even though it was still present in the gap filaments (Locker 1982). The discrepancies between the experiments of King et al. (1981) and those of Locker (1982; and Chapter 1) must be resolved before the gap filament theory of tenderness can be accepted in full.

Experiments presented by Dutson (1983) have demonstrated that the ultrastructural appearance of gap filaments is disrupted by incubation at pH 5.4 and 37°C for 1 hr. Incubation at pH 5.4 and 2°C, at pH 7.4 and 37°C, and at pH 7.4 and 2°C also caused some ultrastructural disruption of gap filaments. However, less disruption of the gap filaments occurred at colder temperatures (2°C) or at pH 7.4. Dutson (1983) did not determine the effect of cooking on gap filaments, so his data cannot be directly compared with those of Locker (1982). However, if the gap filament disruption observed by Dutson (1983) is sufficient to alter its tensile properties, then gap filaments very likely would not have sufficient tensile strength to be a major component of muscle tenderness since muscles normally are at temperatures and pH values in the range used by Dutson (1983) for periods longer than 1 hr before tenderness is measured.

Locker (1982; Chapter 1 this volume) and Locker and Wild (1982A,B) have postulated that the shortening–tenderness relationship in conditioned beef muscle is due to gap filaments and that the improvement in tenderness due to conditioning results from the degradation of gap filaments. The basis of this hypothesis is: As muscles cold shorten, there is an increase in the percentage of fibers that reach a sarcomere length of 1.5 μm or less. At this sarcomere length, the gap filaments would be completely shielded by myosin since myosin filaments from adjacent sarcomeres would be in contact with each other, completely enveloping the gap filaments. Locker (1982; and Chapter 1) suggests that the shielding of gap filaments by myosin prevents their degradation by muscle proteases and that the relationship of percentage shortening to meat tenderness results from gap filament shielding. However, this hypothesis needs to be substantiated by determining the association between the actual number of sarcomeres of 1.5 μm and the degree of tenderness.

The fact remains that gap filaments and connectin are altered by conditioning, particularly at high temperatures (Davey and Graafhuis 1976; Locker et al. 1977; Takahashi and Saito 1979), and this change during conditioning may be related to the tenderness improvement of conditioned muscle (Locker 1982).

Proteolytic Enzyme Activity. The activity of both lysosomal enzymes and calcium-activated factor (CAF) would be reduced as temperatures are lowered from the *in vivo* muscle temperature of 37°C (Weisman 1964; Stagni and deBernard 1968; Dayton et al. 1975; Bird and Carter 1980). Although no direct evidence that these enzymes are responsible for autolytic breakdown of muscle proteins (and resultant muscle tenderness) has been presented, considerable associative evidence is available to show that these protease systems are likely causing many of the tenderness changes during conditioning (Dutson 1983). In fact, when conditions are present in muscle to cause a release of enzymes from the lysosome (lowered muscle pH and increased muscle temperature; Moeller et al. 1976, 1977; Dutson et al. 1980B), alterations associated with increased tenderness also occur (Dutson et al. 1977; Yates et al. 1983; Dutson 1983). Those same conditions have also been shown to increase the proteolytic degradation of muscle proteins (Dutson 1983).

HIGH TEMPERATURE CONDITIONING

The term "high temperature conditioning" is used to refer to the holding of carcasses or muscles at temperatures above those normally used in cold rooms (chillers) of 0° to 5°C, but the term usually implies holding carcasses or muscles at temperatures of 15°C or greater. High temperature conditioning can be applied to carcasses before rigor mortis has occurred (high temperature prerigor conditioning) or after rigor has occurred at cold room temperatures (high temperature postrigor conditioning). The distinction should be made between high temperature prerigor conditioning and high temperature postrigor conditioning since the time at which high temperature conditioning occurs has been shown to affect the amount of tenderization that ensues (Locker and Daines 1975, 1976; Lochner et al. 1980; Marsh et al. 1981).

High Temperature Prerigor Conditioning

High temperature prerigor conditioning of carcasses has been found to be an effective means of improving the tenderness of meat. Smith et al. (1971) have demonstrated an increase in tenderness of more than 40% when carcasses were held at 16°C for 16 to 10 hr as compared with carcasses held at normal chill room temperatures. Bouton et al. (1973C) reported greater than 50% increased tenderness when carcasses held at 15° to 16°C for 2 days were compared with those held at 0° to 1°C for the same period of time. Bouton et al. (1974) showed that tenderness increases were also evident when prerigor carcasses were conditioned at 7° to 8°C, although these tenderness increases were not as great as when similar carcasses were conditioned at 15° to 16°C. Conditioning carcasses at 35° to 37°C causes even greater tenderness improvement than conditioning at 15°C (Locker and Daines 1975, 1976). From the foregoing data (Smith et al. 1971; Bouton et al. 1973C, 1974; Locker and Daines 1975, 1976) and those of others (Locker and Hagyard 1963; Marsh and Leet 1966A,B; McCrae et al. 1971; Harris 1975; Fields et al. 1976; Smith et al. 1976; Davey and Gilbert 1974, 1975B, 1976; Davey et al. 1976; Moeller et al. 1976; Dutson et al. 1977, 1980B; Davey and Niederer 1977; Lochner et al. 1980; Dransfield et al. 1981; Marsh et al. 1981; Calkins 1981), it is clear that increased temperatures during the prerigor period definitely improve meat tenderness and that the higher temperatures (up to 37°C) are more

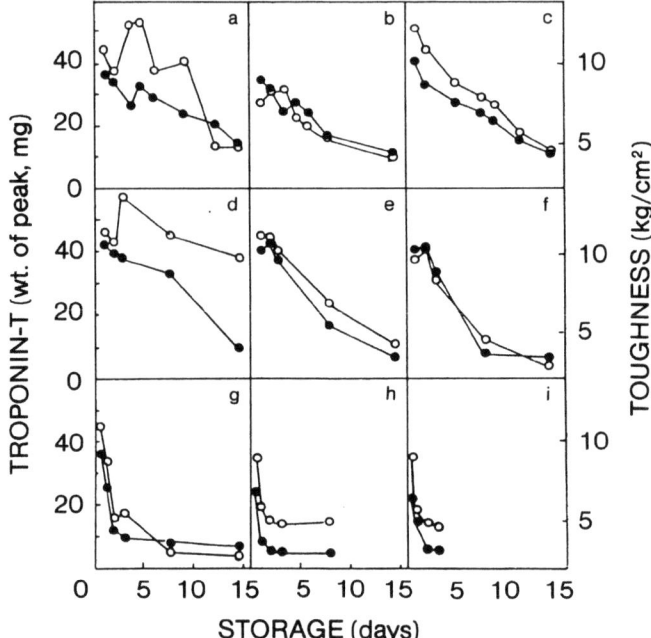

FIG. 2.4. The time course of troponin-T (●) and toughness (○) during conditioning of beef. a, b, and c—M. longissimus dorsi stored at 3°C. d to i—M. semitendinosus stored at: d—0°C; e—5°C; f—10°C; g—15°C; h—25°C; i—35°C.

From Penny and Dransfield (1979) with permission of the authors and publishers.

effective in increasing tenderness. Figure 2.2 shows the relationship of muscle temperature at 2 hr postmortem and muscle tenderness (measured 4 days postmortem). Figure 2.4 demonstrates the effect of increasing prerigor muscle temperature from 0° to 35°C and the resulting tenderness changes that occur in semitendinosus muscles held at those temperatures for up to 15 days (Penny and Dransfield 1979). The rate of troponin-T degradation, a measure of muscle proteolysis (Dutson 1983), is also increased as the muscle temperature is increased (Fig. 2.4). Thus, the data in Fig. 2.4 and data presented previously in this chapter are evidence that high prerigor temperatures not only reduce the toughness associated with cold shortening *per se,* but that they also lessen cold-reduced autolysis.

High Temperature Postrigor Conditioning

Much of the early research on high temperature conditioning was conducted on carcasses or muscles that had previously gone through the rigor process at normal chill room temperatures (Sleeth et al. 1957, 1958; Wilson et al. 1960) and therefore would be classified as high temperature postrigor conditioning. These studies have shown that conditioning postrigor muscles at elevated temperatures increases the tenderness over that of companion muscles held at cooler temperatures for similar periods of time. Other studies (Parrish et al. 1973; Davey and Gilbert 1976; Pierson and Fox 1976; Dransfield et al. 1981) have also shown that high temperature postrigor conditioning increases tenderness up to about 30% over controls held at cooler temperatures (0° to 3°C). Davey and Gilbert (1976) have determined the rate of tenderization at different temperatures of postrigor conditioning and found a continual increase in tenderization rate up to temperatures of 60°C (Fig. 2.5). These authors calculated that the enthalpy of activation was 61.5 kJ for rate of tenderness increase up to 40°C. This enthalpy of activation for postrigor muscle (Davey and Gilbert 1976) is similar to that for prerigor muscle (63 kJ) shown by Penny and Dransfield (1979). This is an indication that the mechanism of tenderiza-

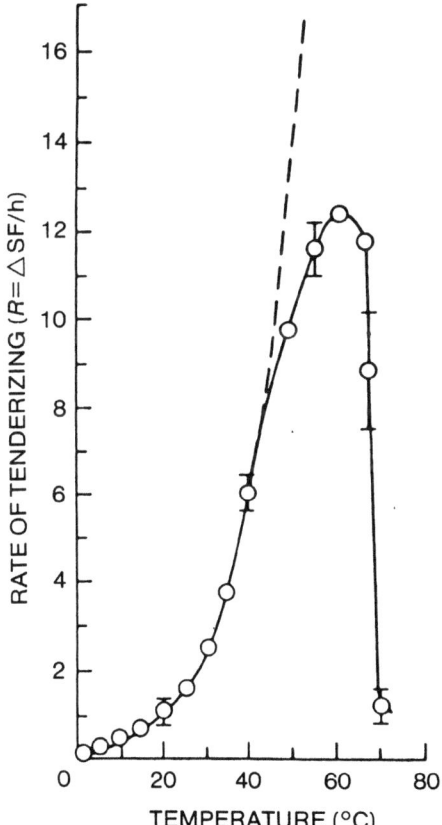

FIG. 2.5. The relationship between rate of tenderizing and holding temperature.
From Davey and Gilbert (1976) with permission of the authors and publishers.

tion due to increased temperatures may be similar in both prerigor and postrigor muscle. However, the data of Locker and Daines (1976) indicate that tenderization is much greater in prerigor muscle than in postrigor muscle when both are held at the same temperature for similar periods of time (37°C for 3 hr).

Recent data presented by Lochner et al. (1980) and Marsh et al. (1981) also indicate that subjecting the carcass to an elevated temperature for the first few hours postmortem is more effective in improving meat tenderness than is the application of high temperature at a later postmortem period (Fig. 2.2).

The mechanism of tenderization of both pre- and postrigor muscle at high temperatures has not been elucidated fully, but the factors responsible for cold-reduced autolysis are probably alleviated under conditions of elevated temperature. This, in turn implies that increased temperatures cause an increase in the rates of proteolysis of the myofibrillar proteins, as mentioned previously herein.

The Effect of High Temperature Conditioning on Collagen

Elevated temperatures may be causing alterations in the connective tissue component of meat, but Pierson and Fox (1976) found no changes in soluble collagen associated with tenderness increases during high temperature postrigor conditioning. Wu et al. (1982), however, did find slight (although not significant) increases in percentage soluble collagen due to extended postmortem time (12 vs 24 hr) and that this increase was enhanced by high temperature prerigor conditioning. These authors also found that there was a significant increase in the amount of small molecular weight collagen components at the later postmortem time, and this postmortem effect was enhanced by high temperature prerigor conditioning. Dutson and Lawrie (1974) found that the amount of collagen that was able to pass through a cheesecloth filter was increased over 100% as a result of cooler conditioning for 14 days, and hypothesized that this increase might be due to activity of lysosomal enzymes which were released by the cooler conditioning treatment. In support of the finding of Dutson and Lawrie (1974), Wu et al. (1981) found that the lysosomal enzymes, β-glucuronidase and β-galactosidase, increased the breakdown of collagen by collagenase, and postulated that these enzymes may cause a release of collagen from the ground substance matrix. Kruggel and Field (1971) and Pfeiffer et al. (1972) also presented evidence for collagen alterations during postmortem conditioning of muscle. Kopp and Valin (1981) have shown that incubation of collagen fibers at 37°C and pH 5.5 in a medium containing lysosomal enzymes significantly affects the thermal stability of muscle collagen by increasing the thermolabile fraction and reducing the fiber strength by 50%. Thus, collagen may be altered during conditioning by the action of lysosomal enzymes.

COOLER CONDITIONING

A large number of research studies have been conducted to evaluate the effects of holding carcasses at 0° to 5°C, particularly as related to tenderness and other palatability traits of meat. Among these studies are reports by Lowe (1948), Partmann (1963), Davey and Gilbert (1966), Dransfield and Rhodes (1975), Cia and Marsh (1976), Davey et al. (1976), Pierson and Fox (1976), Martin et al. (1971, 1977), Bouton et al. (1973B,C, 1976), Hegarty and Allen (1976), Yamamoto et al. (1977), and Smith et al. (1978), all of which demonstrate that meat tenderness increases during cooler conditioning. Reviews including this subject have been published recently by Jeremiah (1978) and by Penny (1980).

The mechanism by which cooler conditioning improves tenderness has also been studied extensively but the conclusions of these studies are unclear. Alterations in Z-lines have been postulated as being instrumental in the tenderization of muscle during cooler conditioning (Davey and Gilbert 1967; 1969; Davey and Dickson 1970; Henderson et al. 1970; Parrish et al. 1973; Penny et al. 1974; Parrish 1978). However, the complete alteration of Z-line structure does not occur until after the maximum tenderness is achieved (Davey and Dickson 1970). In other studies, researchers have found that Z-lines remain quite intact in stretched muscle, even after aging for considerable periods of time (Dutson et al. 1976; Davey and Graafhuis 1976), and that conditions that produce increased muscle fragmentation also preserve the Z-line structure (Dutson and Yates 1978; Yates et al. 1983). Thus, although a reduction in Z-line structure may occur during cooler conditioning, conclusive evidence has not been given that the disruption of the Z-lines is the major factor in tenderness improvement during cooler conditioning (Locker 1982).

Modifications in the protein, troponin-T, have been demonstrated during cooler conditioning with the appearance of a 30,000 dalton protein being the result of this alteration (Hay et al. 1973; Penny and Ferguson-Pryce 1979; Penny 1980). Olson and Parrish (1977), MacBride and Parrish (1977), Penny and Dransfield (1979), and Parrish et al. (1981) have demonstrated an association between meat tenderness and the appearance of the 30,000 dalton protein component; however, troponin-T degradation is likely an indicator of muscle proteolysis rather than actually being involved in the tenderizing process of cooler conditioning (Penny 1980; Yates et al. 1983; Dutson 1983).

Alterations in the myofibrillar ATPase activity of muscle have been found to occur during muscle conditioning (Robson et al. 1967; Arakawa et al. 1976; Cheng and Parrish 1978A,B; Ikeuchi et al. 1978; Penny 1980). The change in myofibrillar ATPase activity has been attributed to alterations in the actin-myosin interaction and, thus, is believed to be a factor in the tenderness changes that occur in conditioned muscle (Goll et al. 1971; Penny 1980; Ouali and Valin 1981).

The discovery of new cytoskeletal proteins in muscle has added a new dimension in the consideration of the mechanisms of tenderization during muscle conditioning. Gap filaments, which are thought to be composed of the proteins connectin and/or titin, have been proposed recently as the major factor causing toughness of meat, with their degradation, therefore, being responsible for the increased tenderness during conditioning (Locker et al. 1977; Locker and Wild 1982A,B). The structure of gap filaments in skeletal muscle and their involvement in toughness of meat has been completely described in Chapter 1. Although the gap filament theory of tenderness is supported by some solid evidence (presented in Chapter 1), there are still some aspects which need clarification since this theory is unable to explain tenderness differences between pre- versus postrigor cooked meat (Davey and Gilbert 1976; Ramsbottom and Strandine 1949; Weidemann et al. 1967). Young et al. (1981) have investigated the relationship of conditioning to alterations in the cytoskeletal proteins connectin and desmin. These authors have postulated that desmin and connectin degradation during conditioning is responsible for the increased myofibrillar fragmentation of conditioned muscle.

ELECTRICAL STIMULATION

The effects of electrical stimulation (ES) on meat quality and the scientific basis of ES are covered in detail in other chapters; however, a few additional comments are in order concerning the similarity of events occurring as a result of the application of ES as compared with high temperature conditioning. As was demonstrated in this chapter and in other reports (Dutson et al. 1977; Yates et al. 1983; Dutson 1983), one of the most important factors for enhancing tenderization (or reducing toughening) during high temperature conditioning of muscle is the production of a lowered pH while the muscle temperature is still relatively high. This not only reduces the amount of shortening but also increases tenderness in the absence of differences in shortening (Dutson et al. 1977; Locker and Daines 1976). The same conditions of lowered pH at an elevated temperature are also present in electrically stimulated muscle (Dutson et al. 1977, 1980B).

Under conditions that would normally produce cold shortening, the application of ES will prevent this shortening from occurring. This is the major reason for the support of research on and the application of ES in New Zealand (see Chapter 3). However, in many of the studies conducted to investigate the effects of ES, no measurable difference was found in sarcomere length between ES and non-ES muscles, whereas a marked improvement was observed in the tenderness of the ES-treated sides (Dutson et al. 1980B). These data led Dutson et al. (1977, 1980B) to postulate that factors other than the prevention of cold shortening were responsible for the increased tenderness of meat from ES carcasses. The "other factors" must be operating by reducing cold toughening through minimization of cold-reduced autolysis (see previous discussion in this chapter).

Various studies have demonstrated that high temperature prerigor conditioning increases the free activity of lysosomal enzymes, probably since the lower pH and higher temperature of this treatment results in the rupturing of the lysosomal membrane (Moeller et al. 1976, 1977; Dutson et al. 1977; Wu et al. 1981). The same increase in free activity of lysosomal enzymes has been observed in ES muscle (Dutson et al. 1980A).

The combination of high temperature and low pH in early postmortem (prerigor) muscle is an occurrence that is almost always associated with increased tenderness of meat (see earlier discussions in this chapter). This low pH, high temperature condition causes many alterations in muscle proteins and muscle structure (such as those observed in myosin, troponin-T, thick filaments, and gap filaments) and increases muscle fragmentation (Dutson and Lawrie 1974; Dutson et al. 1977; Moeller et al. 1977; Wu et al. 1981; Dutson 1983; Yates et al. 1983). Some of these same alterations occur in ES muscle, including increased myofibrillar fragmentation (Sonaiya et al. 1982), myofibril disintegration (Will et al. 1980), and increased free lysosomal enzyme activity (Dutson et al. 1980A); however, the alterations observed in myofibrillar proteins from ES muscle have been inconclusive (Sonaiya et al. 1982).

One study (Marsh et al. 1981) has presented evidence that a high temperature, high pH combination is more effective in producing tenderness than is high temperature, low pH. Muscle shortening or some other factor may have been involved in the study of Marsh et al. (1981) since all of the muscles used (data from only six beef sides were reported) were rather tough (average tenderness ratings of 1.9 to 3.8 on a 7 point scale, with 1 being most tough, 7 most tender) in spite of the high temperatures (37°C for 3 hr) employed. However, insufficient data were presented in this report to determine if shortening, or some other phenomenon, might have influenced their results. In any event, high temperatures early-postmortem seem to be a decisive factor in improved meat tenderness and, in most cases, this high temperature is combined with a lower muscle pH, such as is found in electrically stimulated muscle.

SUMMARY

Postmortem conditioning of meat has been practiced for many years and involves two types of treatments: cooler conditioning, the holding of meat at 0° to 5°C; and high temperature conditioning, holding meat at temperatures above 5°C (usually ranging from 15° to 40°C). Many studies have been conducted on the tenderness effects of conditioning treatments and indicate that conditioning is beneficial to meat tenderness.

High temperature conditioning improves meat tenderness by reducing cold toughening. This is accomplished by reducing cold shortening and/or cold-reduced autolysis, both of which are components of cold toughening. High temperature conditioning can have an effect both pre- and postrigor, with prerigor high temperature conditioning being more effective. Cooler conditioning appears to improve tenderness by allowing more time for muscle autolysis to occur, although autolysis at cold temperatures (0° to 5°C) is not as rapid as at higher temperatures (15° to 40°C).

The mechanism by which a reduction in cold shortening improves tenderness has been widely studied. It is generally accepted that the reduced overlap of the thick and thin filaments produces more tender meat, probably by reducing actomyosin formation. However, maximum toughness occurs when myosin filaments form a continuum from sarcomere to sarcomere, indicating that some factor other than actomyosin formation may be operative. It has been postulated recently that this "other factor" may be the reduced autolysis of gap filaments when they are shielded by a myosin continuum.

The mechanism by which a lessening of cold-reduced autolysis improves tenderness has been less extensively studied; however, the increased degradation of muscle proteins by endogenous enzymes has been identified as an operative factor. It has been demonstrated that muscle proteins are degraded during conditioning, and that high temperature conditioning prerigor increases the free activity of lysosomal enzymes.

The action of endogenous enzymes on muscle proteins appears to be most effectively enhanced by a combination of high temperature and low pH, a condition that is produced by both high temperature conditioning prerigor and electrical stimulation. When elevated temperature in muscle is accompanied by lowered pH, a reduction in cold shortening occurs due to the rapid onset of rigor mortis. Therefore, electrical stimulation may be producing many of its effects by similar mechanisms as are operative during high temperature prerigor conditioning. However, one study has reported that a combination of high muscle pH and high muscle temperature produces the greatest tenderness: this contradictory finding indicates that further investigation is needed.

REFERENCES

ARAKAWA, N., FUJIKI, S., INAGAKI, C. and FUJIMAKI, M. 1976. A catheptic protease active in ultimate pH of muscle. Agric. Biol. Chem. (Tokyo) *40*, 1265.

ASGHAR, A. and YEATES, N.T.M. 1978. The mechanism for the promotion of tenderness in meat during the post-mortem process: A review. CRC Crit. Rev. Food Sci. Nutr., 115.

BENDALL, J.R. 1951. The shortening of rabbit muscle during rigor mortis: Relation to the breakdown of the adenosine triphosphate and creatine phosphate and to muscular contraction. J. Physiol. *114*, 17.

BENDALL, J.R. 1978A. Electrical stimulation of carcasses as a method of avoiding the cold-shortening and toughening of meat. Proc. Eur. Meet. Meat Res. Workers, Kulmbach, W. Germany, *24*, E1, 1.

BENDALL, J.R. 1978B. Variability in rates of pH fall and of lactate production in the muscles on cooling beef carcasses. Meat Sci. *2*, 91.

BIRD, J.W.C. and CARTER, J.H. 1980. Proteolytic enzymes in striated and nonstriated muscle. *In* Degradative Processes in Heart and Skeletal Muscle. p. 51. K. Wildenthal (Editor). Elsevier/North-Holland Biomedical Press, New York.

BOULEY, M. 1874. (Untitled.) C.R. Acad. Sci. Fr. *79*, 739. As cited by Lawrie, R.A. (1979). *In* Developments in Meat Science—1. Applied Science Publishers, London.

BOUTON, P.E., CARROLL, F.D., HARRIS, P.V., and SHORTHOSE, W.R. 1973A. Influence of pH and fiber contraction state upon factors affecting the tenderness of bovine muscles. J. Food Sci. *38*, 404.

BOUTON, P.E., FISHER, A.L., HARRIS, P.V. and BAXTER, R.I. 1973B. A comparison of some post-slaughter treatments on the tenderness of beef. J. Food Technol. *8*, 39.

BOUTON, P.E., HARRIS, P.V., SHORTHOSE, W.R. and BAXTER, R.I. 1973C. A comparison of the effects of aging, conditioning and skeletal restraint on the tenderness of mutton. J. Food Sci. *38*, 932.

BOUTON, P.E., HARRIS, P.V., SHORTHOSE, W.R. and SMITH, M.G. 1974. Evaluation of methods affecting mutton tenderness. J. Food Technol. *9*, 31.

BOUTON, P.E., HARRIS, P.V. and SHORTHOSE, W.R. 1976. Peak shear force values obtained for veal muscle samples cooked at 50 and 60°C: Influence of aging. J. Food Sci. *41*, 197.

BUSCH, W.A., PARRISH, F.C., JR. and GOLL, D.E. 1967. Molecular properties of post-mortem muscle. 4. Effect of temperature on adenosine triphosphate degradation, isometric tension parameters and shear resistance of bovine muscle. J. Food Sci. *32*, 390.

CALKINS, C.R. 1981. Relationships and interrelationships cf selected antemortem and postmortem factors to meat tenderness and palatability. Ph.D. Thesis. Texas A & M University, College Station, TX.

CHENG, C.S. and PARRISH, F.C., JR. 1978A. Effects of postmortem storage conditions on myofibrillar ATPase activity of porcine red and white semitendinosus muscle. J Food Sci. *43*, 17.

CHENG, C.S. and PARRISH, F.C., JR. 1978B. Molecular changes in the salt-soluble myofibrillar proteins of bovine muscle. J. Food Sci. *43*, 461.

CIA, G. and MARSH, B.B. 1976. Properties of beef cooked before rigor onset. J. Food Sci. *41*, 1259.

DAVEY, C.L. and DICKSON, M.R. 1970. Studies in meat tenderness and ultrastructure changes in meat during aging. J. Food. Sci. *35*, 56.

DAVEY, C.L. and GILBERT, K.V. 1966. Studies in meat tenderness: 2. Proteolysis and the aging of beef. J. Food Sci. *31*, 135.

DAVEY, C.L. and GILBERT, K.V. 1967. Structural changes in meat during aging. J. Food Technol. *2*, 57.

DAVEY, C.L. and GILBERT, K.V. 1969. Studies in meat tenderness: 7. Changes in the fine structure of meat during aging. J. Food Sci. *34*, 69.

DAVEY, C.L. and GILBERT, K.V. 1973. The effect of carcass posture on cold, heat and thaw shortening in lamb. J. Food Technol. *81*, 445.

DAVEY, C.L. and GILBERT, K.V. 1974. Carcass posture and tenderness in frozen lamb. J. Sci. Food Agric. *25*, 931.

DAVEY, C.L. and GILBERT, K.V. 1975A. The tenderness of cooked and raw meat from young and old animals. J. Sci. Food Agric. *26*, 953.

DAVEY, C.L. and GILBERT, K.V. 1975B. Cold shortening capacity and beef muscle growth. J. Sci. Food Agric. *26*, 755.

DAVEY, C.L. and GILBERT, K.V. 1976. The temperature coefficient of beef aging. J. Sci. Food Agric. *26*, 244.

DAVEY, C.L. and GRAAFHUIS, A.E. 1976. Structural changes in beef muscle during aging. J. Sci. Food Agric. *26*, 301.

DAVEY, C.L. and NIEDERER, A.F. 1977. Cooking tenderizing in beef. Meat Sci. *1*, 271.

DAVEY, C.L., NIEDERER, A.F. and GRAAFHUIS, A.E. 1976. Effects of ageing and cooking on the tenderness of beef muscle. J. Sci. Food Agric. *26*, 251.

DAYTON, W.R., GOLL, D.E., STROMER, N.H., REVILLE, W.J., ZEECE, M.G. and ROBSON, R.M. 1975. Some properties of a Ca^{2+} activated protease that may be involved in myofibrillar protein turnover. In Proteases and Biological Control. p. 551. E. Reich, D. B. Rifkin and E. Shaw (Editors). Cold Spring Harbor Laboratory, Cold Spring Harbor, NY.

DRANSFIELD, E. 1981. Eating quality of DFD beef. In The Problem of Dark-Cutting in Beef. p. 344. D.E. Hood and P.V. Tarrant (Editors). Martinus Nijhoff Publishers, Boston.

DRANSFIELD, E. and RHODES, D.N. 1975. Texture of beef M. semitendinosus heated before, during and after development of rigor mortis. J. Sci. Food Agric. 26, 483.

DRANSFIELD, E., JONES, R.C.D. and MACFIE, H.J.H. 1981. Quantifying changes in tenderness during storage of beef. Meat Sci. 5, 131.

DUTSON, T.R. 1982. Meat proteolysis. In Food Proteins. pp. 329–336. P.F. Fox and J.J. Condon (Editors). Elsevier Science Publishing Co., New York.

DUTSON, T.R. 1983. Relationship of pH and temperature to disruption of specific muscle proteins and activity of lysosomal enzymes. J. Food Biochem. 7, 223.

DUTSON, T.R. and LAWRIE, R.A. 1974. Release of lysosomal enzymes during postmortem conditioning and their relationship to tenderness. J. Food Technol. 9, 43.

DUTSON, T.R. and YATES, L.D. 1978. Molecular and ulstrastructural alterations in bovine muscle caused by high temperature and low pH incubation. Proc. Annu. Meet. Eur. Meat Res. Workers, Kulmbach, Germany, 24, E6, 3.

DUTSON, T.R., SMITH, G.C., HOSTETLER, R.L. and CARPENTER, Z.L. 1975. Postmortem carcass temperature and beef tenderness. J. Anim. Sci. 41, 289. (Abstract)

DUTSON, T.R., HOSTETLER, R.L. and CARPENTER, Z.L. 1976. Effects of collagen levels and sarcomere shortening on muscle tenderness. J. Food Sci. 41, 863.

DUTSON, T.R., YATES, L.D., SMITH, G.C., CARPENTER, Z.L. and HOSTETLER, R.L. 1977. Rigor onset before chilling. Proc. Recip. Meat Conf. 30, 79.

DUTSON, T.R., SMITH, G.C. and CARPENTER, Z.L. 1980A. Lysosomal enzyme distribution in electrically stimulated ovine muscle. J. Food Sci. 45, 1097.

DUTSON, T.R., SMITH, G.C., SAVELL, J.W. and CARPENTER, Z.L. 1980B. Possible mechanisms by which electrical stimulation improves meat tenderness. Proc. Eur. Meet. Meat Res. Workers, Colorado Springs, CO, 26, II, 84.

FIELDS, P.A., CARPENTER, Z.L. and SMITH, G.C. 1976. Effects of elevated temperature conditioning on youthful and mature beef carcasses. J. Anim. Sci. 42, 72.

GOLL, D.E., STROMER, M.H., ROBSON, R.M., TEMPLE, M.J., EASON, B.A. and BUSCH, W.A. 1971. Tryptic digestion of muscle components stimulates many of the changes caused by postmortem storage. J. Anim. Sci. 33, 963.

HARRIS, P.V. 1975. Meat chilling. CSIRO Food Res. Q. 35, 49.

HARRIS, P.V. and MACFARLANE, J.J. 1971. Effects of some post-slaughter treatments on the mechanical properties of meat. Proc. Eur. Meet. Meat Res. Workers, Bristol. England, 17, 102.

HAY, J.D., CURRIE, R.W., WOLFE, F.H. and SANDERS, E.J. 1973. Effect of postmortem aging on chicken muscle myofibrils. J. Food Sci. 38, 981.

HEGARTY, P.V.J. and ALLEN, C.E. 1976. Comparison of different postmortem temperatures and dissection procedures on shear values of unaged and aged avian and ovine muscles. J. Food Sci. 41, 237.

HENDERSON, D.W., GOLL, D.E. and STROMER, M.H. 1970. A comparison of shortening and Z-line degradation in bovine, porcine and rabbit muscle. Am. J. Anat. 128, 117.

HERRING, H.K., CASSENS, R.G. and BRISKEY, E.J. 1965A. Sarcomere length of free and restrained bovine muscles at low temperatures as related to tenderness. J. Sci. Food Agric. 16, 369.

HERRING, H.K., CASSENS, R.G. and BRISKEY, E.J. 1965B. Further studies on bovine muscle tenderness as influenced by carcass position, sarcomere length and fiber diameter. J. Food Sci. *30,* 1049.

HERRING, H.K., CASSENS, R.G., SUESS, G.G., BRUNGARDT, V.H. and BRISKEY, E.J. 1967. Tenderness and associated characteristics of stretched and contracted bovine muscle. J. Food Sci. *32,* 317.

HOAGLAND, R., McBRYDE, C.N. and POWICK, W.C. 1917. Changes in fresh beef during cold storage above freezing. U.S. Dep. Agric. Bull. *433.*

HOSTETLER, R.L., LANDMANN, W.A., LINK, B.A. and FITZHUGH, H.A. 1970. Influence of carcass position during rigor mortis on tenderness of beef muscles: Comparison of two treatments. J. Anim. Sci. *31,* 47.

HOSTETLER, R.L., LINK, B.A., LANDMANN, W.A. and FITZHUGH, H.A. 1972. Effect of carcass suspension on sarcomere length and shear force of some major bovine muscles. J. Food Sci. *37,* 132.

IKEUCHI, Y., ITO, T. and FUKAZAWA, T. 1978. Change in regulation of myofibrillar ATPase activity by calcium during postmortem storage of muscle. J. Food Sci. *43,* 1338.

JEREMIAH, L.E. 1978. A review of factors affecting meat quality. Agric. Can. Res. Stn., Lacombe, Alberta, Tech. Bull. *1.*

KING, N.L. and HARRIS, P.V. 1982. Heat-induced tenderization of meat by endogenous carboxyl proteases. Meat Sci. *6,* 137.

KING, N.L., KURTH, L. and SHORTHOSE, W.R. 1981. Proteolytic degradation of connectin, a high molecular weight myofibrillar protein during heating of meat. Meat Sci. *5,* 389.

KLOSE, A.A., LUYET, B.J. and MENZ, L.J. 1970. Effect of contraction on tenderness of poultry muscle cooked in the prerigor state. J. Food Sci. *35,* 577.

KOPP, J. and VALIN, C. 1981. Can muscle lysosomal enzymes affect muscle collagen postmortem? A research note. Meat Sci. *5,* 319.

KRUGGEL, W.G. and FIELD, R.A. 1971. Soluble intramuscular collagen characteristics from stretched and aged muscle. J. Food Sci. *36,* 1114.

LAWRIE, R.A. 1979. Meat Science, 3rd Edition. p. 145. Pergamon Press, New York.

LEHMANN, K.B. 1907. Studies of the causes for the toughness in meats. Arch. F. Hyg. 63, 134. As cited by Penny, I.F. 1980. Developments in Meat Science—1. p. 115. Applied Science Publishers, London.

LOCHNER, J.V., KAUFMAN, R.G. and MARSH, B.B. 1980. Early-postmortem cooling rate and beef tenderness. Meat Sci. *4,* 227.

LOCKER, R.H. 1960. Degree of muscle contraction as a factor in tenderness in beef. Food Res. *25,* 304.

LOCKER, R.H. 1982. A new theory of tenderness in meat based on gap filaments. Proc. Recip. Meat Conf. *35,* 92.

LOCKER, R.H. and DAINES, G.J. 1975. Rigor mortis in beef sternomandibularis muscle at 37°C. J. Sci. Food Agric. *25,* 1721.

LOCKER, R.H. and DAINES, G.J. 1976. Tenderness in relation to the temperature of rigor onset in cold shortened beef. J. Sci. Food Agric. *26,* 193.

LOCKER, R.H. and HAGYARD, C.J. 1963. A cold-shortening effect in beef muscles. J. Sci. Food Agric. *14,* 787.

LOCKER, R.H. and LEET, N.G. 1976A. Histology of highly stretched beef muscle. II. Further evidence on the location and nature of gap filaments. J. Ultrastruct. Res. *55,* 157.

LOCKER, R.H. and LEET, N.G. 1976B. Histology of highly stretched beef muscle. IV. Evidence for movement of gap filaments through the Z-line. J. Ultrastruct. Res. *56,* 31.

LOCKER, R.H. and WILD, D.J.C. 1982A. Yield point in raw beef muscle. The effects of ageing, rigor temperature and stretch. Meat Sci. *7,* 93.

LOCKER, R.H. and WILD, D.J.C. 1982B. Myofibrils of cooked meat are a continuum of gap filaments. Meat Sci. *7,* 189.

LOCKER, R.H., DAINES, G.J., CARSE, W.A. and LEET, N.G. 1977. Meat tenderness and the gap filaments. Meat Sci. *1*, 87.
LOWE, B. 1948. Factors affecting the palatability of poultry with emphasis on histological post mortem changes. Adv. Food Res. *1*, 232.
MacBRIDE, M.A. and PARRISH, F.C., JR. 1977. The 30,000 dalton component of tender bovine longissimus muscle. J. Food Sci. *42*, 1627.
MacFARLANE, J.J., HARRIS, P.V. and SHORTHOSE, W.R. 1974. Manipulation of meat quality, particularly tenderness, by the processor. Proc. Aust. Soc. Anim. Prod. *10*, 219.
MARSH, B.B. 1954. Rigor mortis in beef. J. Sci. Food Agric. *5*, 70.
MARSH, B.B. 1977. The nature of tenderness. Proc. Recip. Meat Conf. *30*, 69.
MARSH, B.B. and CARSE, W.A. 1974. Meat tenderness and the sliding filament hypothesis. J. Food Technol. *9*, 129.
MARSH, B.B. and LEET, N.G. 1966A. Studies in meat tenderness. III. The effects of cold-shortening on tenderness. J. Food Sci. *31*, 450.
MARSH, B.B. and LEET, N.G. 1966B. Resistance to shearing of heat-denatured muscle in relation to shortening. Nature *211*, 635.
MARSH, B.B., WOODHAMS, P.R. and LEET, N.G. 1968. Studies in meat tenderness: V. The effect on tenderness of carcass cooling and freezing before completion or rigor mortis. J. Food Sci. *33*, 12.
MARSH, B.B., LEET, N.G. and DICKSON, M.R. 1974. The ultrastructure and tenderness of highly cold-shortened muscle. J. Food Technol. *9*, 141.
MARSH, B.B., LOCHNER, J.V., TAKAHASHI, G. and KRAGNESS, D.D. 1981. Effects of early postmortem pH and temperature on beef tenderness. A research note. Meat Sci. *5*, 479.
MARTIN, A.H., FREDEEN, H.T. and WEISS, G.M. 1971. Tenderness of beef longissimus dorsi muscle from steers, heifers, and bulls as influenced by source, postmortem aging, and carcass characteristics. J. Food Sci. *36*, 619.
MARTIN, A.H., JEREMIAH, L.E., FREDEEN, H.T. and L'HIRONDELLE, P.J. 1977. The influence of pre- and post-rigor muscle changes on intrinsic toughness of beef carcasses. Can. J. Anim. Sci. *57*, 705.
MATSUKURA, U., OKITANI, A., NISHIMURO, T. and KATO, H. 1981. Mode of degradation of myofibrillar proteins by an endogenous protease, cathepsin L. Biochim. Biophys. Acta *662*, 41.
McCRAE, S.E., SECCOMBE, C.G., MARSH, B.B. and CARSE, W.A. 1971. Studies in meat tenderness. IX. The tenderness of various lamb muscles in relation to their skeletal restraint and delay before freezing. J. Food. Sci. *36*, 566.
MOELLER, P.W., FIELDS, P.A., DUTSON, T.R., LANDMANN, W.A. and CARPENTER, Z.L. 1976. Effect of high temperature conditioning on subcellular distribution and levels of lysosomal enzymes. A research note. J. Food Sci. *41*, 216.
MOELLER, P.W., FIELDS, P.A., DUTSON, T.R., LANDMANN, W.A. and CARPENTER, Z.L. 1977. High temperature effects on lysosomal enzyme distribution and fragmentation of bovine muscle. J. Food Sci. *42*, 510.
NEWBOLD, R.P. and HARRIS, P.V. 1972. The effect of pre-rigor changes on meat tenderness. A review. J. Food Sci. *37*, 337.
OLSON, D.G. and PARRISH, F.C., JR. 1977. Relationship of myofibril fragmentation index to measures of beef steak tenderness. J. Food Sci. *42*, 506.
OUALI, A. and VALIN, C. 1981. Effect of muscle lysosomal enzymes and calcium activated neutral proteinase on myofibrillar ATPase activity: Relationship with ageing changes. Meat Sci. *5*, 233.
PARRISH, F.C., JR. 1977. Skeletal muscle tissue disruption. Proc. Recip. Meat Conf. *30*, 87.
PARRISH, F.C., JR. 1978. Changes in myofibrillar proteins during postmortem tenderization of bovine muscle. Proc. Annu. Meet. Eur. Meat Res. Workers, Kulmbach, Germany, *24*, E4, 3.
PARRISH, F.C., JR., GOLL, D.E., NEWCOMB, W.J., II, deLUMEH, B.O., CHAUDHRY, H.M. and KLINE, A.E. 1969. Molecular properties of postmortem muscle.

VII. Changes in nonprotein nitrogen and free amino acids of bovine muscle. J. Food Sci. *34,* 196.
PARRISH, F.C., JR., YOUNG, R.B., MINER, B.E. and ANDERSON, L.D. 1973. Effect of postmortem conditions on certain chemical, morphological and organoleptic properties of bovine muscle. J. Food Sci. *38,* 690.
PARRISH, F.C., JR., SELVIG, C.J., CULLER, R.D. and ZEECE, M.G. 1981. CAF activity, calcium concentration, and the 30,000-dalton component of tough and tender bovine longissimus muscle. A research note. J. Food Sci. *46,* 308.
PARTMANN, W. 1963. Postmortem changes in chilled and frozen muscle. J. Food Sci. *28,* 15.
PENNY, I.F. 1980. The enzymology of conditioning. *In* Developments in Meat Science—1. p. 115. R.A. Lawrie (Editor). Applied Science Publishers, London.
PENNY, I.F. and DRANSFIELD, E. 1979. Relationship between toughness and troponin T in conditioned beef. Meat Sci. *3,* 135.
PENNY, I.F. and FERGUSON-PRYCE, R. 1979. Measurement of autolysis in beef muscle homogenates. Meat Sci. *3,* 121.
PENNY, I.F., VOYLE, C.A. and DRANSFIELD, E. 1974. The tenderizing effect of a muscle proteinase on beef. J. Sci. Food Agric. *25,* 703.
PFEIFFER, N.E., FIELD, R.A., VARNELL, T.R., KRUGGEL, W.G. and KAISER, I.I. 1972. Effects of postmortem aging and stretching on the macro-molecular properties of collagen. J. Food Sci. *37,* 897.
PIERSON, C.J. and FOX, J.D. 1976. Effect of postmortem aging time and temperature on pH, tenderness and soluble collagen fractions in bovine longissimus muscle. J. Anim. Sci. *43,* 1206.
PURCHAS, R.W. and DAVIES, H.L. 1974A. Carcass and meat quality of Friesian steers fed on either pasture or barley. Aust. J. Agric. Res. *25,* 183.
PURCHAS, R.W. and DAVIES, H.L. 1974B. Meat production of Friesian steers: The effect of intramuscular fat on palatability and the effect of growth rates on composition changes. Aust. J. Agric. Res. *25,* 667.
QUARRIER, E., CARPENTER, Z.L. and SMITH, G.C. 1972. A physical method to increase tenderness in lamb carcasses. J. Food Sci. *37,* 130.
RAMSBOTTOM, J.M. and STRANDINE, E.J. 1949. Initial physical and chemical changes in beef as related to tenderness. J. Anim. Sci. *8,* 398.
RHODES, D.N. and DRANSFIELD, E. 1974. Mechanical strength of raw beef from cold-shortened muscles. J. Sci. Food Agric. *25,* 1163.
ROBBINS, F.M., WALKER, J.E., COHEN, S.H. and CHATTERJEE, S. 1979. Action of proteolytic enzymes on bovine myofibrils. J. Food Sci. *44,* 1672.
ROBSON, R.M., GOLL, D.E. and MAIN, M.J. 1967. Molecular properties of postmortem muscle. 5. Nucleoside triphosphate activity of bovine myosin B. J. Food Sci. *32,* 544.
SAMEJIMA, K. and WOLFE, F.H. 1976. Degradation of myofibrillar protein components during postmortem aging of chicken muscle. J. Food Sci. *41,* 250.
SCHWARTZ, W.N. and BIRD, J.W.C. 1977. Degradation of myofibrillar proteins by cathepsins B and D. Biochem. J. *167,* 811.
SLEETH, R.B., HENDRICKSON, R.L. and BRADY, D.E. 1957. Effect of controlling environmental conditions during aging on the quality of beef. Food Technol. *11,* 205.
SLEETH, R.B., KELLEY, G.G. and BRADY, D.E. 1958. Shrinkage and organoleptic characteristics of beef aged in controlled environments. Food Technol. *12,* 86.
SMITH, G.C., ARANGO, T.C. and CARPENTER, Z.L. 1971. Effects of physical and mechanical treatments on the tenderness of the beef longissimus. J. Food Sci. *36,* 445.
SMITH, G.C., DUTSON, T.R., HOSTETLER, R.L. and CARPENTER, Z.L. 1976. Fatness, rate of chilling and tenderness of lamb. J. Food Sci. *41,* 748.
SMITH, G.C., CULP, G.R. and CARPENTER, Z.L. 1978. Postmortem aging of beef carcasses. J. Food Sci. *43,* 823.
SONAIYA, E.B., STOUFFER, J.R. and BEERMAN, D.H. 1982. Electrical stimula-

tion of mature cow carcasses and its effect on tenderness, myofibril protein degradation and fragmentation. J. Food Sci. *47,* 889.

STAGNI, N. and DeBERNARD, B. 1968. Lysosomal enzyme activity in rat and beef skeletal muscle. Biochim. Biophys. Acta *170,* 129.

STOUFFER, J.R. 1977. Post mortem factors affecting tenderness—muscle restraint. Proc. Recip. Meat Conf. *30,* 75.

TAKAHASHI, K. and SAITO, H. 1979. Postmortem changes in skeletal muscle connectin. J. Biochem. *85,* 1539.

VOYLE, C.A. 1969. Some observations on the histology of cold-shortened muscle. J. Food Technol. *4,* 275.

WEIDEMANN, J.F., KAESS, G. and CARRUTHERS, L.D. 1967. The histology of pre-rigor and post-rigor ox muscle before and after cooking and its relation to tenderness. J. Food Sci. *32,* 7.

WEISMAN, G. 1964. Labilization and stabilization of lysosomes. Fed. Proc. Fed. Am. Soc. Exp. Biol. *23,* 1038.

WILL, P.A., OWNBY, C.L. and HENRICKSON, R.L. 1980. Ultrastructural postmortem changes in electrically stimulated bovine muscle. J. Food Sci. *45,* 21.

WILSON, G.D., BROWN, P.D., POHL, C., WEIR, C.E. and CHESBRO, W.R. 1960. A method for the rapid tenderization of beef carcasses. Food Technol. *14,* 186.

WU, J.J., DUTSON, T.R. and CARPENTER, Z.L. 1981. Effect of post-mortem time and temperature on the release of lysosomal enzymes and their possible effect on bovine connective tissue components of muscle. J. Food Sci. *46,* 1132.

WU, J.J., DUTSON, T.R. and CARPENTER, Z.L. 1982. Effect of post-mortem time and temperature on bovine intramuscular collagen. Meat Sci. *7,* 161.

YAMAMOTO, K., SAMEJIMA, K. and YASUI, T. 1977. A comparative study of the changes in hen pectoral muscle during storage at 4°C and −20°C. J. Food Sci. *42,* 1642.

YATES, L.D., DUTSON, T.R., CALDWELL, J. and CARPENTER, Z.L. 1983. Effect of temperature and pH on the postmortem degradation of myofibrillar proteins. Meat Sci. *9,* 157.

YOUNG, O.A., GRAAFHUIS, A.E. and DAVEY, C.L. 1981. Postmortem changes in cytoskeletal proteins of muscle. Meat Sci. *5,* 41.

3

Electrical Stimulation: Its Early Development in New Zealand

B. B. Chrystall[1] and C. E. Devine[1]

Rigor Mortis
Toughness
Electrical Stimulation
Summary
Acknowledgments
References

The current interest in the use of electrical stimulation as a desirable element in meat processing can be traced back to meat science developments in the late 1940s and the 1950s. We believe a debt is owed to the research workers of the time for their concerted efforts in extending the new biochemical and physiological discoveries in muscle function to the events of rigor mortis (Bate-Smith 1948; Bendall 1951; Lawrie 1953; Marsh 1954; Howard and Lawrie 1956). Their efforts have laid the foundations for much of our present knowledge of muscle tissue as a food. Particularly is this the case with developments in electrical stimulation. Comprehensive reviews of the biochemistry and physiology of rigor mortis in striated, skeletal muscle are available (Bendall 1973, 1978; Hamm 1977) as are general reviews of muscle structure (Squire 1981; Franzini-Armstrong and Peachey 1981). In all, they offer the meat scientist a sound understanding of muscle tissue and its conversion through rigor mortis to meat.

Electrical stimulation was frequently used in much earlier times to gain insights into muscle function. Application of bimetallic wire to frog nerve-muscle preparations to show contraction was first reported by Aloysio Galvani in the 1780s (see Needham 1971). From Galvani's time, the interest in using electricity to study muscles increased and during the early 1800s attempts were even made to use electricity to resuscitate corpses. There is one example of reported success in 1818 when Professor James Jeffray of Glasgow University stimulated the corpse of a recently-hanged murderer. The corpse is supposed to have twitched violently and then, to

[1]Meat Industry Research Institute of New Zealand, Inc., P.O. Box 617, Hamilton, New Zealand.

the horror of those present, eventually recovered enough to stand, only to be despatched by more permanent means (MacKenzie 1876; Bonner 1978).

The earliest reported use of electricity on meat animals was the killing of turkeys with electric shock carried out as a party act by Benjamin Franklin in 1749 (see Lopez and Herbert 1975). Although the electrocuted turkeys were "uncommonly tender," we feel it is likely that this was more a consequence of prerigor roasting of the birds than it was of electrical stimulation. Some years later, the French, still pursuing the idea of putting electricity at the service of gastronomy, requested Franklin to kill larger animals with electricity and hasten tenderization. However, Franklin discouraged the idea. So stimulation of larger animals seems to have been bypassed for a further 200 years, although it was suggested around the turn of the century in a New Zealand advertisement (Fig. 3.1).

Then, in the early 1950s, Harsham and Deatherage (1951) and Rentschler (1951), supported by Kroger Co. and Westinghouse Corporation, investigated the use of electrical stimulation as an aid to improving the tenderness of meat. They showed that a brief period of electrical stimulation of beef carcasses helped to tenderize the meat and they suggested that this was due to enhancement of endogenous enzymic activity. No

FIG. 3.1. Electrical stimulation is forecast in this advertisement that appeared in New Zealand around the turn of the century.
From Scott (1973).

commercial use of the process was made after this more recent work which, in consequence, was largely ignored.

Although there was no further work on the use of electrical stimulation to improve tenderness, it was used experimentally to hasten rigor mortis development and to modify parts of the glycolytic pathway in muscle. One study involved stimulation of horse m. sternocephalicus to hasten rigor so that bacterial growth on prerigor and rigor muscles from the same animal could be readily compared (Ingram and Ingram 1955). The effect of electrical stimuli in accelerating steps of the glycolytic pathway was studied by Cori (1945) and later by others, including Karpatkin et al. (1964) and Posner et al. (1965).

The present interest in electrical stimulation as a new element in meat processing lies in the fact that it greatly reduces the risk of cold shortening in muscles by hastening the development of rigor mortis (Chrystall and Hagyard 1976; Davey et al. 1976), increases the rate of tenderization by aging (Savell et al. 1978; George et al. 1980), improves meat color (Savell 1979), reduces "heat ring" (Savell et al. 1978), and prolongs retail shelf-life (Riley et al. 1980).

The use of electrical stimulation to determine the effect of rate of rigor development on tenderness was considered in the excellent work of de Fremery and Pool (1960). They concluded that the rapid rigor induced by electrical stimulation produced toughening, although from later work (de Fremery and Pool 1963) it became evident that this was not so. The toughening that had initially been observed was likely to have developed in muscle which entered rigor while the electrical stimuli were being applied (Chrystall et al. 1982B).

Electrical stimulation of pig muscles in studies of watery pork (Hallund and Bendall 1965) clearly demonstrated that an irreversible acceleration of glycolysis was induced by a 1 to 2 min stimulation of prerigor muscle. A similar irreversible hastening of rigor development can be induced merely by a period of intense nervous stress just prior to slaughter (Forrest and Briskey 1967; McLoughlin 1970). It would appear that because of the risk of inducing the pale, soft exudative (PSE) condition in beef, the studies of Harsham and Deatherage (1951) went unregarded until a specific need arose in New Zealand to overcome the toughening from prerigor chilling and freezing of lamb.

Rediscovery of the use of electrical stimulation for meat animals (Carse 1973) occurred in New Zealand largely as a consequence of the realization that frozen lamb could be tough, and this set the scene for the considerable studies in this country on the effect of refrigeration on the eating quality of lamb. Much of the work has appeared in recent reviews (Locker et al. 1975; Davey and Winger 1980).

Since cold shortening and, indeed, thaw shortening occur only in prerigor muscle, it is evident that these can be prevented by ensuring that rigor mortis is achieved before meat is either chilled or frozen. Rigor development in meat-producing animals can take up to 36 hr, so the aim of the meat scientist has been to hasten rigor development. Early muscle studies indicated that rigor development can be brought forward by antemortem stress (Bate-Smith and Bendall 1949), by maintaining a high carcass temperature (Roschen et al. 1950), or by electrical stimulation of carcasses (Harsham and Deatherage 1951). On humane grounds, the first of these approaches is hardly acceptable. The second has been found to work successfully but is expensive to apply and has not had wide application. Electrical stimulation of carcasses soon after slaughter emerges as the most ingenious and practical approach in inducing early rigor. Not only does it allow meat to be chilled and frozen very soon after slaughter without the damaging toughening of cold and thaw shortening, but it also ensures that further aging will achieve its tenderizing purpose.

This review of research into electrical stimulation is intended to report our experience in New Zealand, a country which is economically dependent on an efficient meat production and export industry. Since more than half its output is sheep meat, which must be frozen within a day of slaughter, the New Zealand meat industry has had an almost exclusive interest in determining the effect of cold on the various qualities of sheep meat, of which tenderness is only one. It is with no parochial intent that most of the work here refers to the New Zealand experience. The host of other studies reported from around the world are acknowledged as essential inspiration in taking our own studies further forward, and in greater depth.

It is understandable in situations where research and related technological transfer are occurring in parallel that a chronological report of developments would become confusing to the reader. For this reason we have ordered our presentation, first by reviewing the basic investigations into electrical stimulation and its effect on rigor development, and second by describing its commercial applications within the New Zealand meat industry, particularly to the lamb processing line.

RIGOR MORTIS

The merit of using electrical stimulation in meat technology is intimately associated with the development of rigor mortis in muscle, with the cold shortening phenomena, with thaw rigor, and with meat aging. Brief attention is therefore given to these topics in preparation for the more extensive discussion on electrical stimulation which follows.

The Development of Rigor Mortis

After an animal dies, its muscles live on in the prerigor state. They are reversibly extensible and can be excited to contract and do work. If starved of oxygen, they enter rigor mortis some minutes or hours postmortem and become nonexcitable and rigid. From extensive studies of rigor development (Bendall 1973), it is evident that the biochemical changes are likely to be the same for all vertebrate species, with the disappearance of ATP and creatine phosphate and, as the product of glycolysis, the appearance of considerable quantities of lactate ions (up to 100 µmol/g muscle). The ultimate pH of muscle is inversely related to the accumulation of these ions, being relatively high (approximately pH 7.0) in excessively exhausted animals and low (approximately pH 5.5) in well fed and rested animals.

Shortening

A prerigor muscle usually shortens slightly on excision to reach its equilibrium length when held horizontally. At temperatures above about 20°C, the muscle also shortens further with rigor development. Such rigor or heat shortening is small at 20°C (less than 5% of the equilibrium length) but rises to as much as 50% at 40°C. Contrary to expectation, muscle from beef animals also shortens by 50% or more prior to and during the development of rigor mortis if held at 2°C (Locker and Hagyard 1963). The magnitude of this so-called cold shortening increases with the fall of temperature toward 0°C. However, it is reduced by increased postmortem delay before chilling. In most respects, the phenomenon is similar to normal contraction involving dephosphorylation of ATP (Newbold 1966).

Shortening also accompanies the development of thaw rigor and is especially evident in unrestrained muscles (Bendall 1973). The shortening occurs on thawing muscles which have been frozen at any point up to almost full rigor development, although as the ATP concentration in the muscle falls, so does the power generated during shortening. Thaw shortening (greater than 50% in strips of excised lamb muscle) still occurs at an ATP concentration of about 0.3 µmol/g muscle (Davey and Gilbert 1976).

To what extent can cold and thaw shortenings occur in a carcass in which most of the skeletal attachments are intact? A few muscles, such as the muscles of the neck, are severed during carcass dressing and are likely to shorten. Some other muscles, notably the m. longissimus, remain attached to the skeleton, but since most of the constituent fibers insert into flexible epimysium, the muscles are able to shorten. Even muscles that are fixed at both ends are capable of cold shortening over a part of their length if they are subjected to a differential chilling rate along their length. This has been demonstrated in restrained ox m. sternomandibularis insulated at both ends and held at 2°C; a zone of marked shortening occurs in the more rapidly chilled, central region with concomitant stretching at the two insulated ends (Marsh and Leet 1966).

TOUGHNESS

Measurement

All shearing measurements reported in this chapter have been obtained with the MIRINZ tenderometer, which measures toughness by applying an increasing load across the muscle fibers of a cooked meat sample at a constant rate (Macfarlane and Marer 1966). The blunt wedge of the tenderometer compresses the meat against a base plate, and, at a certain loading determined by the toughness of the cooked meat, a sharp yielding point is indicated on a load-deformation curve. Shearing forces are measured on an arbitrary scale of 150 force score (FS) units, each being equivalent approximately to a stress of 1.7 kPa.

The relationship between FS values and taste panel score is not a linear one. From many correlations of panel scores and FS values, cooked lamb is regarded as being very tender at shearing values below FS 25, tender from FS 25 to FS 40, and undesirably tough if shearing forces rise much above FS 40. (McCrae et al. 1971). The threshold for beef is considered to be somewhat lower at FS 35.

Toughness and Shortening

The m. sternomandibularis of the ox, running from the sternum to the mandible, is readily available very soon after slaughter. Its uniform, parallel-fibered microanatomy makes it ideal for most of the basic studies reported here. Although it is not part of a commercial cut, having a high connective tissue content, we believe it gives a true picture of the properties of meat tissue. Differences among muscles are of degree rather than of basic characteristics.

Muscle shortening during the early postmortem period affects cooked-meat toughness to a remarkable degree (Locker 1960). The force required to shear the cooked m. sternomandibularis across the fiber direction rises steeply as muscle shortening increases to 40% (Marsh and Leet 1966) (Fig. 3.2), then declines sharply with further shortening toward 60%. Thus, a prominent peak of toughness exists in the cooked muscle with its maximum at about 40% shortening. Cold and thaw shortenings initiated at any stage in the development of rigor produce the same relationship. The peak is encountered in the m. sternomandibularis muscles from animals from all maturities (Davey and Gilbert 1975). This relationship demonstrates above all that the greatest possible cause of toughening in cooked meat is the shortening produced in prerigor muscle; animal age is merely a secondary cause (Locker et al. 1975).

Meat Aging

If rigor muscle is aged for a number of days before cooking, it usually becomes appreciably more tender. It appears that the onset of tenderizing begins at about the time muscles enter rigor mortis, although the point needs further investigation. Force scores of unshortened m. sternomandibularis, for example, held at 15°C, fall from about FS 65 to 25 over a period of 2.5 days, with a slight additional decline to FS 15 on storage for 8 days. The shearing values of muscles shortened by 35%, in contrast, will

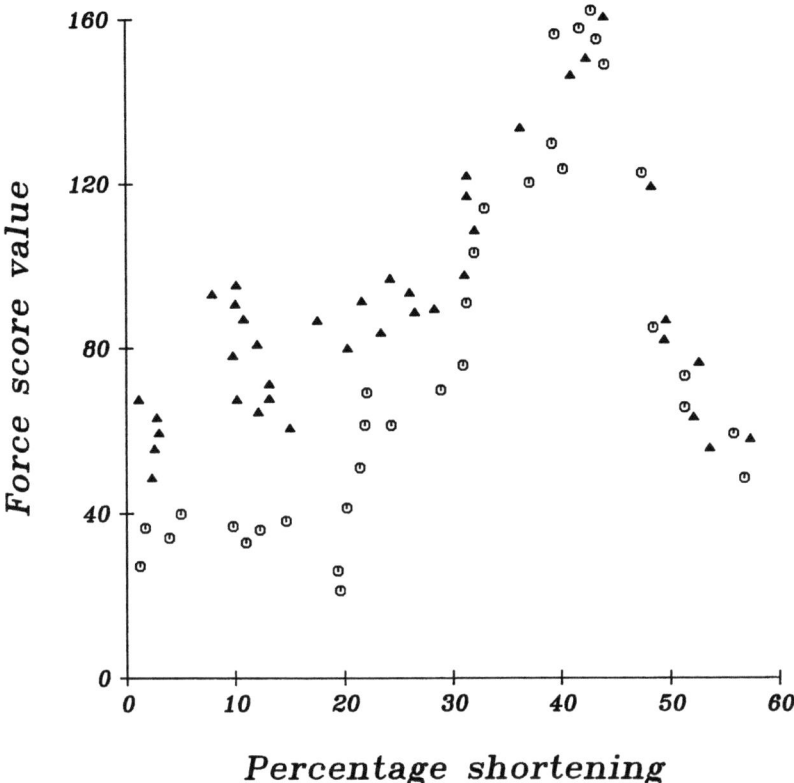

FIG. 3.2. Influence of prerigor muscle shortening on the force score values of unaged and aged beef m. sternomandibularis. ▲ Rigor muscle. ⓘ Muscle aged for 3 days at 15°C after setting in rigor at different degrees of shortening.
After Davey et al. (1967).

similarly fall during 2.5 days storage at 15°C, but in this case from an initial value of FS 95 to a minimum of FS 75 (Davey et al. 1967). Two significant conclusions can be drawn from these studies. First, there is a limit to the tenderizing achieved by aging of m. sternomandibularis, and, second, shortening reduces the degree of tenderizing it is possible to achieve by aging.

The effect of maximal aging on the shortening–toughening relationship is shown in Fig. 3.2. It is stressed that the complete loss of tenderizing capacity at 40% shortening is not a characteristic of all muscles, although high shortenings greatly reduce the tenderizing achieved by aging in all those that have been studied (Herring et al. 1967).

Two quite clear processing requirements arise from our knowledge of rigor mortis, muscle shortening, and meat aging. First, carcasses should

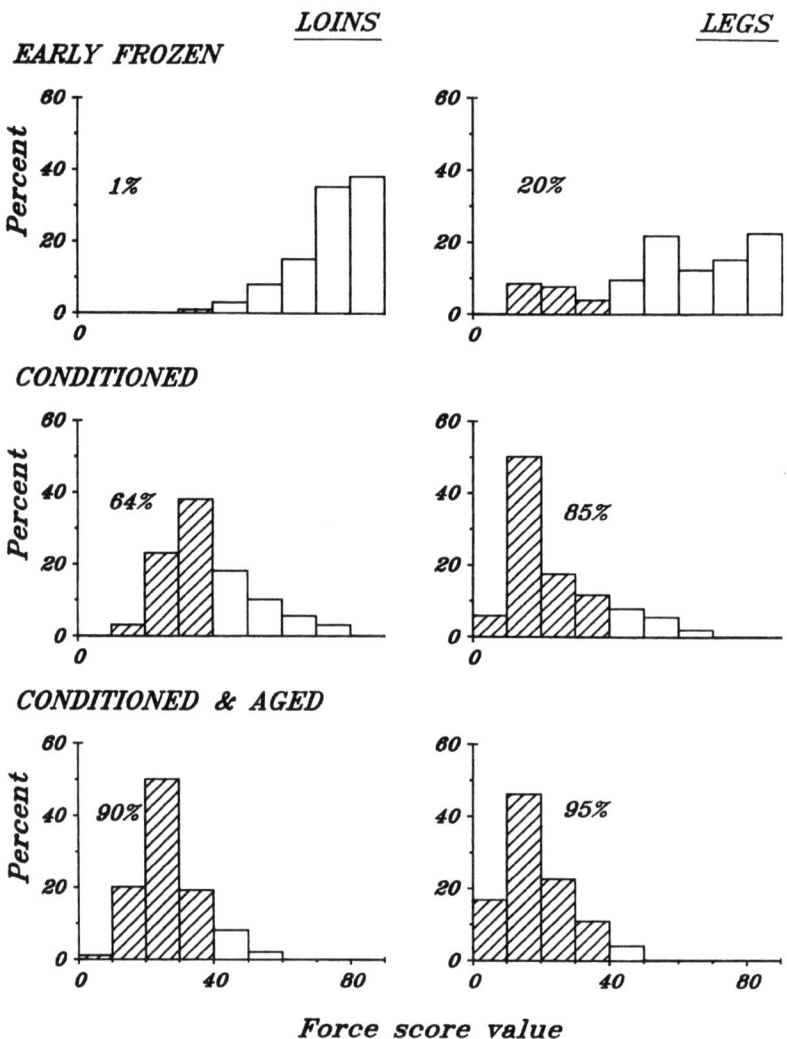

FIG. 3.3. Histograms indicating distribution of force score values for lamb loins and legs cooked from the frozen state after processing differently. In this and other FS histograms, the proportion with FS < 40 is shaded and shown by the figure on the histogram. Early frozen—Lambs were moved into freezers less than 1 hr after slaughter. Conditioned—Lambs were held for 16–24 hr at 15°C before being frozen. Conditioned and aged—Lambs were held at 15°C for 48 hr prior to freezing. The freezing time (to −4°C deep leg) for all carcasses was 12 hr from freezer entry. The storage temperature of −18°C was reached in a further 2 hr.

be "conditioned," (Locker et al. 1975), i.e., allowed to go into rigor mortis before chilling or freezing, to avoid the toughening of cold or thaw shortening. Second, only if shortening is avoided is the full tenderness potential of the meat achieved through aging. It is emphasized that conditioning and aging are quite distinct processes although they may overlap in practice.

Toughness in Conventional Processing Practice

To what extent are cold and thaw shortenings signficant in conventional meat processing practice? Traditionally, in New Zealand, lamb carcasses have been blast frozen very shortly after slaughter. As a consequence, many muscles are toughened appreciably by cold shortening during chilling and, if the cuts are cooked from the frozen state, by thaw shortening during cooking (Marsh et al. 1968). The effect of prefreezing delay on such toughening in lamb is shown in Fig. 3.3. With minimal delay, virtually all

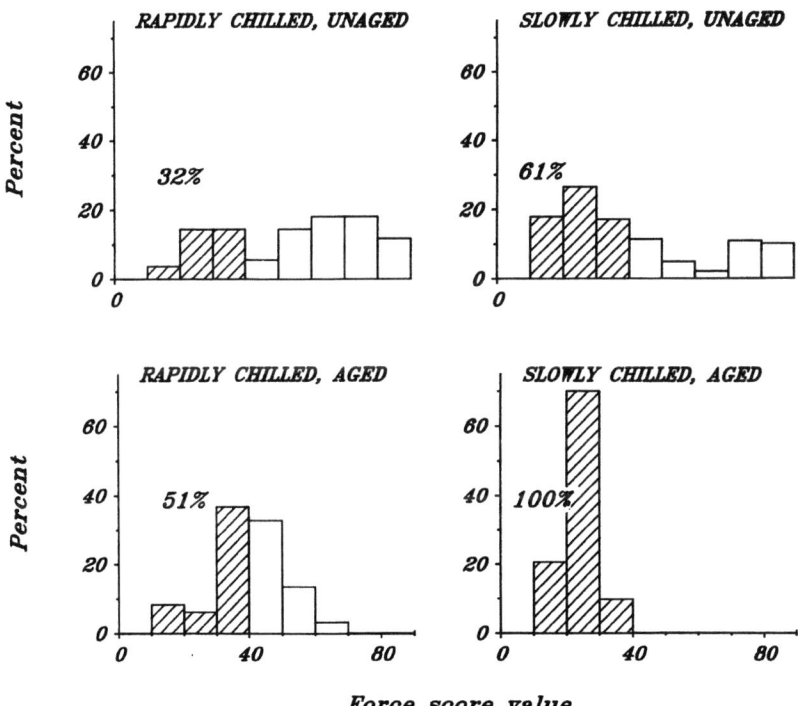

FIG. 3.4. Histograms indicating distribution of force score values for cooked m. longissimus from rapidly and slowly chilled beef sides with and without subsequent aging. Muscles were excised at 18 hr postmortem. Unaged samples were then cooked, whereas the aged samples were held a further 48 hr at 15°C before being cooked.
Redrawn from Davey (1971).

the meat from the loins and 80% of that from the legs was unacceptably tough (FS greater than 40). However, if the carcasses were held above 10°C for 16 to 24 hr to enter rigor mortis, almost 70% of the meat from the loin cuts and 85% of that from the legs was then acceptable. By extending the holding time a further 24 hr, all the meat from both cuts was then acceptably tender.

The chilling or freezing of beef sides can also be sufficiently rapid to toughen the commercially important cuts. A clear demonstration of this has been obtained in a number of large-scale studies. In one (Davey 1971), sides of prime beef carcasses were chilled slowly to a deep leg temperature of 17°C during 18 hr while the other sides were relatively rapidly chilled to 8°C. The effects of the two chilling rates on the tenderness of m. longissimus from the thoracic region are illustrated in Fig. 3.4. Rapid chilling toughened the loin muscle with some 80% of the cooked meat being unacceptably tough (FS greater than 35). The toughened meat did not tenderize to an acceptable level during aging (3 days at 10°C) as shown by the fact that the shearing forces for half the samples remained above FS 35. Very similar results were obtained in experiments using young bulls, cows, and calves; toughness increased with the increased rate of chilling (Buchter 1972). More recently, Martin et al. (1977) has confirmed these studies during the examination of the effect of animal stress and glycolytic rate on toughness development in chilled beef sides.

ELECTRICAL STIMULATION

Measurement of Effects

Rate of Rigor Onset. As with most other muscle research at the Meat Industry Research Institute of New Zealand (MIRINZ), much of the initial experimentation has been carried out with beef m. sternomandibularis. Subsequently, the findings have been extended to other muscles in both sheep and cattle, although in a few instances, the rat has provided tissue for an intermediate experimental step.

The effects of electrical stimulation on hastening rigor development can be assessed in a number of ways. Although an indirect measure, postmortem decline in muscle pH has been used extensively to monitor the development of rigor mortis. Muscle pH values have been determined in iodoacetate/H_2O homogenates at 15°C (Bendall 1973). The fall in pH on stimulation of beef m. sternomandibularis occurs in two stages (Chrystall and Devine 1978) (Fig. 3.5): a fall during the stimulation period itself (ΔpH) at a rate that is 100 to 150 times that of the normal nonstimulated rate; and a fall following cessation of stimulation, the rate (dpH/dt) of which is sustained at 1.5 to 2 times higher than the normal nonstimulated rate. Regardless of stimulation refinements, dpH/dt never seems to be

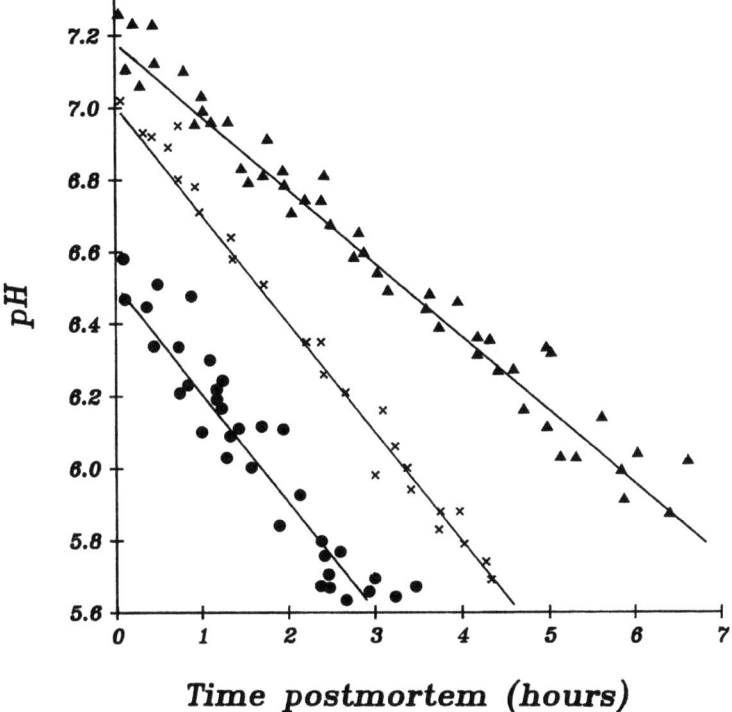

FIG. 3.5. Postmortem acceleration of pH fall in beef m. sternomandibularis caused by electrical stimulation, with 200 V peak, 12.5 pulses/sec. ▲ Unstimulated control stored at 35°C, 0.18 pH units/hr. × Stimulated for 5 sec, then stored at 35°C, 0.30 pH units/hr. ● Stimulated for 120 sec, then stored at 35°C, 0.30 pH units/hr.
From Chrystall and Devine (1978).

more than twice that of nonstimulated muscles. Furthermore, such a rate can sometimes be induced merely by vigorous muscle movement during slaughter (Devine et al. 1979; Chrystall et al. 1980B).

It is the magnitude of ΔpH that largely determines the time delay before muscles lose their capacity to cold shorten. Although Locker and Hagyard (1963) stated that cold shortening does not occur in muscles much below pH 6.3, we have used the threshold figure of pH 6.0 (Bendall 1975) since it offers a greater margin of surety in this respect. The time for muscles to reach pH 6.0 measures the combined effects of ΔpH and dpH/dt and provides a useful measure with which to evaluate the effect of stimulation in inducing early rigor. As an example, the pH of nonstimulated beef m. sternomandibularis at 35°C falls at 0.18 pH units/hr, thus taking nearly 7 hr to reach pH 6.0. In contrast, in maximally stimulated m. sternomandibularis, ΔpH can be as high as 0.7 pH units, and dpH/dt 0.30 pH units/hr, and pH 6.0 is reached in less than 1.5 hr.

Toughness. By inducing early rigor through carcass electrical stimulation, the risks of toughening through cold shortening during chilling and thaw shortening during thawing are remarkably reduced. We believe the advantage of electrical stimulation lies in the reconciliation of the conflicting requirement of early-postmortem refrigeration as an aid to preservation with that of delaying refrigeration until muscles are in rigor to ensure tender meat.

Tenderness measurements provide the only realistic evaluation of a complete process. In New Zealand, tenderness assessments are carried out on meat cooked either as individual muscles or portions in a water bath, or as lamb loin and leg cuts by roasting (McCrae *et al.* 1971), or in the case of beef cuts either by grilling or roasting. Generally, since most of New Zealand meat is exported as frozen product, evaluations have been carried out on meat cooked without prior thawing. We believe that if the meat is tender under these conditions it will be even more tender under all other cooking treatments.

Stimulation Parameters

We have determined the effect of a number of interacting variables on tenderness. Some of the variables relate to the animal, some to the electrical characteristics, and some to the processing conditions following stimulation. In this chapter, unless specified, low voltages refer to those that are below the safety threshold (30–50 V rms) set by electrical supply authorities. High voltages refer to those that are substantially above this safety threshold.

Animal-related Factors. These are the factors that depend on the animal and its interaction with the environment.

Muscle Temperature and Postmortem Delay Before Stimulation. The influence of temperature on the rate of pH fall in muscle has been considered in many publications, notably Bate-Smith and Bendall (1949), Marsh (1954), and Jeacocke (1977). More recently, the temperature dependence of ΔpH in stimulated muscles and dpH/dt in nonstimulated and stimulated muscles have been determined. As the muscle temperature falls, the ΔpH produced by stimulation is reduced. For example, in beef m. sternomandibularis, ΔpH ranges from 0.6 pH units at 35°C to 0.018 units at 15°C. From such determinations of temperature dependence, the energy of activation of ΔpH in stimulated m. sternomandibularis is calculated to be 97 kJ/mol, or very similar to that for calcium-activated actomyosin ATPase (Bendall 1969, 1975; Jeacocke 1977). Indicating a sensitive glycolytic response to temperature, the energy of activation of $d\mathrm{pH}/dt$ in nonstimulated beef m. sternomandibularis is 40–45 kJ/mol. The $d\mathrm{pH}/dt$ of stimulated muscle is even more sensitive to temperature, with the energy of activation approaching 70 kJ/mol. Because of such sensitivity, temperature control is a critical requirement in any useful study designed to optimize stimulation parameters. Of more fundamental interest is the change brought about by stimulation in the energy of activation associated with $d\mathrm{pH}/dt$. It is presumed that the rate-controlling step of glycolysis in nonstimulated muscle is different from that in stimulated muscle. Recent evidence suggests that this change may be related to an alteration in the binding of glycolytic enzymes to actin filaments (Clarke et al. 1980; Walsh et al. 1981).

A postmortem delay in stimulation allows glycolysis to proceed normally. If the delay is protracted, then the muscle pH can have fallen

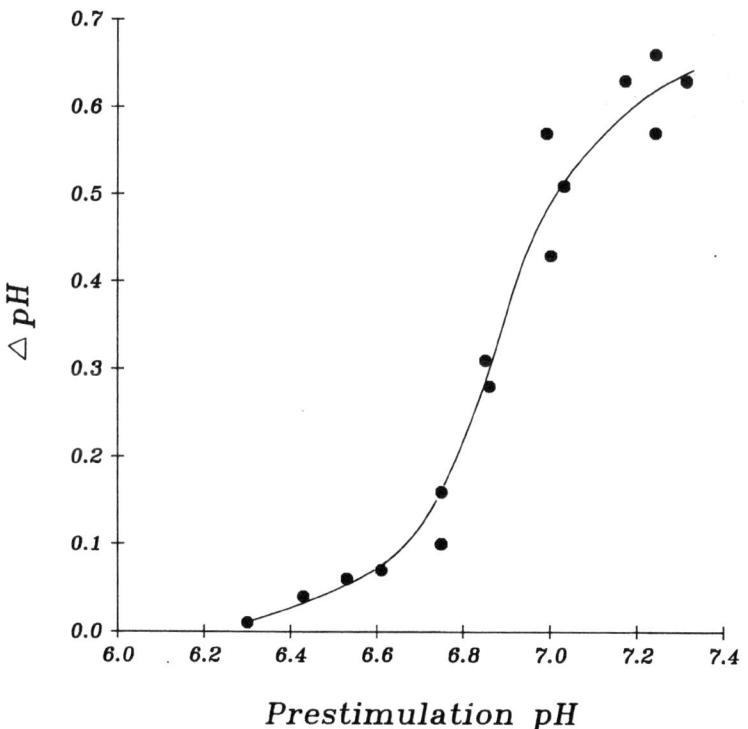

FIG. 3.6. Relationship between prestimulation pH and ΔpH in beef m. sternomandibularis stimulated for 120 sec with 200 V peak, 12.5 pulses/sec at 35°C.
From Chrystall and Devine (1978).

considerably, reducing the magnitude of the ΔpH achieved on eventually applying the stimulus (Chrystall and Devine 1978) (Fig. 3.6). The effect of postmortem delay on muscle response to stimulation is best illustrated in tension traces obtained during low-voltage stimulation of sheep muscles. With short delays (15 to 20 min), the strong response is largely a consequence of nervous recruitment. As the delay increases beyond 30 min, the elicited response is then largely due to direct muscle stimulation (Fig. 3.7) and requires higher voltages to produce a maximal response. Further evidence of the decay of nervous effect has been obtained in studies of pH fall in muscles of isolated lamb legs held at constant temperature (Chrystall et al. 1980A,B) (Fig. 3.8).

Studies using the neuromuscular blocker, curare, support the view that a functional nervous system is essential if low voltage stimulation is to be effective (Devine et al. 1979), but it is less essential at high voltages.

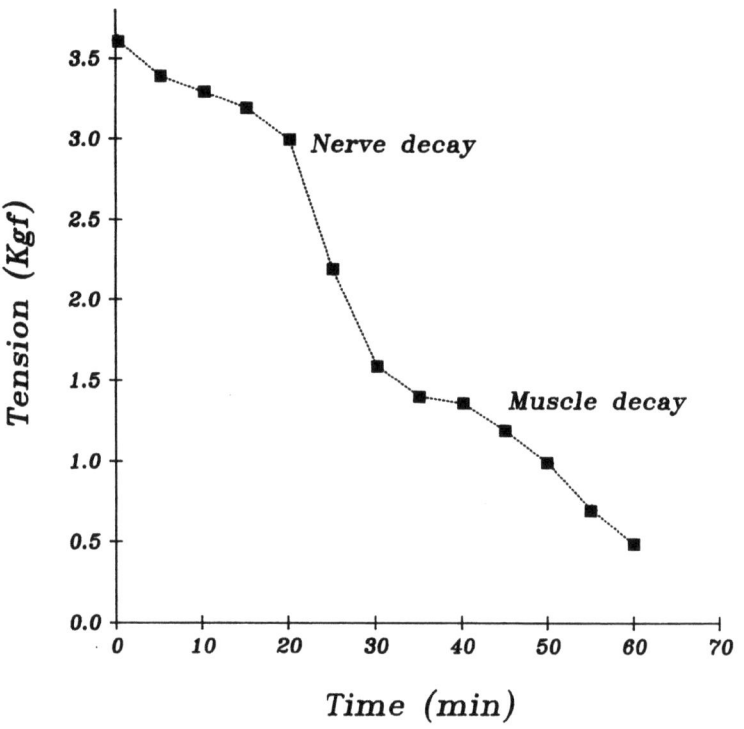

FIG. 3.7. Graph of peak tension produced by the m. gastrocnemius of a lamb stimulated periodically with 1-sec bursts of 15 V peak square wave pulses at 14.28/sec. Zero time is time of slaughter.

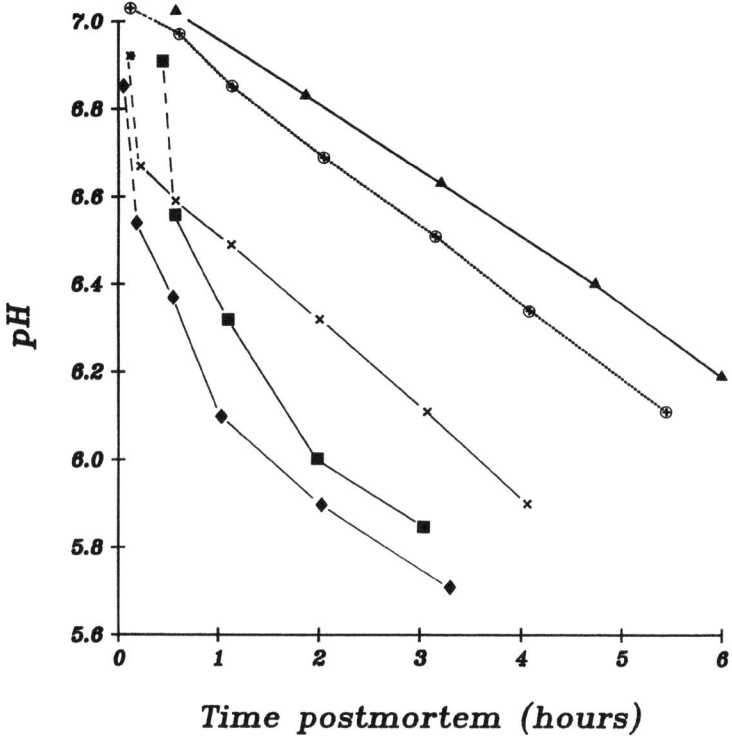

FIG. 3.8. pH–time curves for m. biceps femoris in isolated lamb legs held at 30°C. ▲ Control. ⊗ Stimulated via nerves with 12 V peak at 30 min postmortem. × Stimulated via nerves with 12 V peak at 5 min postmortem. ■ Stimulated directly with 200 V peak, 30 min postmortem. ♦ Stimulated directly with 200 V peak, 5 min postmortem.
After Chrystall et al. (1980A).

Muscles. As various muscles in the body have different roles to play during life, it is not surprising that they behave differently toward electrical stimulation after slaughter. Excellent reviews of the different muscle fiber types and their metabolic differences are available (Beatty and Bocek 1970; Gauthier 1970). We have studied the responses of a limited number of beef muscles to electrical stimulation. In the fast twitch m. cutaneus trunci, largely composed of white-muscle fibers, the values for ΔpH and dpH/dt are large, indicating a distinct response to stimulation. In contrast, in the slow twitch m. masseter, there is neither a distinct ΔpH nor an acceleration of dpH/dt, which in this muscle is naturally rapid (0.4 pH units/hr).

The m. sternomandibularis and m. longissimus are composed of a mix-

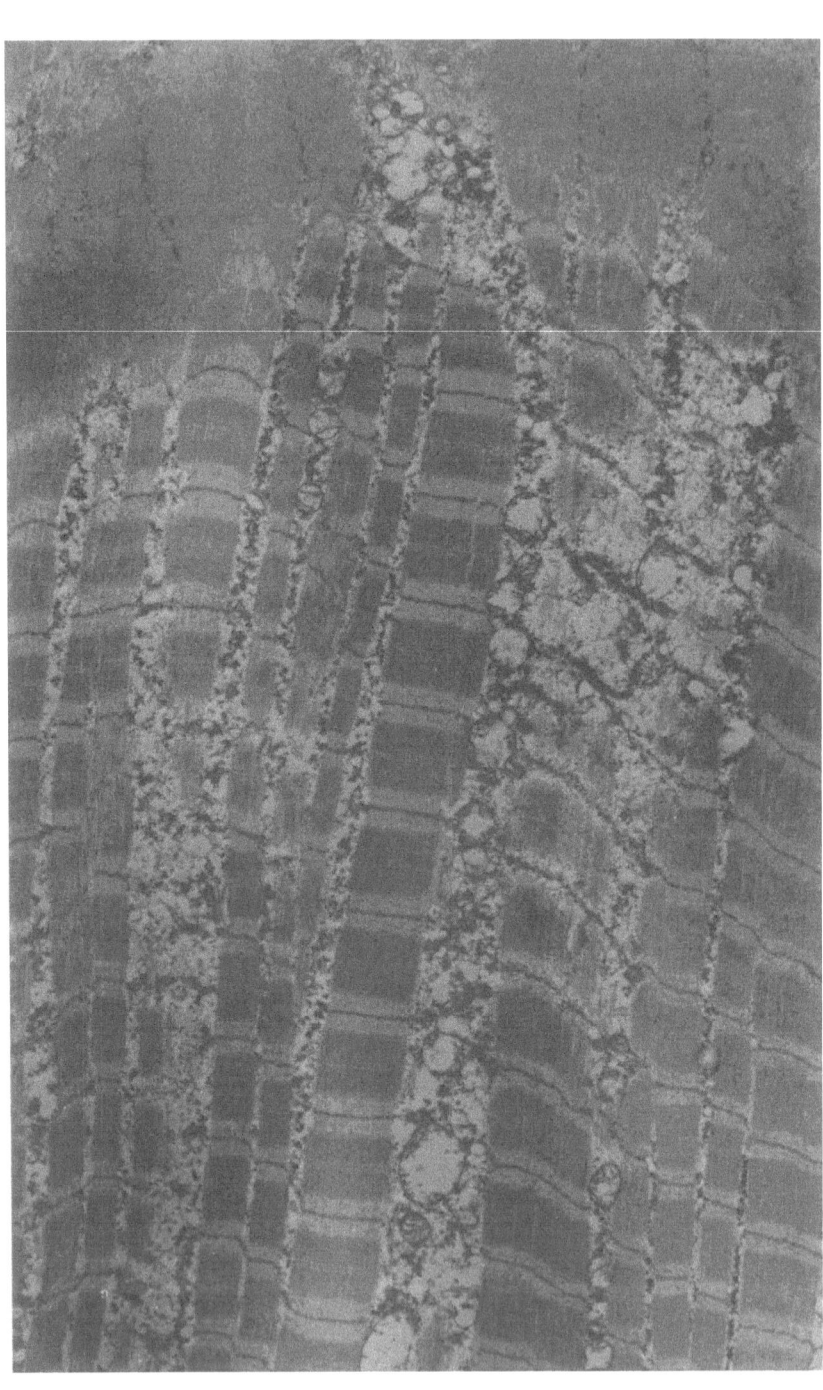

ture of fast- and slow-twitch fibers, and changes in ΔpH and dpH/dt are intermediate to those of the m. cutaneus trunci and m. masseter. Houlier et al. (1980) have also shown that ΔpH is minimal in predominantly slow-twitch muscles while applying low-voltage gradients. Muscles are likely to differ in their optimum response frequency in much the same way as they differ in twitch characteristics. Beef m. longissimus is more responsive to a frequency of 14.28 pulses/sec than it is to 40 pulses/sec. It has been reported that the converse is true for the m. semimembranosus (Bouton *et al.* 1980). Again, rat muscles, which have a much shorter twitch time than most beef muscles, have an optimum response at higher frequencies (33–50 pulses/sec) (Devine and Chrystall 1982). There is some evidence that cold-shortening ability also varies among muscles. For example the m. cutaneus trunci does not cold shorten unless its temperature has fallen to 0°C, whereas the m. sternomandibularis and m. longissimus begin to shorten at 10°C.

Ultrastructural studies have shown that two distinct changes occur after stimulation, mitochondrial swelling and muscle fiber supercontracture. Mitochondria swell in muscles that show the most change in glycolytic rate on stimulation, e.g., m. cutaneus trunci, m. longissimus, and m. sternomandibularis, but they do not swell markedly in muscles such as the m. masseter. Supercontracture is not always a consequence of stimulation, being rarely found in the m. sternomandibularis or m. cutaneus trunci, but it occurs more in the m. longissimus. In contrast, supercontracture always seems to be present in the m. masseter even though this muscle is otherwise little affected by stimulation (Fig. 3.9).

Spread of Response. The degree to which the development of rigor mortis is accelerated in different parts of a carcass is of some practical importance. The spread of response depends not only on the predominant fiber type within a muscle but also on the current density in, or voltage gradient across, the muscles.

Attempts have been made to determine the current density in different parts of a carcass during stimulation by measuring either voltage gradients with Langmuir probes or current densities indirectly through resistive heating (Chalcroft and Chrystall 1974; Gilbert 1980). A clear demonstration of the nonuniformity of current spread was encountered in lamb carcasses stimulated with high voltages (3600 V peak) while suspended from an earthed rail by shackles on the right hind legs. The directly stimulated legs were tender, whereas the free legs were tough (Chrystall and Hagyard 1975). Confirmatory evidence has been provided

FIG. 3.9. Low power electron micrograph of stimulated lamb m. longissimus showing a region of supercontracture, depleted glycogen content, and swollen mitochondria. The region adjacent to the supercontracture is not disrupted, but the sarcomeres are stretched. The muscle was stimulated on the carcass and fixed 10 min after stimulation ($\times$6875).

by measuring the value of ΔpH in, and tension developed by, the m. gastrocnemius of rats stimulated with low voltages. Both ΔpH and the developed tension were markedly lower in the leg not in the direct electrical pathway (Chrystall and Devine 1983) (Fig. 3.10). The nonuniformity of response in terms of rigor development and tenderness throughout the beef carcass has also been shown in studies in which low-voltage stimulation was applied immediately after slaughter via a rectal probe (Kerr et al. 1979). In this case, significant forequarter muscles remained unaffected by the rectal stimulus and took as long to enter rigor mortis as those from nonstimulated carcasses.

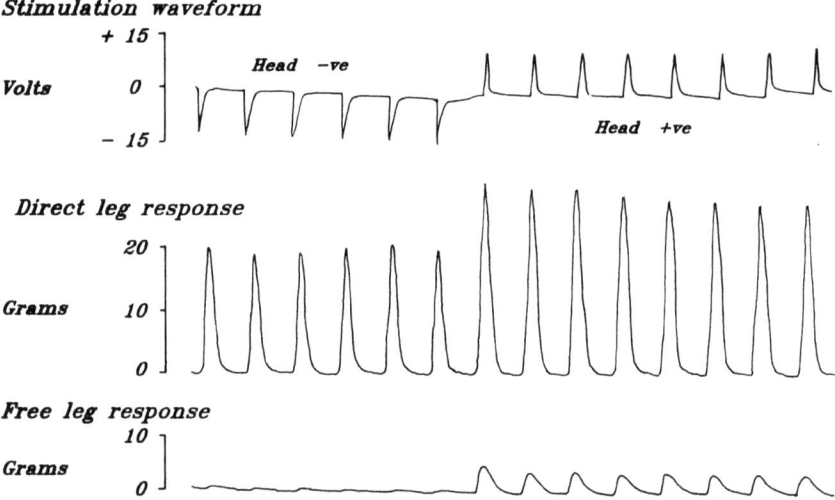

FIG. 3.10. Tension response and the applied stimulation waveform recorded from m. gastrocnemius of a freshly slaughtered rat when stimulated from the nose to the direct leg. The free leg was stimulated only via the nervous system. Pulse repetition rate 14.28 pulses/sec.

Animal Stress. If animals are stressed, they may come to slaughter with partially depleted glycogen reserves and therefore enter rigor mortis much sooner than would normally be the case. Under these circumstances, electrical stimulation may bring forward rigor to such a degree that it develops close to or during the period of stimulation itself. Stimulation in such cases may induce toughening due to irreversible contracture of the muscles. This view is supported by tenderness studies in lamb (Chrystall et al. 1982B). The legs of stimulated, exercise-stressed lambs were significantly tougher than those from nonstimulated lambs (FS 31 cf FS 20), but the difference was less pronounced in the loins. In much the same way as the physiological (tension) and biochemical (ΔpH and dpH/dt) responses to stimulation vary within an animal, so the responses are likely to vary for the same muscle but between species.

Our view is emphasized at this point that the overriding merit of electrical stimulation is to overcome cold and thaw shortenings with their attendant toughening. Unless the prerigor musculature of a carcass is likely to be exposed to conditions that could give rise to cold shortening or thaw shortening, we do not advocate electrical stimulation as a means of ensuring meat tenderness.

Stimulation Parameters. In optimizing the stimulation conditions, it is well to remember that, in addition to animal factors already described, a number of other interacting variables have to be considered. Among these we have the stimulation period and the electrical parameters of pulse shape, wave form, polarity, frequency, current, and voltage.

Period of Stimulation. Increasing the period for which isolated muscles are stimulated increases the magnitude of ΔpH up to a limit of about 0.7 pH units (Chrystall and Devine 1978; Devine 1976). In contrast, the maximum achievable values of dpH/dt are achieved with no more than a 5–10 sec period (Chrystall and Devine 1978).

Although Carse (1973) in his studies stimulated lamb carcasses for 30 min to achieve pH 6 in the musculature in 3 hr, it was soon evident that most of the effect of stimulation in inducing early rigor was achieved in the first few minutes (Chrystall and Hagyard 1975). This was even more the case if stimulation was applied less than 5 min after slaughter. By 60 min postmortem, the pH values of the m. longissimus of lamb carcasses stimulated with 15 pulses/sec, 2 amp peak, for 20, 30, or 60 sec, were pH 6.3, 6.1, and 5.9, respectively. If the delay before stimulation was increased to 30 min, a 90 sec stimulation period was needed to achieve the same pH value at 90 min poststimulation as was achieved by a 45 sec period applied within 5 min of slaughter (Chrystall 1978; Hagyard et al. 1980).

The stimulation waveform frequency and voltage can influence the choice of the period of stimulation. For example, lamb carcasses suspended from a gambrel between the hind legs can only be stimulated with 50 Hz, 800 V (rms) for 45 sec before the Achilles tendons melt. Despite a 2 hr delay before freezer entry, the legs and loins of such carcasses are tough. The choice of an effective stimulation period for beef carcasses or sides also depends on the extent of postmortem delay. For delays up to 20 min, a 90 sec period is recommended, but with longer delays, up to 60 min postmortem, the period is 120 sec (MIRINZ 1978A). Australian results (Kerr et al. 1979) also demonstrate the need to extend stimulation periods with longer postmortem delays.

Pulse Shape, Wave Form, and Polarity. It has been possible to study only a limited variety of wave forms. However, many trials have been undertaken with half sine-wave pulse trains derived from the 50 Hz AC mains supply, with square-wave pulses from laboratory stimulation units, and spike pulses from capacitive discharge units. It would appear that pulse shape does not greatly determine the extent to which rigor development is hastened. In one comparison with beef m. sternomandibularis, half sine-wave pulses of 10 msec duration delivered at 12.5 pulses/sec (derived by using every eighth pulse of the 50 Hz supply), and square waves of 10 msec duration, both at 200 V peak, produced virtually the same ΔpH values of 0.68 and 0.63, respectively (Chrystall and Devine 1978). Physiologically, maximum contractile response is dependent on both pulse voltage and pulse duration; however, the voltages considered here are supramaximal, maximum responses in terms of tension and rigor development are achieved over a wide range of pulse durations. Thus, halving the duration of square-wave pulses to halve the energy input did not significantly reduce ΔpH values in beef m. sternomandibularis from their maximum value (0.7 pH units).

The minor dependence on pulse shape was also shown in detailed studies with rat m. gastrocnemius. Square-wave pulses (5 msec duration) and half sine-wave pulses (10 msec duration) of the same peak voltage produced almost identical responses measured either as ΔpH or as muscle tension development. Because of the small influence of pulse shape in inducing early rigor, most of our subsequent studies have used trains of 10 msec half sine-wave pulses derived from a 50 Hz AC electrical supply. By suitable chopping and rectification, either unidirectional or bidirectional pulse trains can be obtained.

At high stimulating voltages, the values of pH in exercised beef m. sternomandibularis were virtually unaffected by pulse polarity (Chrystall and Devine 1978). However, the bidirectional pulse trains were markedly superior in achieving uniform tenderness in frozen lamb (Hagyard *et al.* 1980). A different result was obtained in studies comparing trains of unidirectional and alternating-polarity spike pulses from a high-voltage capacitive discharge unit. Loins from lamb carcasses stimulated at less than 5 min after slaughter with unidirectional pulses and then moved to freezing at 60 min postmortem were reasonably tender. In contrast, those from carcasses stimulated with a alternating polarity were tough (Chrystall and Hagyard 1975). The lesson from such observations clearly is that the values of ΔpH and dpH/dt are not the sole indices of effective stimulation. The ultimate index is the tenderness of the meat.

Visual evidence of the influence of pulse polarity on the contractile response was obtained during trials with low-voltage stimulation (80 V peak) of beef carcasses. When a bidirectional pulse train was used, the muscles of the carcass twitched at half the stimulation frequency. With a unidirectional pulse train, however, the muscles twitched with every stimulus. It is well recognized that nerves and muscles only react to cathodic stimuli favoring the use of unidirectional pulses of appropriate polarity. This is well exemplified in a detailed study on rats (Fig. 3.11). The re-

Stimulation waveform

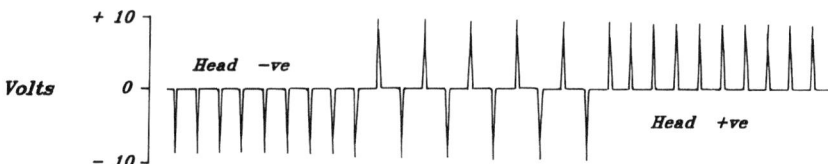

Direct leg response

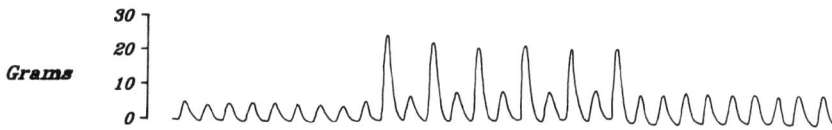

FIG. 3.11. Tension response to the applied stimulation waveform recorded from the directly stimulated m. gastrocnemius from a freshly slaughtered rat. Pulse repetition rate was 14.28 pulses/sec.

sponse illustrated by the tension records was reflected in a reduction of ΔpH values with bidirectional pulse trains. The merit of using unidirectional cathodic pulse trains of low voltage for sheep and cattle is now recognized. With high voltages, alternating polarity becomes important because of the effects of electrode polarization and transformer core saturation.

Frequency. The magnitude of ΔpH largely determines the time for muscles of stimulated carcasses to reach pH 6.0. The frequency of the applied voltage is an important determining factor. Frequencies between 9 and 16 pulses/sec give the greatest ΔpH in beef m. sternomandibularis (Fig. 3.12). At frequencies between 10 and 20 pulses/sec, ΔpH values are, respectively, 40 and 75% greater than at 50 and 100 pulses/sec (Chrystall and Devine 1978). The frequency optimum of around 9–16 pulses/sec seems to hold for most muscles of sheep and beef carcasses. In animals where the metabolic rate is higher, such as in rats (Krebs 1950), muscle twitch times are shorter and maximum values for ΔpH and sustained tension occur at 33 pulses/sec. It is to be expected, therefore, that optimum pulse frequencies will vary among species and may be higher for rabbits and poultry than for beef and sheep.

When considering such complex interactions, a clearcut assembly of the influences of stimulation parameters is not to be expected; thus, with short stimulation periods, high frequencies (50 to 100 pulses/sec) give greater ΔpH values in beef m. sternomandibularis than those produced by lower frequencies. If longer stimulation periods can be used (120 sec), then 9–16

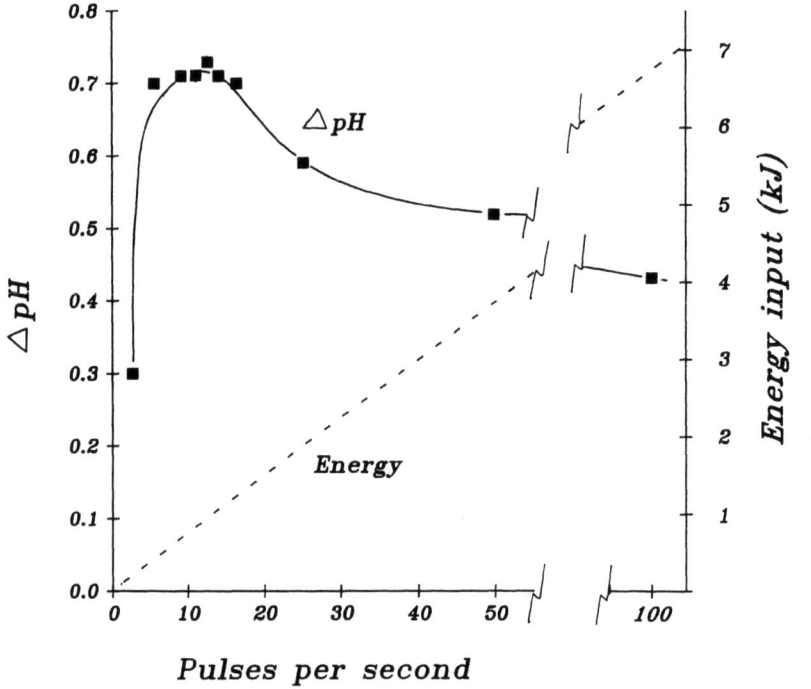

FIG. 3.12. Effect of pulse frequency on pH fall during stimulation (ΔpH) of beef m. sternomandibularis with 200 V peak pulses for 120 sec.
After Chrystall and Devine (1978).

pulses/sec give the greatest ΔpH (Chrystall and Devine 1978). The lower frequencies produce a slightly lower peak tetanic tension than the higher frequencies but maintain their peak tension for considerably longer (Fig. 3.13). For this reason, electrical pulses at 14.28 pulses/sec are now being used too maintain tension of the back muscles during downward hide-pulling of cattle, greatly reducing the incidence of broken backs (Gilbert 1979; Richardson 1980).

An advantage of lower stimulation frequencies is the lowered energy input. For example, at 14.28 pulses/sec, the energy input is only one-seventh of that at 100 pulses/sec (Fig. 3.12). This significantly reduces heating at the electrode contacts and in the musculature.

Voltage and Current. Harsham and Deatherage (1951) used voltages ranging from less than 50 to greater than 3000 V peak, the latter giving better current distribution throughout the beef carcasses. Based on the time to reach pH 6, Carse (1973) showed that this time decreased markedly as the voltage increased up to 250 V. In experiments using excised beef m. sternomandibularis, at all frequencies, with a 30 sec stimulation period, an increase in stimulation voltage from 50 to 320 V increased the value of ΔpH produced by over 60% (Fig. 3.14). The difference decreased as the stimulation period was increased, but even with a period of 120 sec, it was still significant (Chrystall and Devine 1978).

In processing practice, is it worthwhile using high stimulation voltages merely to make small additional reductions in the time it takes rigor mortis to develop? As already noted, the final criterion of the merit of refining the stimulation parameters is the tenderness of the cooked meat. The influence of stimulation voltage on tenderness is shown by the FS values of cooked legs and loins from lambs (Fig. 3.15). With 14.28 pulses/sec derived from 400 V rms (566 V peak) electrical supply, the loins were considered acceptably tender but the legs were moderately tough. With similar pulses derived from the 800 V rms (1130 V peak) electrical supply, there was a further slight improvement in the loins and a major improvement in the legs (Chrystall 1978).

The stimulation parameters giving the greatest reduction in the time for rigor to develop are the same for both lamb and beef carcasses. As with lamb, the stimulating voltage is critical. Based on pH determinations, values close to the ultimate are obtained after 8 hr in m. longissimus, m.

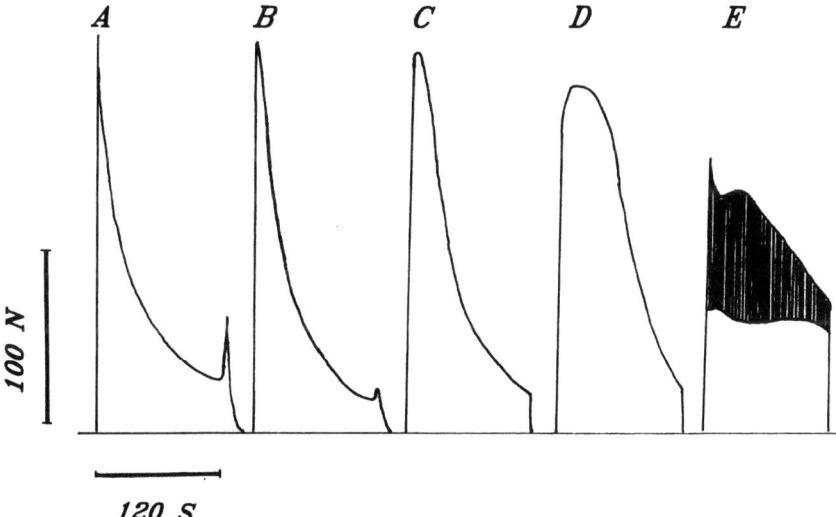

FIG. 3.13. Tension-time curves recorded from similar sized (0.001 m²) beef m. sternomandibularis stimulated with 200 V peak unidirectional pulses at the following frequencies: A—100 pulses/sec. B—50 pulses/sec. C—25 pulses/sec. D—12.5 pulses/sec. E—2.5 pulses/sec.
After Chrystall and Devine (1978).

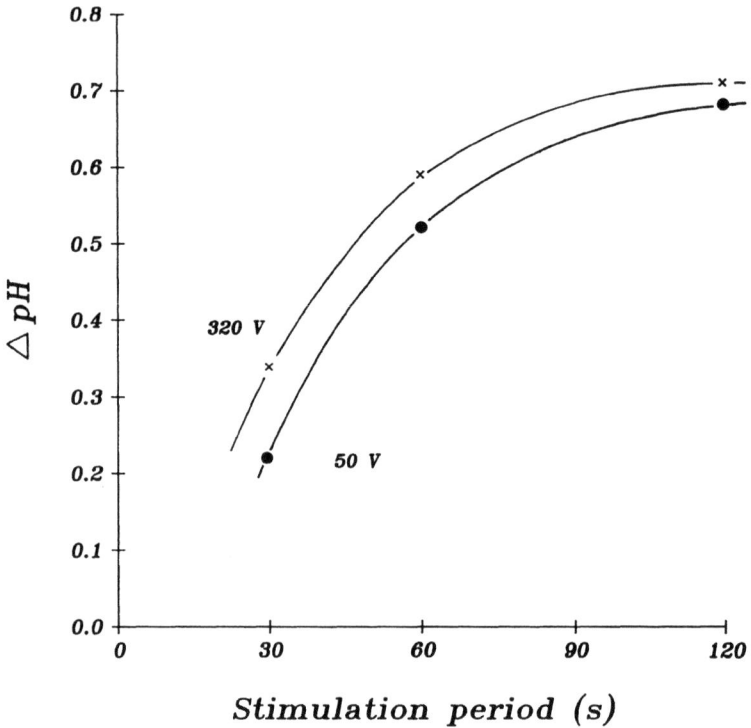

FIG. 3.14. The effect of stimulation period on the magnitude of ΔpH produced at two voltages.
After Chrystall and Devine (1978).

semimembranosus, and m. triceps brachii of beef carcasses stimulated for 120 sec with 400 V (rms) pulses (14.28 per sec). In contrast, a stimulation voltage of 800 V (rms) was required to achieve 8-hr pH values approaching their ultimate in muscles such as m. rectus abdominus, m. brachiocephalicus, and m. vastus lateralis (Gilbert and Davey 1980, unpublished data). If maximal rates of rigor development are essential to the meat processor, then high stimulating voltages (800 V rms or above) should be used. This view is supported by Bendall (1980) and Bendall et al. (1976). However, in circumstances where maximal rates of rigor development are not an essential processing demand, much lower stimulation voltages can be used with the advantage that stringent safety requirements do not then have to be met.

Swedish researchers (Fabiannson et al. 1979; Ruderus and Berquist 1980; Jonsson et al. 1978) and Australian researchers (Shaw and Walker

1977; Walker et al. 1977; Bouton et al. 1978) have been strong advocates of low-voltage beef stimulation. The Swedish group have used 85 V peak, unidirectional square-wave pulses (5 msec duration) at 14 Hz applied less than 5 min after slaughter. The Australian group have used unidirectional square-wave pulses (2 msec duration) at 40 Hz applied as stepwise increases of voltage from 20 to 110 V (peak) (Bouton et al. 1978). Using beef m. longissimus, the pH, after low-voltage stimulation, was reduced to 6.0 in 3–5 hr (Bouton et al. 1978; Taylor and Marshall 1980; Ruderus and Fabiansson 1980).

Low-voltage stimulation has been developed as a useful element in beef processing in circumstances where carcasses are not subjected to hot deboning, rapid chilling, or freezing. While recognizing the critical need for high voltages to stimulate lambs subjected to rapid chilling or freezing soon after slaughter, we believe there could be situations, especially in small plants, where low-voltage stimulation is an appropriate alternative.

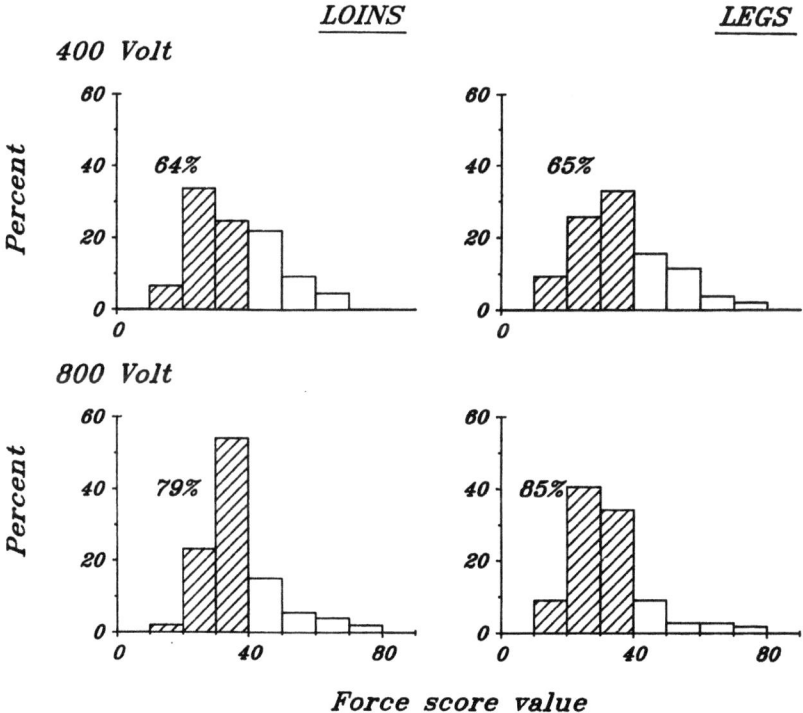

FIG. 3.15. Histograms of force score values for lamb loins and legs cooked from the frozen state after the carcasses had been stimulated at 30 min postmortem for 90 sec at 400 or 800 V (rms), then moved into a freezer at 2 hr postmortem. Deep leg temperatures reached −4°C 14 hr postmortem.

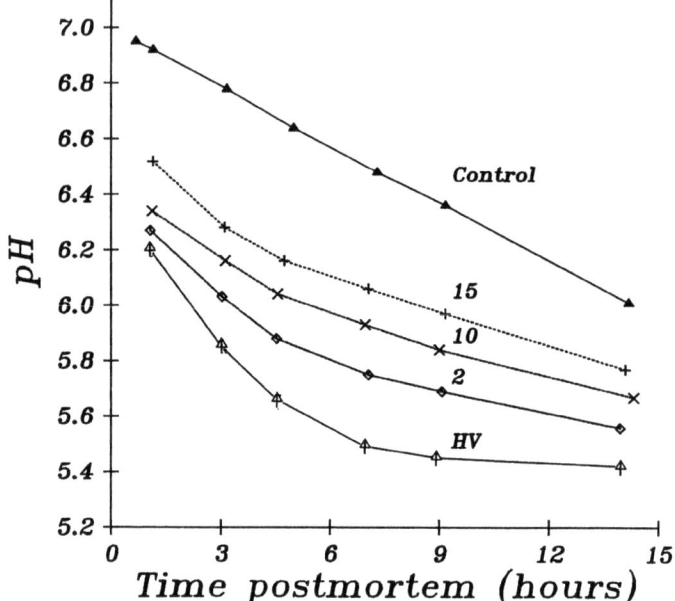

FIG. 3.16. pH–time curves for m. longissimus of nonstimulated and stimulated lamb carcasses at 10°C. ▲ Nonstimulated, controls. + Low voltage stimulated, 120 sec period at 15 min postmortem. × Low voltage stimulated, 120 sec period at 10 min postmortem. ◇ Low voltage stimulated, 120 sec period at 2 min postmortem. △ High voltage for 90 sec period at 30 min postmortem.

The laboratory rat has served as a useful model, allowing us to extend our knowledge of low-voltage stimulation (Chrystall and Devine 1983). Based on tension development, very low voltages (0.1 V) produce maximal tension in the leg muscles of the rat when applied indirectly through the sciatic nerve, but much higher voltages (1.0 V) are needed, however, if these are applied directly to the muscle. Higher voltages again are required if the stimulation is applied across the whole body.

Extending the studies to lamb, the pH fall in the m. longissimus was followed in carcasses held at 10°C after stimulation with 80 V peak, unidirectional pulses at 14.28/sec applied between nose and both hind legs. A number of stimulation periods were used, but maximum acceleration of rigor development was achieved with a 120-sec period. The fall of pH in carcasses stimulated at different times postmortem is compared with those in nonstimulated carcasses and in carcasses stimulated with high voltages (Fig. 3.16) (Chrystall *et al.* 1982A). The time taken for the pH to reach 6 provides an indication of the effectiveness of the different stimulation treatments in advancing rigor mortis development. At the high voltage, pH 6 was reached in less than 2 hr. In contrast, at low voltage, even if applied within 2 min after slaughter, pH 6 was reached in something over 3.5 hr. As the figure clearly shows, postmortem delay before stimulation greatly increases the time required for the m. longissimus pH to fall to 6.

Poststimulation Processing

Electrical stimulation alone does not ensure a tender product. However, by hastening rigor development, it reduces the need for protracted holding periods prior to chilling or freezing, an especially important requirement in meat plants with large throughputs.

The Processing of Lambs. *High- and Low-voltage Stimulation.* In New Zealand, most lamb carcasses are frozen on the day of slaughter, and the major proportion of the carcass musculature is therefore particularly

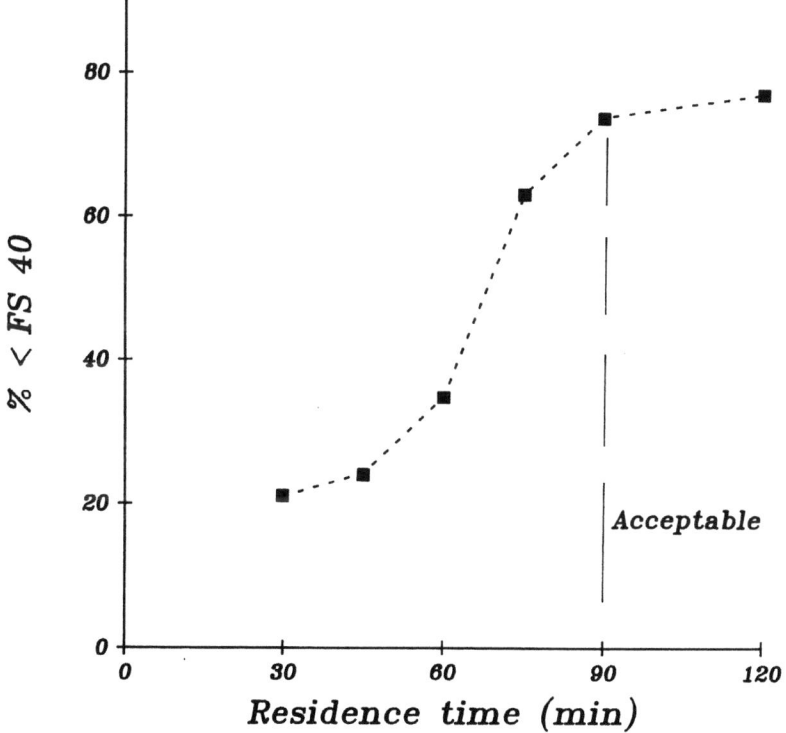

FIG. 3.17. The effect of time that lamb carcasses spend above 6°C after stimulation on the tenderness of loins roasted from the frozen state. Ninety minutes or more are necessary for the loins to be acceptably close to "conditioned" tenderness.

prone to cold and thaw shortening. The pH values of major muscles of lamb fall to pH 6 in 45 min and to pH 5.8 in 70 min when high voltages (1130 V, peak) are applied to carcasses for 60 sec immediately after slaughter. The dressed carcasses can then be frozen without a major risk of cold or thaw shortening. In practice, the stimulation period needs to be reduced to 45 sec to avoid excessive carcass stiffening during dressing operations. Under these circumstances, major muscles reach pH 6 in an hour and pH 5.8 in an hour and a half. To avoid cold and thaw shortenings completely, a holding period at 6°–10°C is required before carcasses are transferred to freezing.

If the stimulation is applied to the dressed carcass, up to 30 min after slaughter, the stimulation period should be extended to 90 sec and the carcasses should then be held for 90 min above 6°C before transfer to

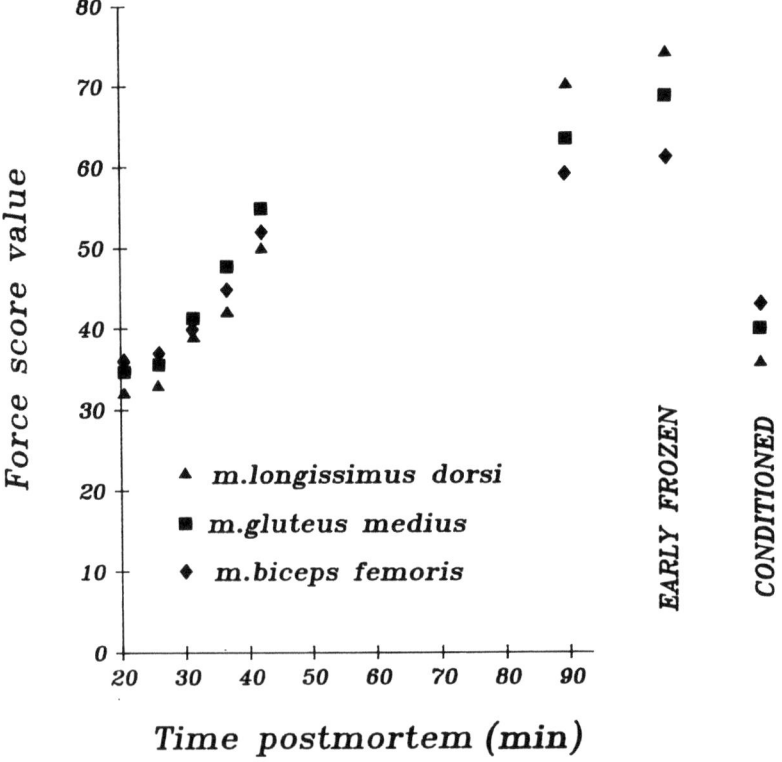

FIG. 3.18. The effect of time of stimulation with 1130 V peak, 14.28 pulses/sec on the tenderness of lamb m. longissimus, m. gluteus medius, and m. biceps femoris. All carcasses were frozen 90 min after stimulation. The tenderness of muscles for early frozen and conditioned lambs is shown for comparison.

freezing (Fig. 3.17). Electrical stimulation of carcasses much beyond 30 min is ineffectual in inducing early rigor, and muscles of these carcasses can shorten and are tough (Fig. 3.18).

The values of ΔpH and dpH/dt in the major muscles of lamb carcasses stimulated with low voltages (80 V peak) are less than those obtained with high voltages. In this respect, with 1130 V (peak) applied 30 min after slaughter, pH 6 was reached in the m. longissimus in under 2 hr, whereas with the lower voltage, applied less than 5 min from slaughter, a period of 3 hr was required (Chrystall et al. 1982A).

If required, stimulated lamb carcasses can be held for a period to age before being frozen. Very tender meat can be produced by holding the carcasses for up to 8 hr at not less than 6°C before freezing (Fig. 3.19). Other time–temperature combinations are possible. In one, the carcasses are held for 10 hr at 10°C to eliminate the need for a controlled, slow freeze (Hagyard 1979). In contrast to requirements for conventional processing, these aging times may appear to be very short. However, it is to be remembered that the carcasses enter rigor mortis at about 30°C, so that much of the aging occurs rapidly at a relatively high temperature.

Freezing Conditions for Lamb Carcasses. Under the prescribed stimulation conditions, full rigor is rarely achieved at the time of freezer entry. While cold shortening is avoided, thaw shortening with its attendant toughening remains a likely hazard. In order to maintain practicable holding times before freezer entry, a controlled freeze is specified, allowing full rigor development during the chilling phase of freezing. If the deep leg temperature falls below -4°C in less than 14 hr from slaughter (12 hr from freezer entry), the tenderness of legs and loins is less than that considered desirable (Chrystall and Hagyard 1975; Chrystall 1980) (Fig. 3.20).

Frozen Storage of Lamb. If freezing is too rapid, there is a danger that full rigor will not be achieved throughout the musculature of lamb carcasses and thaw shortening is then a concern. Fortunately, during storage at temperature above about -14°C, frozen carcasses progress slowly into a state of rigor mortis through slow dephosphorylation of residual ATP. At -12°C, for instance, the time for this to occur is 20 to 30 days (Davey and Gilbert 1976). The tenderness of lamb stored at -12°C improves as a consequence of the loss of ATP (Davey and Foster 1977). With the trend to much lower storage temperatures (-25°C), the advantage of this so-called subzero conditioning is largely lost. Strict adherence to specified stimulation parameters, poststimulation holding, and freezing times is a more reliable way to achieve uniform tenderness. Subzero conditioning is merely a fortuitous additional dividend.

The Processing of Beef Carcasses and Sides. In contrast to lamb carcasses, beef sides in New Zealand are not frozen on the day of slaughter but are merely chilled for up to 2 days before boning. Thaw shortening

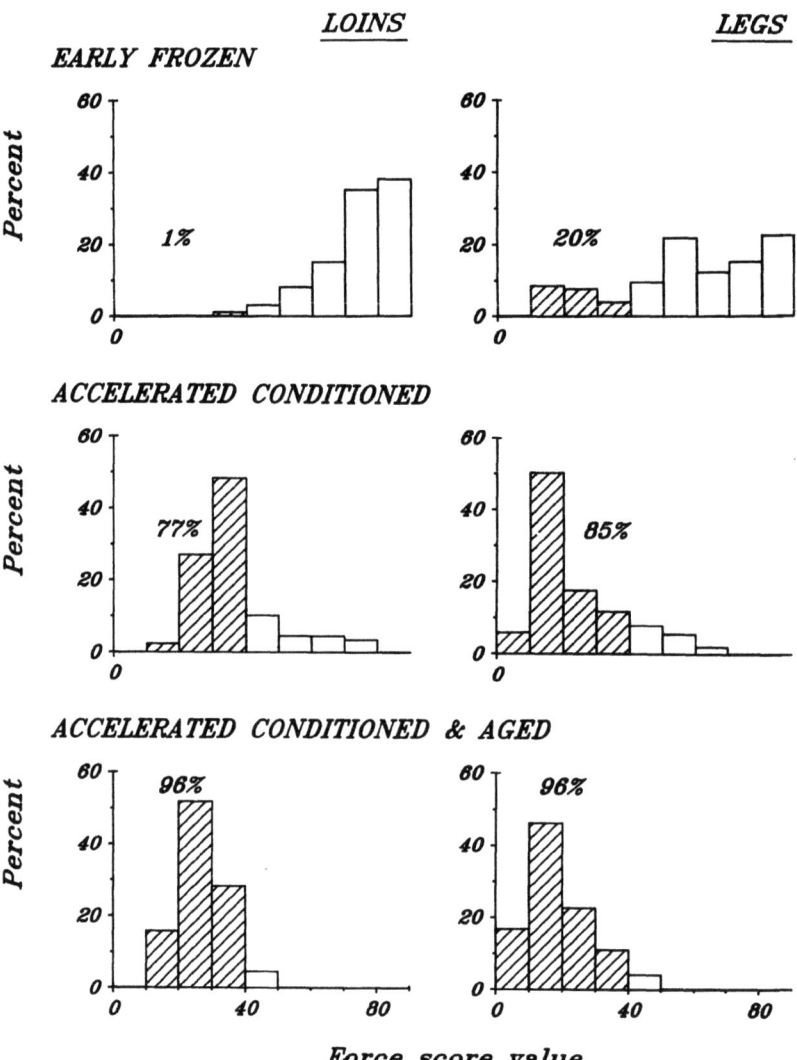

FIG. 3.19. Histograms of force score values for lamb loins and legs roasted from the frozen state after various treatments. The freezing time (to −4°C deep leg) for all carcasses was 12 hr from freezer entry. The storage temperature of −18°C was reached in a further 2 hr. Early frozen (as for Fig. 3.3)—Moved into freezer 2 hr postmortem. Accelerated conditioned (AC)—Processed according to AC specifications (Table 3.1). Accelerated conditioned & aged—As for accelerated conditioned (AC), except held at 7°C till 8 hr postmortem, then frozen.

does not present a processing hazard, although cold shortening may do so if the sides are chilled rapidly to deep-muscle temperatures below 10°C in 24 hr or less.

When high voltages (1130 V peak) are applied to beef sides for 120 sec at 60 min postmortem (Davey *et al.* 1976), the pH values of major muscles fall to pH 6 in about 3.5 hr. Even rapid chilling presents no threat to the tenderness of major muscles. The early setting of the muscles can facilitate later processing steps. Low voltages (50–80 V peak) have been used to stimulate beef carcasses. Based on our own studies, the value of ΔpH and dpH/dt in major muscles are not as large as those achieved by high-voltage stimulation (1130 V peak), and the time for the pH of most muscles to fall to pH 6 is considerably longer. However, chilling conditions for beef are never as severe as those for lamb carcasses and the somewhat longer time required for rigor development is hardly a disadvantage. The choice

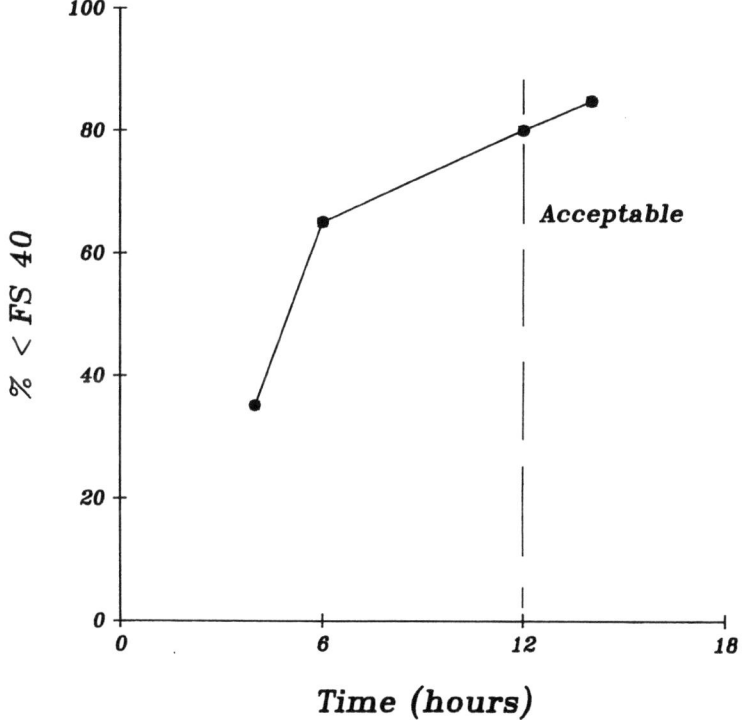

FIG. 3.20. The effect of freezing time (time for deep leg to reach −4°C) on the tenderness of stimulated lamb loins roasted from the frozen state. A freezing time of 12 hr is needed to produce "conditioned" quality tenderness.

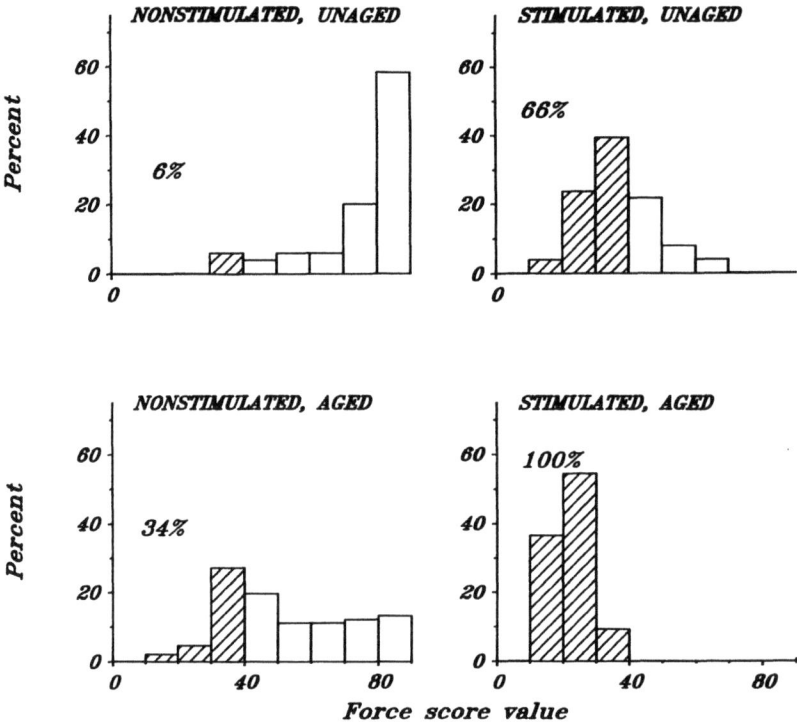

FIG. 3.21. Histograms of force score values for cooked beef m. longissimus from nonstimulated and stimulated carcasses chilled at a rate sufficient to induce toughening through cold shortening. Unaged samples were excised and cooked at 18–24 hr postmortem. Aged samples were excised and cooked 66 hr postmortem.
Redrawn from Davey et al. (1976).

of the stimulating voltage becomes largely a matter of plant layout and throughput.

Rapid Chilling and Freezing of Beef. In contrast to lamb, beef sides traditionally have been slowly chilled with little risk of cold shortening of muscles in the primal cuts. However, with the move to more rapid chilling to satisfy regulatory demand, there is a likelihood of cold shortening occurring, especially in lighter, leaner sides of beef. Electrical stimulation then becomes an unquestionable need.

In one laboratory study, beef sides were subjected to 1–2 min of high voltage (3600 V peak) electrical stimulation immediately after carcass dressing (Davey *et al.* 1976). The time for rigor setting was reduced from 24 hr to about 5 hr. Even with rapid chilling (deep leg temperature <8°C at 24 hr), the stimulated carcasses were still warm as rigor developed so that cold shortening could not occur. The meat was reasonably tender and could be aged to a high degree of uniform tenderness (Fig. 3.21). In other words, by the use of electrical stimulation, the tenderness of slowly chilled beef is achieved despite rapid chilling. Similar results have been obtained in experiments using beef carcasses stimulated for 1 min with low voltage pulses (80 V peak) applied immediately after slaughter (Fig. 3.22) (Gilbert and Chrystall 1982).

Early Boning. In an extension of this study (Gilbert and Davey 1976), stimulated beef sides were boned out 5 hr after slaughter since by then

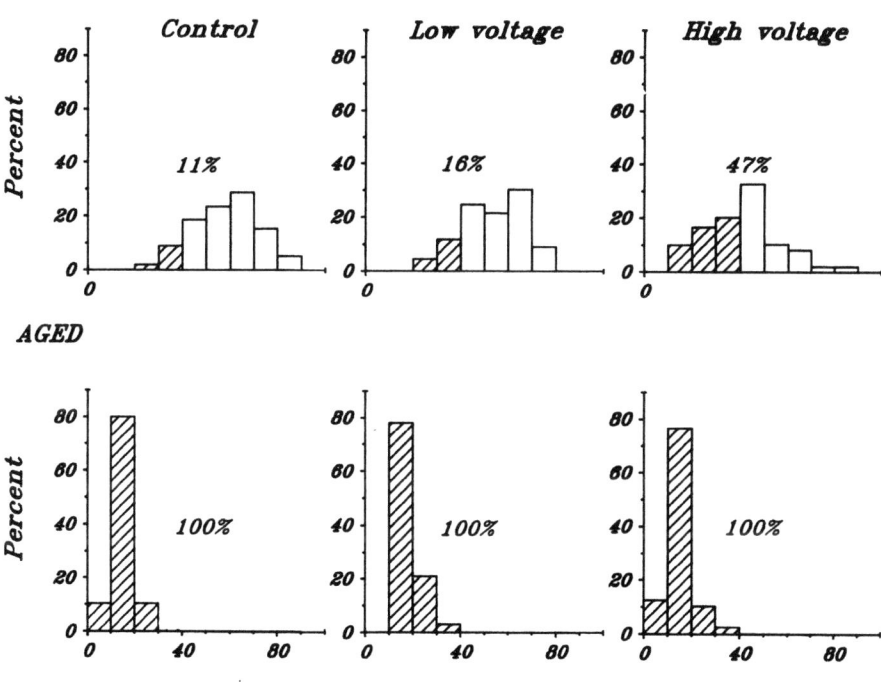

FIG. 3.22. Histograms of force score values for cooked m. longissimus from control, low voltage, and high voltage stimulated beef subjected to rapid chilling. Low voltage stimulation was for a 60 sec period with 80 V peak unidirectional pulses at 14.28/sec. High voltage stimulation was with 1130 V peak pulses of alternating polarity at 14.28 pulses/sec applied to sides 60 min postmortem for 120 sec.

they had entered rigor mortis. Rapid chilling during the early postmortem period has hardened the surface fat and this aided conventional boning. As the meat was set in rigor mortis, there was no risk of toughening from cold shortening in the boned out cuts. The meat was as acceptably tender as that from unstimulated sides that had been slowly chilled and boned out at 24 hr. The tenderness was further improved by aging. There was no evidence of increased microbial spoilage from boning while the meat was still warm. By such early boning, chilling space and processing time are saved and the handling of cartoned meat through aging and freezing can be carried out mechanically.

Hot Boning. Yet another processing variant is that of hot boning the electrically stimulated (high-voltage 3600 V peak) sides of beef. In one study (Gilbert et al. 1977), the primal cuts were transferred to chilling or freezing at 2 hr postmortem. Despite rapid freezing (7–10 hr), rigor was achieved before the deboned meat temperature had fallen below 8°C. The meat was as tender as that boned from slowly chilled sides deboned at 24 hr. If the cuts were chilled to 5°C over a period of 46 hr before freezing, shortening was avoided and the meat aged to a highly acceptable and uniform level of tenderness. Although bacterial proliferation was limited by the much shorter processing time, further studies will be necessary to determine whether selective growth occurs in the packaged, hot-boned meat.

The tenderness achieved in early-frozen, hot-boned meat is influenced by the stimulation conditions. Meat from nonstimulated carcasses and from carcasses stimulated for 60 sec immediately after slaughter with low-voltage (80 V peak, unidirectional, half sine-wave pulses, 10 msec duration, 14.28 pulses/sec) was significantly less tender than that from sides stimulated for 120 sec, 60 min after slaughter with high-voltage pulses (1130 V peak, alternating, half sine-wave pulses, 10 msec duration, 14.28 pulses/sec) (Fig. 3.23) (Gilbert and Chrystall 1982).

Processing Applications

Introduction to Processing Lines. Research defining the limits for effective stimulation has been translated into practical processing. The availability of sites within a plant suitable for the effective stimulation of lamb and beef carcasses is determined by time constraints and existing plant layout. For sheep carcasses and for beef carcasses or sides, we have given most attention to locating stimulation systems at the beginning and end of processing lines. This does not mean that incorporation at other locations is not feasible.

Sheep Processing Line. *High Voltage.* High-voltage stimulation immediately after slaughter was initially favored since the body temperature

UNAGED

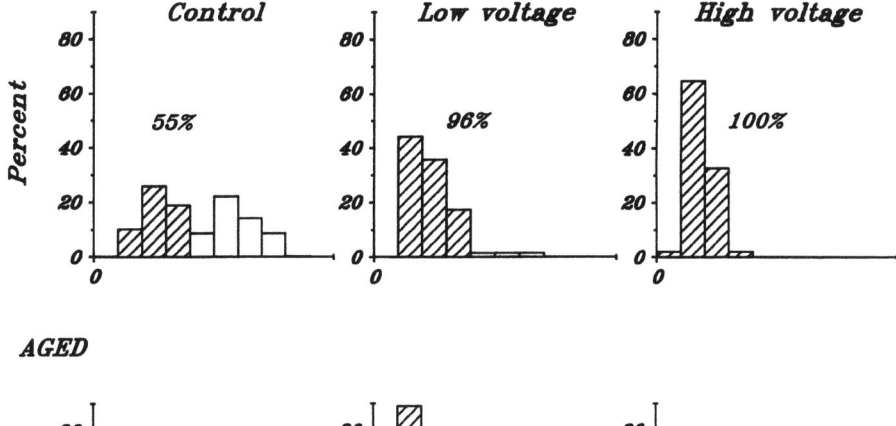

AGED

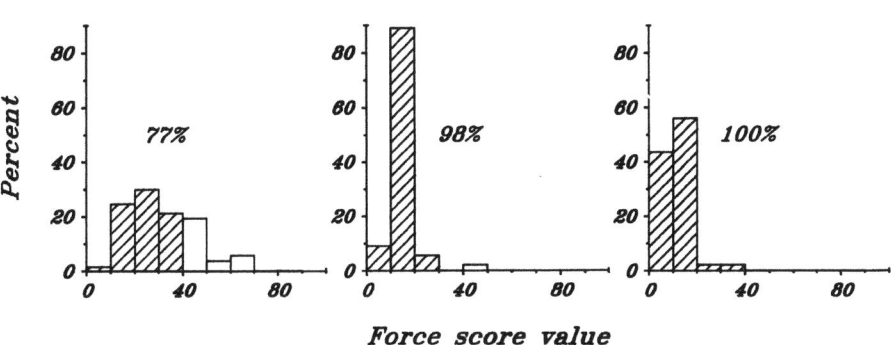

Force score value

FIG. 3.23. Histograms of force score values for cooked m. longissimus hot boned from control, low voltage, and high voltage stimulated beef. The unaged samples were moved into the freezer 2 hr postmortem, whereas the aged samples were held 12 hr at 10°C before freezing. Stimulation conditions were as in Fig. 3.22.

of the freshly slaughtered animals is high and the nervous system is still vital. Furthermore, carcass dressing times can be incorporated into the specified holding times to allow rigor to be achieved before freezing.

In an early on-line system (Chrystall and Hagyard 1975), each pelt-on lamb carcass was suspended by a shackle on its right hind leg from an earthed conveyor rail. The carcass moved across a series of electrodes, rubbing contact being made across the back just below the shoulders (Fig. 3.24). Each electrode was supplied with high-voltage pulses of electricity (3600 V peak at 15 pulses/sec), which collapsed under load to 1600 V (peak) while supplying 2 amp (peak). High-pressure jets were used to apply water to the rubbing contact area to prevent sparking which could scorch the wool and pelt. Since the carcass was suspended by only one of its

FIG. 3.24. Photo of early predressing stimulation of lamb.

hind legs, the other leg was much less tender (Chrystall and Hagyard 1975). The carcasses were also prone to stiffen before completion of dressing. In later developments, a mechanical claw was devised to capture the free leg and provide an electrical contact. However, unless the contact resistance at the claw and shackle were essentially identical, an unequal split of current occurred between the two legs. In this respect, at least 80% of the total current flowed through the gripped leg if the claw was 20 cm nearer the crotch than the shackle was on the other (Chrystall and Hagyard 1975).

Stimulation immediately after dressing (Fig. 3.25) has become the more generally accepted alternative. The carcasses suspended by skids and gambrels move at about one per sec along an earthed conveyor rail and rub

3 ELECTRICAL STIMULATION: ITS EARLY DEVELOPMENT

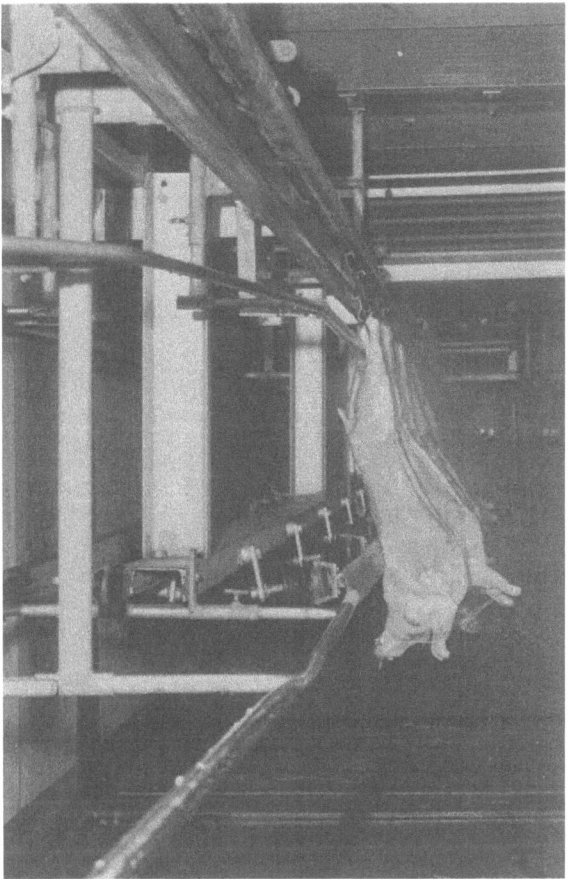

FIG. 3.25. Photo of post scales stimulation of lamb.

at shoulder level across an electrode supplied with high-voltage pulses (1100 V peak, half sine-wave pulses, 10 msec duration, 14.28 pulses/sec). The skid and gambrel provide good electrical contact for both hind legs, and at peak voltage the current approaches 2 amp. The resistances from rail to skid and from skid to gambrel are minimal; that between gambrel and carcass is about 300 ohms and that between carcass and the rubbing electrode about 100 ohms (Loeffen and Hagyard 1979). Since the electrical resistance was remarkably constant among carcasses, a single, unsegmented electrode could be used to stimulate them as they moved through the tunnel.

Low Voltage. Even if lamb carcasses are stimulated for 2 min immediately after slaughter with low-voltage pulses (80 V peak, unidirectional

FIG. 3.26. Low voltage stimulation of lamb. The voltage (80 V peak half sine-wave unidirectional 14.28 pulses/sec) is applied between the front and hind legs. The lambs have been electrically stunned before being hung up as shown. Sticking, opening up of the neck, and tying the esophagus take place while the current is flowing with no discomfort to the worker. The animal is completely still during and after stimulation.
Photo courtesy of Pacific Freezing N.Z., Ltd., Oringi works.

half sine-wave pulses, 10 msec duration, 14.28 pulses/sec the development of rigor is slower than if high voltages are used at 30 min after slaughter. To ensure that "conditioned" quality tenderness is produced, holding times at 6°–10°C before freezer entry must be almost twice those specified for lambs stimulated at high voltage. Automatic low-voltage stimulation, with current flow from front to hind legs, is in commercial operation in one New Zealand plant (Fig. 3.26). One advantage of low-voltage stimulation systems is that dressing work can continue on the carcass while the stimulation current is flowing.

Beef Processing Line. High Voltage. As with sheep processing, the beginning and the end of the beef processing line are ideal locations for carrying out the high-voltage stimulation of beef carcasses. Whole bodies can be stimulated (1130 V peak, half sine-wave pulses, 10 msec duration, 14.28 pulses/sec) either before or after dehiding using a rubbing electrode system. To ensure consistently good contact through both hind legs, stimulation should occur after the legging operation while the carcass is suspended from the earthed rail by hooks through the hind hocks. If stimulation occurs before dehiding, water sprays should be used to improve electrical contact and provide cooling to avoid hide scorching. Downward hide-pulling after high- and low-voltage stimulation can lead to an increased incidence of broken backs because the muscles do not then respond to the electrical pulses applied to stiffen the muscles during the pulling operation (Richardson 1980). If stimulation is carried out after dehiding, sterilization of the electrode between successive carcasses may be required to satisfy hygiene regulations.

It is difficult to stimulate individual sides with high-voltage pulses applied with simple rubbing electrodes since the side curls, laterally raising the neck more than 1 m above its rest position. An ingenious electrode system of hanging chains has been commercially developed (Fig. 3.27) (MIRINZ 1981). The electrode system moves along with the sides and, despite their considerable movement, consistent electrical contact is achieved. The chains are cleaned and sterilized before contacting the next side.

Low Voltage. Low-voltage stimulation of beef carcasses can be successful provided that stimulation occurs soon after or immediately prior to slaughter. Stimulation is carried out while each carcass is hanging by both hind legs or by one hind leg with the free leg electrically connected to it by a second clip from an earthed rail. The live electrode is attached manually to the nose or jaw of the carcass. All electrical contacts must be clean since with low-voltage stimulation, any reduction in effective voltage has a marked influence on response. Some New Zealand systems electrically check that the total resistance is below a critical threshold before stimulation commences, and if the resistance is too high, a signal is given so that contacts can be cleaned before stimulation is undertaken. If stimulation is applied between stunning and sticking, the delay from cessation of stimulation to sticking must be very short so that there is little chance for the muscles to either replenish their energy supply or to get rid of the lactate built up during stimulation.

Specifications. Meat tenderness is not determined solely by electrical stimulation, but downstream processing may be very important, especially in the New Zealand circumstance where lamb carcasses must be frozen on the day of slaughter. Processing specifications must therefore incorporate not only the stimulation requirement but all aspects of prior and subsequent processing that may affect the final tenderness of the meat. The process, termed "accelerated conditioning," includes a period of electrical stimulation and, in the case of lamb, prefreezing delays and

FIG. 3.27. Photo of on-line automatic high voltage stimulation of beef carcass sides. The hanging chains provide the electrical contact. After the 120 sec stimulation period, the chain electrodes pass through special cleaning and sterilizing baths before coming in contact with another side.
Photo courtesy of The Canterbury Frozen Meat Company Limited.

freezing times. Specifications have been drawn up for the industry so that they have a set of rules that can be followed to ensure that at least a basic level of tenderness is always achieved.

Specification for the accelerated conditioning of lamb either at predressing or postdressing sites are given in Table 3.1 (MIRINZ 1976, 1977). If a higher level of tenderness is required, it can be achieved by following the specifications for producing the production of accelerated-conditioned and aged lamb (MIRINZ 1978B). These require that stimulation be followed by holding at a temperature of not less than 6°C for 8 hr prior to freezing. The tenderness of lamb treated in this way approaches that of conditioned and aged lamb (see Fig. 3.3). Specifications are not yet available for low-voltage stimulation of lamb.

A specification describing the requirements for the high-voltage stimulation of beef is also given in Table 3.1. The strict time–temperature requirements that are part of the lamb specifications are not as essential since the sides of beef are chilled so slowly that the meat enters rigor mortis well before it is of sufficiently low temperature to cold shorten.

Quality Assurance. If meat of the required tenderness is to be achieved, the electrical parameters and the time–temperature requirements of the specifications must be strictly adhered to. Because of the difficulty in identifying product treated according to the specification, a dual quality assurance program is advocated. It is essential that there be a regular and comprehensive program for monitoring various electrical parameters used for stimulating carcasses or sides. It is critical also that there be no departures from the well defined time requirements of the specification, especially in the allowed delay from slaughter to stimulation and the stimulation period itself. The time–temperature requirements for holding lamb carcasses after stimulation and the rate at which they are frozen are also critical points in the quality assurance program.

The final criterion of the effectiveness of the specifications is the tenderness of the meat itself. As part of quality monitoring, a statistically-based program is used to sample loin chops from frozen carcasses. The loin chops

TABLE 3.1. Specifications for Accelerated Conditioning of Lamb and Beef

	Lamb		Beef	
	Predressing	Postdressing	Predressing	Postdressing
Time for slaughter	less than 5 min	less than 30 min	less than 20 min	less than 60 min
Minimum stimulation period	45 sec	90 sec	90 sec	120 sec
Prefreezing delay above 6°C	90 min		not required	
Freezing time (time for deep leg to reach −4°C)	not less than 12 hr from freezer entry		not required	
Electrical specifications for high voltage systems Pulse characteristics:				
shape	half sine-wave			
duration	10 msec			
frequency	14.28 pulses/sec			
polarity	alternating			
peak voltage under load[a]	1130 V			

[a] A much higher open circuit voltage (3600 V peak) is used for predressing lamb stimulation to establish current flow through the high resistance offered by wool and skin.

must be cooked according to closely defined schedules and the tenderness measured. Although this is the true measure of effectiveness, the process monitoring of the process steps provides the immediate quality assurance information. An important element in the national quality assurance program is that occasional surveys be carried out, especially of the critical points of the accelerated conditioning process. In addition, plants must undertake their own routine monitoring.

Safety. Operator safety is an overriding consideration in the design of all electrical stimulation systems. Safety is especially necessary with high-voltage systems. In conjunction with the national electrical supply authority, MIRINZ has drawn up safety guidelines (MIRINZ 1978C) that are the minimum safety requirements for systems in New Zealand. In essence, the guidelines require that the electrodes for high-voltage systems must be enclosed in a tunnel such that the entry of people is impossible while the electrodes are alive. Pressure-sensitive footplates, located at the tunnel entrance and exit, shut off electrical power to the system. Photoelectric sensors provide a second line of safety protection. Access is of course required for maintenance and cleaning. To protect the staff responsible, strategically placed key switches must be activated in a set sequence and within a prescribed time to activate the system. All safety circuits must be tested in both open and closed positions before the high-voltage supply can be energized. If any of the safety devices is tripped, the power is shut off and the startup procedure must be repeated. Although the guidelines err on the side of caution, compliance ensures that high-voltage stimulation systems for lamb and beef will be completely safe.

SUMMARY

Acceptance of the role of cold and thaw shortening in causing toughness in meat led to research into methods that could be used to overcome or eliminate the processing-induced toughness. Holding at moderate temperatures to allow development of rigor mortis before chilling or freezing avoids the ravages of cold and/or thaw shortening, but for high throughputs, very large, expensive holding areas are required. Electrical stimulation is one means of advancing the development of rigor mortis so that the period of risk is reduced.

Although the effects of electrical stimulation can be most easily determined by the fall of pH during stimulation and the subsequent increased rate of pH fall, the ultimate measure of its effectiveness is the tenderness of the final product. Poststimulation processing interacts with stimulation conditions to determine the final tenderness.

Because the rate of pH fall after stimulation is largely independent of stimulation conditions, the fall in pH during stimulation determines differences in the acceleration of rigor mortis development obtained by different stimulation conditions. The magnitude of the drop is affected by animal-related factors, such as muscle temperature, postmortem delay, muscle type, and animal condition, as well as by the characteristics of the stimulating waveform. The range of possible stimulating conditions is almost limitless, but as a result of extensive trials, we have found two types that are appropriate under New Zealand conditions. High-voltage stimulation [1130 V peak, half sine-wave pulses (10 msec duration), 14.28 pulses/sec of alternating polarity] applied for appropriate periods is recommended when large numbers of carcasses are being treated automatically or when stimulation takes place more than 5 min after slaughter. High-voltage stimulation is especially recommended for situations when carcasses or meat cuts are to be rapidly chilled or frozen soon after slaughter of the animal. Low-voltage stimulation [80 V peak, unidirectional half sine-wave pulses (10 msec duration) or square-wave pulses (5 msec duration) at 14.28 pulses/sec] is recommended for smaller manual operations and for situations when the carcasses are larger and will not be cooled very rapidly or very soon after slaughter. Low-voltage stimulation has therefore been recommended for beef and deer processing.

Specifications for processes incorporating electrical stimulation have been developed for both lambs and beef and quality assurance programs prepared to ensure that the desired tenderness levels are being achieved and maintained.

The methods of applying high- and low-voltage stimulation to sheep and beef carcasses have been considered fully, with special attention paid to safety of operators.

Electrical stimulation has become a useful aid in the processing of sheep, cattle, deer, and goats to achieve their inherent level of tenderness.

ACKNOWLEDGMENTS

Among the many contributions of the staff of the New Zealand Meat Industry Research Institute to the field, the authors would especially like to recognize the long support of Cedric Hagyard, Kevin Gilbert, Bert Tavener, Dick Hand, and Peter Macfarlane.

REFERENCES

BATE-SMITH, E.C. 1948. The physiology and chemistry of rigor mortis, with special reference to the aging of beef. Adv. Food Res. *1*, 1.

BATE-SMITH, E.C. and BENDALL, J. 1949. Factors determining the time course of rigor mortis. J. Physiol. *110*, 47.

BEATTY, C.H. and BOCEK, R.H. 1970. Biochemistry of red and white muscles. In Physiology and Biochemistry of Muscle as a Food, Vol. 2. p. 155. E.J. Briskey, R.G. Cassens and B.B. Marsh (Editors). University of Wisconsin Press, Madison.

BENDALL, J.R. 1951. The shortening of rabbit muscles during rigor mortis: It's

relation to the breakdown of adenosine triphosphate and creatine and to muscular contraction. J. Physiol. *114*, 71.
BENDALL, J.R. 1969. Muscles, Molecules & Movement. Heinemann, London.
BENDALL, J.R. 1973. Postmortem changes in muscle. In The Structure and Function of Muscle, Vol. 3. p. 227. G.H. Bourne (Editor). Academic Press, New York.
BENDALL, J.R. 1975. Cold-contracture and ATP-turnover in the red and white musculature of the pig, post-mortem. J. Sci. Food Agric. *26*, 55.
BENDALL, J.R. 1978. Variability in rates of pH fall and of lactate production in the muscles on cooling beef carcasses. Meat Sci. *2*, 91.
BENDALL, J.R. 1980. The electrical stimulation of carcasses of beef animals. In Developments in Meat Science, Vol. 1. p. 37. R.A. Lawrie (Editor). Applied Science, London.
BENDALL, J.R., KETTERIDGE, C.C. and GEORGE, A.R. 1976. The electrical stimulation of beef carcasses. J. Sci. Food Agric. *27*, 1123.
BONNER, J. 1978. The man who lived twice. Scots Mag., Dundee (Mar.) 618.
BOUTON, P.E., FORD, A.L., HARRIS, P.V. and SHAW, F.D. 1978. Effect of low voltage stimulation of beef carcasses on muscle tenderness and pH. J. Food Sci. *43*, 1392.
BOUTON, P.E., WESTE, R.R. and SHAW, F.D. 1980. Electrical stimulation of calf carcasses; Response of various muscles to different waveforms. J. Food Sci. *45*, 148.
BUCHTER, L. 1972. The influence of chilling temperature on the toughness of loin muscles from calves and young bulls. In Chilling; Why and How, Symp. 2. p. 4.51. Meat Research Inst., Langford, England.
CARSE, W.A. 1973. Meat quality and the acceleration of post-mortem glycolysis by electrical stimulation. J. Food Technol. *8*, 163.
CHALCROFT, J.P. and CHRYSTALL, B.B. 1974. Current distribution in carcasses during electrical stimulation. In Meat Industry Research Institute of New Zealand Annu. Rep. 1973–74, 27. MIRINZ, Hamilton, N.Z.
CHRYSTALL, B.B. 1978. Electrical stimulation, refrigeration and subsequent meat quality. 24th Eur. Meat Res. Workers Conf., Kulmbach, W. Germany, Pap. *E7.3*.
CHRYSTALL, B.B. 1980. Accelerated conditioning of lamb: Problems of implementation. 21st Meat Ind. Res. Conf., Hamilton, New Zealand, 9.
CHRYSTALL, B.B. and DEVINE, C.E. 1978. Electrical stimulation, muscle tension and glycolysis in bovine sternomandibularis. Meat Sci. *2*, 49–58.
CHRYSTALL, B.B. and DEVINE, C.E. 1983. A rat model for electrical stimulation studies. Meat Sci. *9*, 33.
CHRYSTALL, B.B. and HAGYARD, C.J. 1975. Accelerated conditioning of lamb. Meat Ind. Res. Inst. N.Z. Publ. *470*.
CHRYSTALL, B.B. and HAGYARD, C.J. 1976. Electrical stimulation and lamb tenderness. N.Z. J. Agric. Res. *19*, 7.
CHRYSTALL, B.B., DEVINE, C.E. and DAVEY, C.L. 1980A. Studies in electrical stimulation: Post mortem decline in nervous response in lambs. Meat Sci. *4*, 69–78.
CHRYSTALL, B.B., DEVINE, C.E. and DAVEY, C.L. 1980B. Electrical stimulation: Influence of post mortem delay and curare on nervous responses in lamb. In Fibrous Proteins: Scientific, Industrial and Medical Aspects, Vol. 2. p. 67. D.A.D. Parry and L.K. Creamer (Editors). Academic Press, London.
CHRYSTALL, B.B., DEVINE, C.E., ELLERY, S. and WADE, L. 1982A. Low voltage electrical stimulation of lamb. Meat Sci. (submitted)
CHRYSTALL, B.B., DEVINE, C.E., SNODGRASS, M. and ELLERY, S. 1982B. Tenderness of excerise-stressed lambs. N.Z. J. Agric. Res. *25*, 331.
CLARKE, F.M., SHAW, F.D. and MORTON, D.J. 1980. Effect of electrical stimulation post mortem of bovine muscle on the binding of glycolytic enzymes. Biochem. J. *186*, 105.
CORI, G.T. 1945. The effect of stimulation and recovery on the phosphorylase a content of muscle. J. Biol. Chem. *158*, 333.

DAVEY, C.L. 1971. Beef processing and aging. Food Technol. N.Z. 6 (5) 31.
DAVEY, C.L. and FOSTER, M. 1977. Freezing rate, frozen storage and lamb tenderness. Meat Sci. 1, 157.
DAVEY, C.L. and GILBERT, K.V. 1975. The tenderness of cooked and raw meat from young and old beef animals. J. Sci. Food Agric. 26, 953.
DAVEY, C.L. and GILBERT, K.V. 1976. Thaw contracture and the disappearance of adenosinetriphosphate in frozen lamb. J. Sci. Food Agric. 27, 1085.
DAVEY, C.L. and WINGER, R.J. 1980. The structure of skeletal muscle and meat toughness. In Fibrous Proteins: Scientific, Industrial and Medical Aspects, Vol. 1. p. 97. D.A.D. Parry and L.K. Creamer (Editors). Academic Press, London.
DAVEY, C.L., KUTTEL, H. and GILBERT, K.V. 1967. Shortening as a factor in meat aging. J. Food Technol. 2, 53–56.
DAVEY, C.L., GILBERT, K.V. and CARSE, W.A. 1976. Carcass electrical stimulation to prevent cold shortening toughness in beef. N.Z. J. Agric. Res. 19, 13.
deFREMERY, D. and POOL, M.F. 1960. Biochemistry of chicken muscle as related to rigor mortis and tenderization. Food Res. 25, 73.
deFREMERY, D. and POOL, M.F. 1963. The influence of post mortem glycolysis on poultry tenderness. J. Food Sci. 28, 173.
DEVINE, C.E. 1976. Accelerated conditioning of meat. Proc. 18th N.Z. Meat Ind. Res. Conf., 1976, Rotorua, N.Z., 10.
DEVINE, C.E. and CHRYSTALL, B.B. 1982. Unpublished results. Meat Industry Research Inst. of New Zealand, Hamilton, N.Z.
DEVINE, C.E., CHRYSTALL, B.B. and DAVEY, C.L. 1979. Studies in electrical stimulation: Effect of neuromuscular blocking agents in lamb. J. Sci. Food Agric. 30, 1007.
FABIANSSON, S., JONSSON, G. AND RUDERUS, H. 1979. Optimum conditions for low voltage electrical stimulation of beef carcasses. 25th Eur. Meat Res. Workers Conf., 1979, Budapest, Pap. 2.3.
FORREST, J.C. and BRISKEY, E.J. 1967. Response of striated muscle to electrical stimulation. J. Food Sci. 32, 483.
FRANZINI-ARMSTRONG, C. and PEACHEY, L.D. 1981. Striated muscle—Contractile and control mechanisms. J. Cell Biol. 91 (3.2) 166s.
GAUTHIER, G.F. 1970. The ultrastructure of three fiber types in mammalian skeletal muscle. In Physiology and Biochemistry of Muscle as a Food, Vol. 2. p. 103. E.J. Briskey, R.G. Cassens and B.B. Marsh (Editors). University of Wisconsin Press, Madison.
GEORGE, A.R., BENDALL, J.R. and JONES, R.C.D. 1980. The tenderizing effect of electrical stimulation of beef carcasses. Meat Sci. 4, 51.
GILBERT, K.V. 1979. Electrical stimulation during hide pulling. 29th Annu. Conf. N.Z. Leather and Shoe Res. Inst., 1979, Palmerston North, New Zealand.
GILBERT, K.V. 1980. Personal communication. Meat Industry Research Institute of New Zealand, Hamilton, N.Z.
GILBERT, K.V. and CHRYSTALL, B.B. 1982. Unpublished. Meat Industry Research Institute of New Zealand, Hamilton, N.Z.
GILBERT, K.V. and DAVEY, C.L. 1976. Carcass electrical stimulation and early boning of beef. N.Z. J. Agric. Res. 19, 429.
GILBERT, K.V. and DAVEY, C.L. 1980. Unpublished. Meat Industry Research Institute of New Zealand, Hamilton, N.Z.
GILBERT, K.V., DAVEY, C.L. and NEWTON, K.G. 1977. Electrical stimulation and the hot boning of beef. N.Z. J. Agric. Res. 20, 139.
HAGYARD, C.J. 1979. Aging regimes for electrically stimulated lamb. Meat Ind. Res. Inst. N.Z. Publ. RM88.
HAGYARD, C.J., HAND, R.J. and GILBERT, K.V. 1980. Lamb tenderness and electrical stimulation of dressed carcasses. N.Z. J. Agric. Res. 23, 27.
HALLUND, O. and BENDALL, J.R. 1965. The long-term effect of electrical stimula-

tion on the post-mortem fall of pH in the muscles of Landrace pigs. J. Food Sci. 30, 296.
HAMM, R. 1977. Post mortem breakdown of ATP and glycogen in ground muscle: A review. Meat Sci. 1, 15.
HARSHAM, A. and DEATHERAGE, F.E. 1951. Tenderization of meat. U.S. Pat. 2,544,681. Mar. 13.
HERRING, H.K., CASSENS, R.G., SUESS, G.G., BRUNGARDT, V.H. and BRISKEY, E.J. 1967. Tenderness and associated characteristics of stretched and contracted bovine muscles. J. Food Sci. 32, 317.
HOULIER, B., VALIN, C., MONIN, G. and SALE, P. 1980. Is electrical stimulation efficiency muscle dependent? Proc. 24th Eur. Meat Res. Workers Conf., Colorado Springs, Pap. J5.
HOWARD, A. and LAWRIE, R.A. 1956. Studies on beef quality. Part II. Physiological and biochemical effects of various pre-slaughter treatments. Dep. Sci. Ind. Res. (DSIR) Spec. Rep. Food Invest. Board, London, 63.
INGRAM, M. and INGRAM, G.C. 1955. The growth of bacteria on horse muscle, in relation to the changes after death leading to rigor mortis. J. Sci. Food Agric. 6, 602.
JEACOCKE, R.E. 1977. The temperature dependence of an anaerobic glycolysis in beef muscle held in a linear temperature gradient. J. Sci. Food Agric. 28, 551.
JONSSON, G., FABIANSSON, S. and NILSSON, H. 1978. Experiments to avoid cold shortening in beef. Proc. 24th Eur. Meat Res. Workers Conf. 1978, Kulmbach, W. Germany, E10:1.
KARPATKIN, S., HELMREICH, E. and CORI, C.F. 1964. Regulation of glycolysis in muscle. II. Effect of stimulation and epinephrine in isolated frog sartorius muscle. J. Biol. Chem. 239, 3139.
KERR, D.T., BOUTON, P.E., HARRIS, P.V. and SHAW, F.D. 1979. Extra low voltage stimulation of beef carcasses using a rectal probe. CSIRO Meat Res. Rep. 8/79.
KREBS, H.A. 1950. Body size and tissue respiration. Biochim. Biophys. Acta 4, 249.
LAWRIE, R.A. 1953. The onset of rigor mortis in various muscles of the draught horse. J. Physiol. 121, 275.
LOCKER, R.H. 1960. Degree of muscular contraction as a factor in tenderness of beef. Food Res. 25, 304.
LOCKER, R.H. and HAGYARD, C.J. 1963. A cold shortening effect in beef muscles. J. Sci. Food Agric. 14, 787.
LOCKER, R.H., DAVEY, C.L., NOTTINGHAM, P.M., HAUGHEY, D.P. and LAW, N.H. 1975. New concepts in meat processing. Adv. Food. Res. 21, 157.
LOEFFEN, M.P.F. and HAGYARD, C.J. 1979. Unpublished observations. Meat Industry Research Inst. of Hamilton, N.Z., New Zealand.
LOPEZ, C.A. and HERBERT, E.W. 1975. The Private Franklin: the Man and His Family, 1st Edition. p. 44. W.W. Norton and Co., New York.
MACFARLANE, P.G. and MARER, J.M. 1966. An apparatus for determining the tenderness of meat. Food Technol. 20, 838.
MacKENZIE, P. 1876. Reminiscences of Glasgow and the West of Scotland. J. Tweed, Glasgow.
MARSH, B.B. 1954. Rigor mortis in beef. J. Sci. Food Agric. 5, 70.
MARSH, B.B. and LEET, N.G. 1966. Studies in meat tenderness. III. The effects of cold shortening on tenderness. J. Food Sci. 31, 450.
MARSH, B.B., WOODHAMS, P.R. and LEET, N.G. 1968. Studies in meat tenderness. 5. The effects of carcass cooling and freezing before the completion of rigor mortis. J. Food Sci. 33, 12.
MARTIN, A.H., JEREMIAH, J.E., FREDEEN, H.T. and L'HIRONDELLE, P.J. 1977. Influence of pre- and post-rigor muscle change on intrinsic toughness of beef muscle. Can. J. Anim. Sci. 57, 705.
McCRAE, S.E., SECCOMBE, C.G., MARSH, B.B. and CARSE, W.A. 1971. Studies in

meat tenderness. 9. The tenderness of various lamb muscles in relation to their skeletal restraint and delay before freezing. J. Food. Sci. *36,* 566.
McLOUGHLIN, J.V. 1970. Muscle contraction and postmortem pH changes in pig skeletal muscle. J. Food Sci. *35,* 717.
MIRINZ. 1976. Specification for AC of lambs before dressing. Meat Ind. Res. Inst. N.Z. Publ. *525.*
MIRINZ. 1977. Specification for AC of lambs after dressing. Meat Ind. Res. Inst. N.Z. Publ. *RM54.*
MIRINZ. 1978A. Specification for accelerated conditioning of beef. Meat Ind. Res. Inst. N.Z. Publ. *RM73.*
MIRINZ. 1978B. Specification for AC & A lamb. Meat Ind. Res. Inst. N.Z. Publ. *RM70.*
MIRINZ. 1978C. Safety requirements for the operation of accelerated conditioning systems. Meat Ind. Res. Inst. N.Z. Publ. *643.*
MIRINZ. 1981. Meat Industry Research Institute of New Zealand Annual Report. MIRINZ, Hamilton, N.Z.
NEEDHAM, D.M. 1971. Machina Carnis. Cambridge University Press, Cambridge.
NEWBOLD, R.P. 1966. Changes associated with rigor mortis. *In* The Physiology and Biochemistry of Muscle as a Food. p. 213. E.J. Briskey, R.G. Cassens and J.G. Trautman (Editors). University of Wisconsin Press, Madison.
POSNER, J.B., STERN, R. and KREBS, E.G. 1965. Effects of electrical stimulation and epinephrine on muscle phosphorylase, phosphorylase b kinase and adenosine 3',5'-phosphate. J. Biol. Chem. *240,* 982.
RENTSCHLER, H.C. 1951. Apparatus and method for the tenderization of meat. U.S. Pat. 2,544, 724. Mar. 13.
RICHARDSON, J.C. 1980. Innovations in downward hide pulling. Proc. 21st Meat Ind. Res. Conf., 1980, Hamilton, N.Z., 25.
RILEY, R.R., SAVELL, J.W., SMITH, G.C. and SHELTON, M. 1980. Quality, appearance and tenderness of electrically stimulated lamb. J. Food Sci. *45,* 119.
ROSCHEN, H.L., ORFSCHEID, B.J. and RAMSBOTTOM, J.H. 1950. Tenderizing meats. U.S. Pat. 2,519,931. Aug. 22.
RUDERUS, H. and BERQUIST, A. 1980. Industrial application of low voltage electrical stimulation in Sweden. Ann. Technol. Agric. *29,* 659.
RUDERUS, H. and FABIANSSON, S. 1980. Research on low voltage electrical stimulation of beef carcasse in Sweden. *Ann. Tech. Agric. 29,* 581.
SAVELL, J.W. 1979. Electrical stimulation of meat—Other aspects. *In* Conf. Proc. Electr. Stimulation for Improving Meat Qual., Texas A&M Univ., College Station, 26–31.
SAVELL, J.W., SMITH, G.C. and CARPENTER, Z.L. 1978. Effect of electrical stimulation on quality and palatability of light-weight beef carcasses. J. Anim. Sci. *46,* 1221.
SCOTT, D. 1973. Stock in Trade: Hellaby's First 100 Years 1873–1973. Southern Cross Books, Auckland.
SHAW, F.D. and WALKER, D.J. 1977. Effect of low voltage stimulation of beef carcasses on muscle pH. J. Food Sci. *42,* 1140.
SQUIRE, J. 1981. The Structural Basis of Muscular Contraction. Plenum Press, New York.
TAYLOR, D.G. and MARSHALL, A.R. 1980. Low voltage electrical stimulation of beef carcasses. J. Food Sci. *45,* 144.
WALKER, D.J., HARRIS, P.V. and SHAW, F.D. 1977. Accelerated processing of beef. Food Technol. Aust. *29,* 504.
WALSH, T.P., MASTERS, C.J., MORTON, D.J. and CLARKE, F.M. 1981. The reversible binding of glycolytic enzymes in ovine skeletal muscle in response to tetanic stimulation. Biochim. Biophys. Acta *675,* 29.

4

Effects of Electrical Stimulation on Meat Quality, Color, Grade, Heat Ring, and Palatability

G. C. Smith[1]

Tenderness
Flavor
Lean Color
Heat Ring Prevention
Marbling and Quality Grade
Retail Caselife
Processing Properties
Low- vs High-voltage Electrical Stimulation
Implementation by the Packing Industry
References

In the late 1960s, scientists at Texas A&M University (TAMU) developed a process for tenderization of carcasses that was ultimately called "Texas A&M Tenderstretch" (Orts *et al.* 1971). Despite extensive research documenting its effectiveness in increasing tenderness of beef and lamb, the "Tenderstretch" process was not implemented by packers in the United States. The process is used commercially by a few packers in North America and some packers in South America and it has been used extensively in Australia and New Zealand. It was the use of "Tenderstretch" to prevent toughness in rapidly processed (chilling and/or freezing) lamb carcasses in New Zealand that led, ultimately, to TAMU investigations of electrical stimulation.

TAMU scientists first learned of electrical stimulation from a former student upon his return from a 1974 visit to New Zealand; he called to report that scientists there were experimenting with a new process for "conditioning" (treating carcasses in a manner that minimizes the toughening of muscles during chilling and/or freezing) lambs. Historically, New Zealand packers have frozen lambs intended for the export trade as quickly as possible after slaughter. About 20 years ago, New Zealand scientists

---

[1]Professor and Head, Department of Animal Science, Texas A&M University, College Station, TX.

determined that early postmortem freezing causes "cold shortening" of muscle fibers and consequent toughening of the meat. "Cold shortening," it was later discovered, could be prevented by holding lambs at high (about 62°F) temperatures for the first 14 to 18 hr after slaughter or by hanging the lamb carcasses by the "Tenderstretch" (they call it "Squat-Posture") method of suspension. It was the threatened discontinuation of the use of "Tenderstretch" to minimize "processing toughness" in lambs intended for export to North America that aroused our interest in investigating electrical stimulation. Use of electrical stimulation in New Zealand was not intended to tenderize conventionally handled carcasses; rather, it was used to prevent "processing toughness" so that rapidly processed lamb would have tenderness equivalent to that which it would have had if it had been conventionally chilled.

The information that TAMU scientists had regarding electrical stimulation was quite sparse (as it turned out, it was also inaccurate). Beginning with the information that the process involved 2 min of stimulation with 36,000 V (TAMU's source was, in fact, incorrect on this aspect; the New Zealand scientists were actually using 3600 V) and with the intent of comparing its effectiveness with that of "Tenderstretch," TAMU scientists sought assistance from others in academia and in industry and from meat scientists throughout the United States that would allow simulation of the process. Initial progress was deterred by the unusually high voltage requirements, but TAMU scientists ultimately resorted to using the only source of electrical impulse then available to accomplish reasonable simulation—a commercially available hog stunner.

A search of the literature ultimately revealed that the use of electricity to increase meat tenderness was not new; Benjamin Franklin had suggested its use for that purpose in 1749. In a biography of Franklin, it was reported that "killing turkeys electrically, with the pleasant side effect that it made them uncommonly tender, was the first practical application that had been found for electricity" (Lopez and Herbert 1975). According to Lemisch (1961), Franklin stated the following in a 1773 letter to two French scientists:

It has been observed that lightning, by rarefying and reducing into vapour the moisture contained in solid wood, in an oak, for instance, has forcibly separated its fibres, and broken it into small splinters; that, by penetrating intimately the hardest metals, as iron, it has separated the parts in an instant, so as to convert a perfect solid into a state of fluidity; it is not then improbable, that the same subtle matter, passing through the bodies of animals with rapidity, should possess sufficient force to produce an effect nearly similar. The flesh of animals, fresh killed in the usual manner, is firm, hard, and not in a very eatable state, because the particles adhere too forcibly to each other. At a certain period, the cohesion is weakened, and, in its progress towards putrefaction, which tends to produce a total separation, the flesh becomes what we call tender, or is in that state most proper to be used as our food. It has frequently been remarked, that animals killed by lightning putrefy immediately. It is not unreasonable to presume, that, between the period of their death and that of their putrefaction, a time intervened in which the flesh might be only tender, and only sufficiently so to be served at table. Add to this that persons, who have eaten of fowls killed by our feeble imitation of lightning (electricity), and dressed immediately, have asserted, that the flesh was remarkably tender. Experience alone will inform us of the requisite proportions (of electricity) for animals of different forms and ages. Probably not less will be required to render a small bird, which is very old, tender, than for a larger one, which is young.

In 1951, over two centuries after Franklin's observation, patents were issued for electrical stimulation processes to Harsham and Deatherage (1951) and Rentschler (1951); no reports of the research studies upon which those patents were based were ever published in the scientific or popular literature. It is now known (Savell 1979) that patent rights for those processes were sold to Kroger and Westinghouse, respectively, but neither company implemented the processes commercially. Whatever the reason, no commercial use was made of the processes and the patents expired in 1968.

The first report of scientific research on electrical stimulation and meat tenderness in New Zealand was released in 1973 (Carse 1973); research on electrical stimulation and meat tenderness was initiated at the University of Florida and at Texas A&M University in 1975 and each institution gave its first report of scientific research in 1976 (Grusby et al. 1976; Savell et al. 1976). The probable sequence of events (ideas and research) in development of electrical stimulation and the first research report from each of several countries is presented in Table 4.1.

Several of the events listed in Table 4.1 did not result in publications in the refereed scientific literature. It now seems quite likely that had the

TABLE 4.1. Probable Sequence of Events (Ideas and Research) in Development and Use of Electrical Stimulation

Year	Country	Event	Reference
1749	United States	Ben Franklin's observation	Lopez and Herbert (1975)
1943	United States	Kroger-Westinghouse research	Deatherage (1980)
1951	United States	Process patent for electrical stimulation	Harsham and Deatherage (1951)
1951	United States	Process patent for electrical stimulation	Rentschler (1951)
1964	England	Research with pork	Hallund and Bendall (1965)
1966	United States	Research with pork	Forrest and Briskey (1967)
1972	New Zealand	Research with lamb	Carse (1973)
1975	United States	Research with beef	Grusby et al. (1976)
1975	United States	Research with goat, lamb, and beef	Savell et al. (1976)
1975	England	Research with rabbit and lamb	Bendall (1976)
1976	Australia	Research with beef	Shaw and Walker (1977)
1976	Canada	Research with beef	Swatland (1977)
1978	Denmark	Research with pork	Braathen (1979)
1978	Norway	Research with beef	Braathen (1979)
1979	France	Research with beef	Houlier et al. (1980)
1979	Sweden	Research with beef	Ruderus and Fabiansson (1980)
1979	Belgium	Research with beef	Demeyer and Vandendriessche (1980)

electrical stimulation process not been tied up in patents and had results of the Harsham and Deatherage (1951) or the Rentschler (1951) studies been made available to the scientific community, the process might have been adopted by commerical industry at a much earlier date. Several of the events listed in Table 4.1 did not relate electrical stimulation to meat palatability. Listed in Table 4.2, in chronological order, are scientific reports that related electrical stimulation to changes in meat palatability.

Research on electrical stimulation by TAMU scientists initially concentrated on determining its effectiveness in increasing tenderness; subsequent studies identified a number of benefits and/or advantages for electrical stimulation that were not expected and/or that had not been previously reported. As benefits (Table 4.3) continued to accrue, it became more and more apparent that commercialization of the electrical stimulation process was possible in the United States.

As scientific research was being conducted, attention was also paid by TAMU personnel to developmental efforts, to in-plant testing, and to parameters related to possible industry implementation of the process. The progression from the Best and Donovan Hog Stunner to the LeFiell-USDA Test Unit to the LeFiell Commercial Unit and the developmental effort in methods of applying electrical impulses are outlined in Table 4.4. In-plant tests by TAMU personnel, initiated in 1976 at Bristol, VA, are listed in Table 4.5. In-plant tests were conducted (1) to gather research data, (2) to identify conditions necessary to facilitate implementation by industry, and (3) to demonstrate the procedure to industry.

TAMU studies of electrical stimulation involving field research and the development and implementation of specifics for its commercial use are detailed in Table 4.6. Among developmental efforts, the chief advancements were made after TAMU scientists were able to very precisely identi-

TABLE 4.2. Sequence of Scientific Reports Describing Effects of Electrical Stimulation on Meat Palatability

Year	Country	Relationship	Reference
1973	New Zealand	Tenderness of lamb	Carse (1973)
1976	United States	Tenderness of beef	Grusby et al. (1976)
1976	United States	Tenderness of goat, lamb, beef	Savell et al. (1976)
1976	New Zealand	Tenderness of lamb	Chrystall and Hagyard (1976)
1976	New Zealand	Tenderness of beef	Davey et al. (1976)
1976	England	Tenderness of beef	Bendall and Rhodes (1976)
1978	Australia	Tenderness of beef	Bouton et al. (1978)
1978	United States	Tenderness of pork	Westervelt and Stouffer (1978)
1980	France	Tenderness of beef	Houlier et al. (1980)
1980	Sweden	Tenderness of beef	Ruderus and Fabiansson (1980)
1980	Belgium	Tenderness of beef	Demeyer and Vandendriessche (1980)
1980	Canada	Tenderness of beef	Jeremiah and Martin (1980)

TABLE 4.3. Principal Findings of Research on Electrical Stimulation by TAMU Scientists

Year	Species	Finding	Reference
1975	Beef	Increased tenderness	Savell et al. (1976)
1976	Lamb, goat	Increased tenderness	Savell et al. (1977)
1976	Beef	Reduced "heat ring"	Smith et al. (1977)
1977	Beef	Mode of action; structural damage	Savell et al. (1978A)
1977	Beef	Brighter lean color	Savell et al. (1978B)
1977	Lamb	Faster pH decline	Bowling et al. (1978)
1977	Beef	Improved marbling and grade	Savell et al. (1978C)
1977	Beef	Reduced need for aging	Savell et al. (1978C)
1978	Beef	Facilitates hot boning	Seidman et al. (1979)
1978	Beef	Improved flavor	Savell and Smith (1979)
1978	Beef	Faster pH decline	Savell et al. (1979)
1978	Calf	Increased tenderness	Smith et al. (1979A)
1978	Goat	No effect of stage-in-slaughter	McKeith et al. (1979)
1978	Forage-fed beef	Complementarity in tenderization	Smith et al. (1979B)
1979	Lamb	No effect of stage-in-slaughter	Riley et al. (1980A)
1979	Beef	Facilitates earlier grading	Calkins et al. (1980)
1979	Lamb	Enhances retail caselife	Riley et al. (1980B)
1979	Mature cows	Improved tenderness and grade	McKeith et al. (1980B)
1979	Beef	Same effect—Sides or carcasses	McKeith et al. (1980A)
1979	Beef	No effect on protein quality	Dutson et al. (1980A)
1979	Lamb	Mode of action; lysosomal enzymes	Dutson et al. (1980B)
1979	Beef	More rapid rate of chilling	Dutson et al. (1980D)
1979	Beef	Enhances retail appearance	Hall et al. (1980)
1979	Rabbit, pork, lamb, beef	No effect on bacteria on meat	Mrigadat et al. (1980)
1980	Beef	Mechanism of brighter lean color	Orcutt et al. (1981)
1980	Beef	No effect on bacterial growth in beef	Butler et al. (1981)
1980	Beef	Identified muscles tenderized	McKeith et al. (1981A)
1980	Beef	No effect of stage-in-slaughter	McKeith et al. (1981B)
1980	Beef	Parameters: 550 V, 1 min	McKeith et al. (1981B)
1980	Beef	No effect on moisture, pH of beef roasts	Terrell et al. (1981)
1980	Beef	Greatest impact if aged short time	Savell et al. (1981)
1981	Ram lambs	Increased tenderness	Riley et al. (1981)
1981	Veal	Improved color, firmness of lean	McKeith et al. (1982)
1981	Pork	Little effect on processing properties	Swasdee et al. (1982)
1981	Mature cows	Little effect on processing properties	Terrell et al. (1982A)
1981	Bullock beef	Little effect on processing properties	Terrell et al. (1982B)
1981	Beef	Little effect on processing properties	Terrell et al. (1982C)
1981	Beef	Mode of action; flavor	Calkins et al. (1982)
1981	Beef	Cannot prevent "dark cutters"	Dutson et al. (1982)
1981	Pork	Minimal effects on quality-palatability	Johnson et al. (1982)
1981	Beef	Complementary with cooking-start temperature	Hostetler et al. (1982)
1981	Beef	Minimal effects on retail display traits	Riley et al. (1982)
1981	Bullock beef	Improved lean color	Savell et al. (1982)
1982	Slaughter calf	Improved palatability	Rouquette et al. (1983)
1982	Bullock beef	Improved tenderness	Riley et al. (1983A,B)

TABLE 4.4. History; Methods of Stimulation, Texas A&M University

Year	Method	Machine
1975	Discs on bar	B and D Hog Stunner
1975	Cables, two clamps	B and D Hog Stunner
1976	Cables, two probes	B and D Hog Stunner
1977	Cable, one probe	LeFiell-USDA Test Unit
1978	Cable, contact bar	LeFiell-USDA Test Unit
1978	Rub bar	LeFiell Commercial Unit

Source: Smith (1979A).

fy the proper voltage, impulse duration, and number of impulses needed to effect desired changes in tenderness and in meat quality indicators. The most useful results of field research were those that allowed identification of locations on the kill floor for possible installation of the electrical stimulator in an in-line, hand-operated system for all carcasses at H and H Meat Products in Mercedes, TX. The H and H Meat Products company has successfully employed the hand-operated system for electrical stimulation since April 1978. From an implementation standpoint, the TAMU research program received its greatest impetus when the LeFiell Company developed a commercial stimulation unit. This unit was installed at Sam Kane Beef Processors in Corpus Christi, TX. The Sam Kane Beef Processors company installation was the first in-line, fully automated system for electrical stimulation of beef carcasses in the United States.

The role of Texas A&M University in use of electrical stimulation was that of demonstrating the commercial feasibility of the process and encouraging its implementation by the packing industry of the United States. As the system evolved, at least with respect to TAMU research and development efforts, the voltage used, the fact that impulses should be intermittent rather than continuous, and the expectations in terms of its benefits were decidedly different (Table 4.7) from those in the two other countries where electrical stimulation had been researched and/or used in the 1970s.

TABLE 4.5. History; In-plant Tests, Texas A&M University

Year	Location	Species
1976	Bristol, VA	Calves
1977	Corpus Christi, TX	Beef
1977	Mercedes, TX	Beef
1978	Mercedes, TX	Beef
1978	Greeley, CO	Beef
1978	Denver, CO	Beef
1978	Houston, TX	Beef
1978	Abilene, TX	Beef

Source: Smith (1979A).

TABLE 4.6. History; Field Research, Development and Implementation of Electrical Stimulation, Texas A&M University

Year	Species	Event
1976	Beef	Development; number of impulses
1977	Lamb	Research; comparison of United States and New Zealand systems
1977	Beef	Development; LeFiell-USDA Test Unit
1978	Beef	Research; location on kill floor
1978	Beef	Research; impulse specifications
1978	Beef	Implementation; in-line, hand-operated system, all carcasses (H and H Meat Products)
1978	Lamb	Research; rapid chilling
1978	Beef	Research; grade factors
1978	Beef	Development; LeFiell Commercial Unit
1978	Beef	Implementation; in-line, fully automated system, all carcasses (Sam Kane Beef Processors)

Source: Smith (1979A).

During a period of eight years, a series of studies was conducted by researchers at Texas A&M University to determine effects of electrical stimulation on quality and palatability of beef, pork, lamb, and goat meat. Of interest in those studies were advantages that might accrue to the packer, retailer, purveyor, restaurateur, and/or consumer as a result of implementation of electrical stimulation technology by the slaughter-dressing industry. Improving the tenderness of beef, pork, lamb, and goat meat has been a major goal of TAMU scientists for more than two decades; during studies intended to quantitate the tenderizing effect of electrical stimulation (ES), it became apparent that the ES process also had quality enhancement capabilities. As a result, research studies ensued that had the objective of documenting response in quality indicators as well as in palatability attributes to the electrical stimulation of beef, pork, lamb, and goat carcasses. Later, it was also important to determine effects of electrical stimulation on shrinkage loss and appearance of wholesale and retail cuts and on processing characteristics of muscles from electrically stimulated carcasses. In the following sections, effects of electrical stimulation on specific carcass, cut, or muscle traits will be discussed.

TABLE 4.7. Comparison of Electrical Stimulation Methodology Used in Three Countries

Country	Voltage	Impulses	Purpose
New Zealand	3600	1 for 2 min	Prevention of cold shortening in frozen lamb
England	750	1 for 2 min	Prevention of cold shortening in rapidly chilled beef
United States	100 to 800, depends on purpose	8 to 50, varies but is intermittent	Tenderization; prevention of "heat ring"; improvement of muscle color; more rapid development of grade factors

Source: Smith (1979A).

TENDERNESS

Possibilities of use of electrical stimulation to tenderize meat were brought to the attention of TAMU scientists by research in New Zealand (as was mentioned earlier), but our intent in using the process was decidedly different from theirs. In its original application in New Zealand, electrical stimulation was used to prevent meat from becoming tough; in the United States, electrical stimulation was (and still is) used to make meat more tender. If a lamb carcass with inherently "tough" muscle is split and one side (side A) is treated by use of the New Zealand process (stimulated with a single impulse of high voltage-low amperage electricity, then rapidly chilled and frozen) and the other side (side B) is treated in the manner identified by the TAMU research group (stimulated by use of repetitive electrical impulses of intermediate voltage and intermediate amperage, then conventionally chilled), muscle from side A will still be "tough" whereas muscle from side B will be "tender."

Support for the latter statement is provided in a statement by Bendall (1976), who referred to a then-unpublished study in which electrical stimulation was being investigated by researchers in New Zealand. Bendall (1976) stated that, "The method has been further developed by New Zealand workers for use in lamb slaughtering abattoirs. This enables lamb carcasses to be frozen much sooner than normal after slaughter, without danger of contracture and toughening of meat on thawing." Further support for the difference in objectives between United States and New Zealand studies of electrical stimulation is provided in Table 4.8 although conditions were not exactly equivalent to those used in the two countries. Data in Table 4.8 compare the tenderization achieved in lamb loin chops by using the Gallagher Energizer (an electric fence-charger from New Zealand that was used in some electrical stimulation studies in that country) and the Best and Donovan "Electro-Sting" Hog Stunner (the machine used in early studies of electrical stimulation at Texas A&M University).

TABLE 4.8. Comparison of Effects of the Gallagher Energizer and the Best and Donovan Hog Stunner in Improving Tenderness of Lamb Loin Chops

Comparison	Number of electrical stimulation impulses	Tenderness changes (%)	
		Sensory[a]	Shear[b]
Control vs Gallagher Energizer	1	−8	−2
Control vs Gallagher Energizer	25	−8	−8
Control vs Best and Donovan Stunner	1	+25	−32
Control vs Best and Donovan Stunner	25	+27	−46
Gallagher Energizer vs Best and Donovan Stunner	1	+27	−33
Gallagher Energizer vs Best and Donovan Stunner	25	+42	−28

Source: Smith (1979B).
[a] Sensory = tenderness ratings by a trained taste panel.
[b] Shear = Warner-Bratzler shear force determinations.

TABLE 4.9. Summary of Effects of Electrical Stimulation on Tenderness of Beef, Pork, Lamb, and Goat Meat

Meat	Shear force value[a]		Sensory panel ratings[b]	
	Number of carcasses or samples	Percentage improvement	Number of carcasses or samples	Percentage improvement
Beef	656	23	452	26
Pork	90	9	180	3
Lamb	137	24	109	12
Goat	731	29	229	32

Source: Smith et al. (1980).
[a] The force required to shear a 1.27 cm core of cooked beef, pork, lamb, or goat muscle.
[b] Ratings of comparative tenderness–toughness of cooked product by an experienced panel of 8 to 10 members.

Use of the Gallagher Energizer changed tenderness by −8 to +8% over that of untreated controls; use of the Best and Donovan Hog Stunner increased tenderness by 25 to 46%. Direct comparison of the two instruments revealed that use of the Best and Donovan Hog Stunner tenderized lamb by 27 to 42% more than that achieved by the Gallagher Energizer.

Further verification that the New Zealanders were pursuing use of electrical stimulation for preventing "processing toughness" is provided in the first scientific report of their research in a paper by Carse (1973). Carse (1973) concluded that electrically stimulated lamb carcasses put into a blast freezer at 5 hr postmortem were as tender as untreated (not electrically stimulated) carcasses put into a blast freezer at 20 hr postmortem that had been held for 16 of those 20 hr at elevated temperature to avoid toughness caused by postmortem shortening. Chrystall and Hagyard (1976) were also studying "processing toughness" when they determined that electrical stimulation prior to freezing of the lamb carcass greatly improved tenderness in comparison with freezing without having electrically stimulated the carcass. However, Davey et al. (1976) used electrical stimulation so that "even with rapid chilling the stimulated carcasses are still warm at rigor entry, so that cold shortening with toughening does not develop and the meat can be aged to a high degree of uniform tenderness;" they achieved substantial improvements in tenderness by electrically stimulating beef carcasses that were then chilled and never frozen.

Research at Texas A&M University has demonstrated that sizable tenderness improvements result from use of electrical stimulation (Table 4.9). Although most of the TAMU research has involved beef, efforts have also been made to determine the influence of ES on pork, lamb, and goat. On the average, tenderness, measured by reduced resistance to shear force, was increased approximately 23, 9, 24, and 29% for beef, pork, lamb, and goat carcasses, respectively. Sensory panel evaluations indicated that steaks or chops from electrically stimulated sides were, on the average, 26,

TABLE 4.10. Summary of Electrical Stimulation Effects on Tenderness of Steaks from Bovine Animals of Differing Kinds and Ages

Description of animals	Percentage change in tenderness values[a]	
	Sensory rating[b]	Shear force[b]
29 forage-fed steers	+24	−25
30 grain-fed heifers	+21	−13
9 grain-fed steers	—	−46
6 forage-fed steers	+55	−41
30 hot-skinned calves	+16	−22
30 cold-skinned calves	+18	−17
5 grain-fed steers	+27	−24
30 grain-fed heifers	+11	−15
12 grain-fed steers	+27	−20
60 calves	+21	−20
50 grain-fed heifers	—	−24
40 aged cows	—	−26
331 cattle[c]	+19.7	−21.6

Source: Stiffler et al. (1982).
[a] Calculated by comparing electrically stimulated and not-stimulated sides of the same animal.
[b] An increase in sensory panel ratings and a decrease in shear force values indicate more tender meat.
[c] Average percentage improvement in meat tenderness.

3, 12, and 32% more tender than steaks or chops from counterpart untreated sides for beef, pork, lamb, and goats, respectively (Table 4.9).

Within the bovine species, not all meat is affected by electrical stimulation—with regard to its tenderization—to the same qualitative or quantitative degree. On the average, ES improved tenderness approximately 21% (Table 4.10) in TAMU studies when data were composited across age groups and feeding-management regimens. Smith (1979B) reported that electrical stimulation increased tenderness of forage-fed beef by 24%, of calf by 18%, of aged-cow beef by 26%, and of grain-fed beef by 21%. Savell et al. (1979) concluded that electrical stimulation is as effective in improving palatability of heavyweight, grain-fed beef as previous research has shown it to be in improving characteristics in lightweight, short-fed, or forage-fed beef. West (1982) reported (Table 4.11) that: (1) shear force value was decreased by electrical stimulation for loin steaks from Choice, Good, and Standard—but not from Prime—carcasses; (2) in cases where the average shear force value indicated borderline acceptability in tenderness—as was the case for steaks from untreated sides of Standard grade carcasses—electrical stimulation brought the average shear force values within an acceptable range in tenderness, (3) electrical stimulation tended to reduce the percentage of beef in a population that was unacceptable in tenderness; but, if the beef was inherently tender, little change in tenderness was effected by electrical stimulation, and (4) beef that had been electrically stimulated reached its ultimate tenderness at an earlier time postmortem than did nonstimulated beef and, thus, aging time could be reduced if beef had been electrically stimulated.

That the inherent tenderness of beef affects its response, in terms of tenderization, to electrical stimulation was observed in the earliest TAMU studies. Smith et al. (1977), summarizing research conducted on goat, lamb, beef, and calf, reported that the largest change in shear force associated with the use of electrical stimulation was 4.6 kg for grain-fed beef and that the smallest reduction in shear force was 0.9 kg for grain-fed lambs. They further reported that reductions in shear force for treated samples were generally large when control samples (from untreated sides) had high shear force requirements and small when control samples (from untreated sides) had low shear force requirements, suggesting that electrical stimulation is of greatest benefit for carcasses that would produce less-tender meat if untreated (Smith et al. 1977). Judge et al. (1980) also reported that sensory panel tenderness ratings and shear force values demonstrated that long-fed beef (beef that should be more tender at the outset) was improved less in tenderness than short-fed beef or beef fed no grain (beef that should be less tender at the outset) when electrically stimulated.

Savell (1982) surmised that the increase in tenderness observed in meat from electrically stimulated carcasses appears to be different for different animals. He had previously observed that carcasses that would otherwise produce tough meat benefited greatly from electrical stimulation while those carcasses that would otherwise produce tender meat did not benefit from electrical stimulation (Savell 1979; Savell et al. 1981).

In the Savell et al. (1981) research, two studies (study 1 = 23 forage-fed steers; study 2 = 20 grain-fed steers and heifers) were conducted to determine relationships of electrical stimulation (ES) and/or postmortem aging (PA) to tenderness of beef. In an attempt to determine the effects of ES and PA on beef of differing inherent levels of tenderness, within each study two groups were created by Savell et al. (1981) based on the median tend-

TABLE 4.11. Shear Force Values for Loin Steaks from Beef Carcasses of Different USDA Carcass Grades

USDA quality grade of carcass	Number[a]	Shear force value (kg) of loin steaks			Statistical significance of difference
		From not-electrically stimulated side	From electrically stimulated side	Difference	
Prime	7	3.7	3.4	0.3	$P > 0.05$
Choice	126	4.7	3.9	0.8	$P < 0.001$
Good	150	5.1	4.0	1.1	$P < 0.001$
Standard	92	5.2	4.0	1.2	$P < 0.001$

Source: West (1982).
[a] Studies involved paired sides of 375 carcasses using voltages of 300 to 600 with both continuous and pulsatory stimulations and variable times postmortem involved.

erness rating and the median Warner-Bratzler shear (WBS) value for not-ES steaks from all cattle; these groups were referred to as "tough" (group 1) and "tender" (group 2), and results of subsequent analyses are graphically illustrated in Fig. 4.1, 4.2, and 4.3. The most striking feature in these figures is that the lines for not-ES values have greatly different slopes for steaks in group 1 as compared with the lines in group 2; conversely, the lines for the ES values are essentially the same for group 1 and group 2. Furthermore, the relationship between treatments (ES vs not-ES) for each group is interesting—for study 1 (Fig. 4.1), there is a large difference in WBS values between ES and not-ES samples, regardless of the group even at day 14, whereas for study 2 (Fig 4.2 and 4.3), there is a large difference between ES and not-ES values only until day 5 or 8 for group 1 steaks and little difference between ES and not-ES values at days 11 or 14 for group 1 steaks and at any PA period for group 2 steaks (Savell *et al.* 1981).

Salm *et al.* (1981) found a differential response in tenderness to electrical stimulation and time-on-feed; carcasses from animals fed longer times responded less to electrical stimulation than carcasses from animals

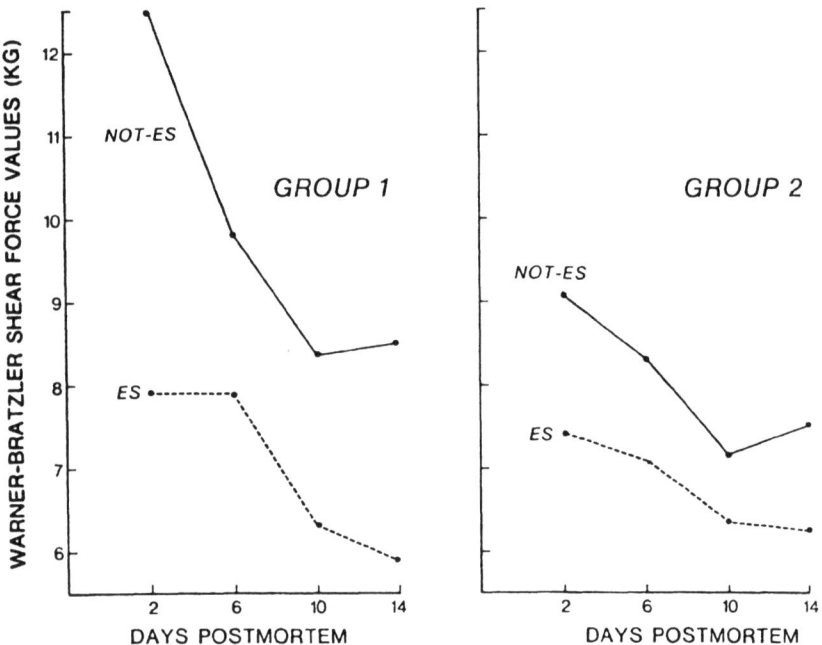

FIG. 4.1. Comparison of effects of electrical stimulation and postmortem aging time on Warner-Bratzler shear force (WBS) values for "tough" (Group 1) and "tender" (Group 2) steaks from carcasses in Study 1.
From Savell et al. (1981). Reprinted from J. Food Sci. 1981. 46:1780. Copyright © by Institute of Food Technologists.

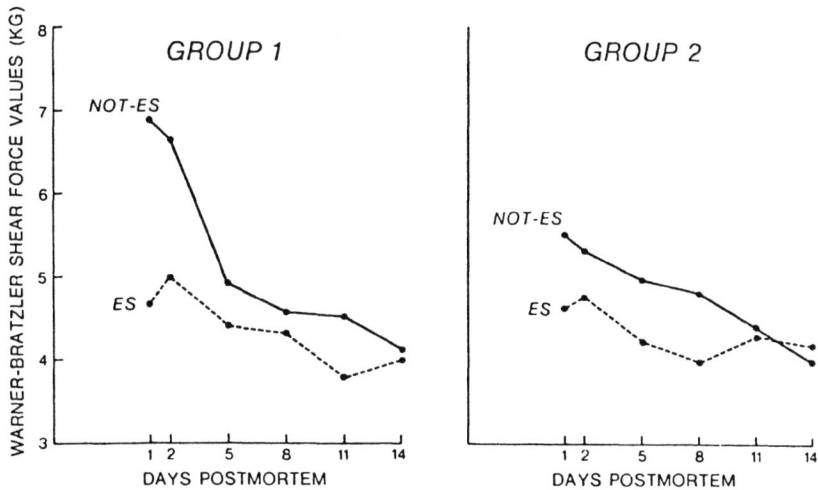

FIG. 4.2. Comparison of effects of electrical stimulation and postmortem aging time on WBS values for "tough" (Group 1) and "tender" (Group 2) steaks from carcasses in Study 2.
From Savell et al. (1981). Reprinted from J. Food Sci. 1981. 46:1780. Copyright © by Institute of Food Technologists.

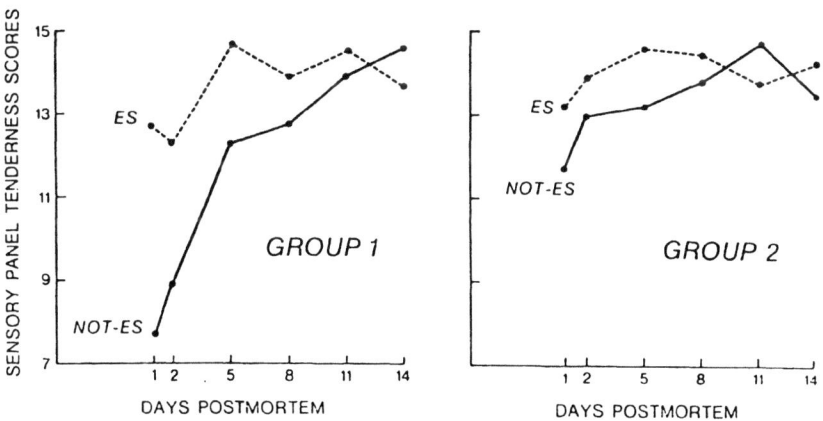

FIG. 4.3. Comparison of effects of electrical stimulation and postmortem aging time on sensory tenderness ratings for "tough" (Group 1) and "tender" (Group 2) steaks from carcasses in Study 2.
From Savell et al. (1981). Reprinted from J. Food Sci. 1981. 46:1780. Copyright © by Institute of Food Technologists.

TABLE 4.12. Changes in Shear Force Values for Beef[a] Aged for Different Periods of Time

Aging period (days)	Change in shear force value (%)	
	Electrically stimulated beef sides	Not-electrically stimulated beef sides
2	−29	None
6	−34	−16
10	−40	−22
14	−44	−26

Source: Smith (1979B).
[a] Beef carcasses were of Choice, Good, or Standard grade. Percentages are increases compared with that of steaks from not-ES sides aged for 2 days.

fed shorter times. It is not clearly understood why this phenomenon occurs, but the differential response in tenderness to electrical stimulation is probably related to prevention of cold shortening; some carcasses undoubtedly will undergo more cold shortening than others and benefit more correspondingly, from electrical stimulation than carcasses that undergo less cold shortening of muscle fibers.

Two other studies have been conducted by TAMU researchers to determine the extent to which aging negates or complements tenderness differences affected by electrical stimulation. Results of one of these studies (Table 4.12) revealed that aging of ES loin steaks for 14 days increased their tenderness by 15% over that (29%) achieved by electrical stimulation, whereas 14 days of aging of not-ES loin steaks increased their tenderness by 26%. The latter results suggest that aging complements the tenderness achieved by electrically stimulating beef. Results of the other aging study (Table 4.13) reveal that loins from electrically stimulated carcasses, aged for 7 days, were about 10% more tender than loins from untreated (not-ES) carcasses that were aged for 21 days.

Most of the early TAMU studies of electrical stimulation were performed using split carcasses; one side was electrically stimulated, the other side served as the untreated (not-ES) control. Because the ES pro-

TABLE 4.13. Changes in Sensory Panel Ratings and Shear Force Values for Beef Aged for Different Periods of Time

Comparison (days)		Advantage for electrical stimulation (%)	
Control (not-ES)	Electrically stimulated	Sensory panel rating	Shear force value
7	7	+9	−7
21	21	+18	−22
21	7	+5	−15

Source: Savell et al. (1978C).

cedure involves massive contractions of the major muscles, there was considerable uncertainty that the same tenderization results could be achieved by electrically stimulating unsplit (intact) carcasses. Studies of goat carcasses, youthful beef carcasses, and mature cow carcasses (Table 4.14), electrically stimulated as sides or as intact carcasses, convinced TAMU researchers that splitting is not necessary to achieve the desired response in tenderness. The latter conclusion was also supported by the research of McKeith et al. (1980A).

TABLE 4.14. Comparison of Effectiveness of Electrically Stimulating Intact Carcasses vs Sides

Treatment	Improvement over not-ES samples (%)	
	Sensory panel rating	Shear force value
ES as sides (goats)	+50	−27
ES as intact carcasses (goats)	+28	−29
ES as sides (beef)	+56	−33
ES as intact carcasses (beef)	+51	−32
ES as sides (mature cow beef)	—	−26
ES as intact carcasses (mature cow beef)	—	−22
ES as sides (beef)	+28	−30
ES as intact carcasses (beef)	+31	−25

Source: McKeith et al. (1979, 1980B, 1981B); Smith (1979B).

Because early TAMU studies used sides, electrical stimulation was, by necessity, performed immediately prior to placement of carcasses in the chill cooler. It was of consequence with regard to progressing toward industry implementation to identify the site on the kill floor that was best for achieving the desired level of tenderization; results of one such study of this kind that was performed are presented in Table 4.15. Stimulation of intact goat carcasses after evisceration increased the tenderness of muscles from the loin and leg by 35.7%; stimulation after bleeding, after pelting, or after splitting increased the tenderness of these same muscles by 27.7, 28.3, and 27.0%, respectively. The latter study and a similar study performed by McKeith et al. (1981B) with beef carcasses (Table 4.16) suggest that electrical stimulation can be performed at any of a number of

TABLE 4.15. Effect of Location on Kill Floor at Which Goat Carcasses Were Electrically Stimulated on the Tenderness of Loin and Leg Muscles

Time of electrical stimulation	Change from control (not-ES) (%)		
	Loin	Inside leg	Outside leg
After bleeding	−21	−27	−35
After pelting	−31	−16	−38
After evisceration	−36	−27	−41
After splitting	−35	−19	−27

Source: McKeith et al. (1979).

TABLE 4.16. Effect of Location on Kill Floor at Which Beef Carcasses Were Electrically Stimulated on the Tenderness of Rib Steaks

Time of electrical stimulation	Change from control (not-ES) (%)	
	Sensory panel rating	Shear force value
Before evisceration	+15	−22
After evisceration	+31	−25
After splitting	+28	−30

Source: McKeith et al. (1981B).

locations on the kill floor with approximately the same end result in terms of tenderness. Available space on the slaughter–dressing floor and safety (to workmen) are probably as important as tenderization efficacy in selecting the kill-floor site for electrical stimulation.

Of further concern to TAMU scientists were direct comparisons of the amount of tenderization accomplished by electrical stimulation with that achieved by use of high temperature conditioning, TAMU Tenderstretch, cooler aging, and/or blade tenderization (Smith et al. 1979B). Studies were conducted to make such comparisons; results of these tests are presented in Tables 4.17 and 4.18. Electrical stimulation was more effective than high temperature chilling or TAMU Tenderstretch in increasing tenderness of loin steaks from forage-fed beef (Table 4.17); aging (14 days at 34°F) resulted in a further increase in tenderness (across all treatments) of 16%. TAMU Tenderstretch was slightly more effective than electrical stimulation in increasing tenderness of round steaks in samples that were not blade tenderized (Table 4.18), but differences in tenderness achieved by use of these two techniques after blade tenderization were negligible. Blade tenderization increased tenderness by 27% over that achieved by the singular use of electrical stimulation, high temperature chilling, or TAMU Tenderstretch. Muscles in the round contain much more connective tissue than does the ribeye muscle; as a result, tenderization techniques that affect connective tissue (e.g., blade tenderization via cutting of connective tissue strands; TAMU Tenderstretch by stretching connective

TABLE 4.17. Comparison of Tenderization Methods and Tenderness of Loin Steaks[a]

Method	Change from control (%)	
	Sensory panel rating	Shear force value
Electrical stimulation (ES)	+55	−41
High temperature chilling (HT)	+11	−16
TAMU Tenderstretch (TS)	+47	−27
ES plus aging	+74	−54
HT plus aging	+25	−32
TS plus aging	+66	−39

Source: Smith et al. (1979B).
[a] Forage-fed beef.

tissue proteins, and/or making them more susceptible to breakdown by the heat of cooking) would be expected to increase tenderness more than would a procedure like electrical stimulation, which acts primarily on the muscle fibers. Savell et al. (1982) determined effects of electrical stimulation, postmortem aging, and blade tenderization (BT) on the palatability attributes of beef from young bulls. They found that ES significantly improved ($P < 0.05$) palatability traits in 2 of 24 comparisons; BT significantly improved palatability traits in 12 of 24 comparisons, and 18-day postmortem aging significantly improved palatability traits in 7 of 12 comparisons. No significant reductions ($P > 0.05$) in shear force values were observed for steaks from ES vs not-ES sides, whereas significant reductions ($P < 0.05$) were observed for steaks from BT vs not-BT cuts (4 of 6 comparisons) and for steaks from cuts aged for 18 vs 4 days (10 of 12 comparisons). BT and 18-day postmortem aging were more effective for increasing palatability or for decreasing shear force requirements than was ES.

Several studies have been conducted to determine the efficacy with which electrical stimulation complements other postmortem tenderization treatments. Studies by Elgasim et al. (1981), Smith et al. (1979B), Savell et al. (1981, 1982), and Sonaiya and Stouffer (1982) have reported the effects of electrical stimulation in conjunction with other postmortem tenderization treatments and have found that the impact of electrical stimulation on tenderness can be lessened if electrical stimulation is used in conjunction with tenderization treatments that are severe in their action (like blade tenderization). While Savell et al. (1981) found that postmortem aging will lessen the impact of electrical stimulation, especially in populations of "tender" carcasses, electrical stimulation used in conjunction with postmortem aging can greatly decrease the percentage of steaks that are "tough," thereby helping to ensure that steaks purchased by consumers are usually "acceptable" in tenderness.

Electrical stimulation is also successful in improving the tenderness of meats less-traditionally used for the block-beef trade; these include mature cows (McKeith et al. 1980B), forage-fed steers (Smith et al. 1979B;

TABLE 4.18. Comparison of Tenderization Methods and Tenderness of Round Steaks[a]

Method	Change from control (%)	
	Sensory panel rating	Shear force value
Electrical stimulation (ES)	+6	−16
High temperature chilling (HT)	+6	−5
TAMU Tenderstretch (TS)	+26	−11
ES plus blade tenderization	+49	−29
HT plus blade tenderization	+55	−18
TS plus blade tenderization	+47	−34

Source: Smith et al. (1979B).
[a] Forage-fed beef.

Davis et al. 1981), weanling calves (Rouquette et al. 1983), and young bulls (Savell et al. 1981; Riley et al. 1983A,B). ES also improves tenderness of cuts from young rams (Riley et al. 1981). With the use of a postmortem process like electrical stimulation, more parts of the carcass could be used for steaks, roasts, or chops rather than limiting their use to ground or comminuted products.

Many of the major muscles of the beef carcass are tenderized by use of electrical stimulation. McKeith et al. (1981B) found that in the chuck only some of the muscles of the seven-bone steak were improved in tenderness by electrical stimulation whereas none of the muscles of arm and blade steaks was significantly affected by ES. With the exception of the semitendinosus muscle, however, all of the muscles of round steaks and all of the major muscles of top sirloin steaks, rib steaks, and loin steaks were improved by electrical stimulation. Such findings are important because steaks from the round, sirloin, loin, and rib are usually cooked with "dry heat," whereas steaks and roasts from the chuck are usually cooked with "moist heat." Since moist-heat cookery is more conducive to tenderization of muscles than is dry-heat cookery, it is much more important that the round, loin, sirloin, and rib be tenderized by means other than cookery than those cuts (e.g., arm steak, blade steak) that are normally cooked with moist heat.

Of particular interest in the future in the area of electrical stimulation is the need to ensure that all muscles of the carcass be affected by electrical stimulation. Based on the research available, electrical stimulation appears to be most effective for improving the tenderness of the longissimus dorsi muscle, partially effective in improving the tenderness of some of the muscles of the sirloin and round, and relatively ineffective for improving the tenderness of almost all of the muscles of the chuck. Identifying effective means for tenderizing all of the muscles in the carcass appears to be an important area for future electrical stimulation research.

Some research has been conducted to determine effects of electrical stimulation on the tenderness of pork. Westervelt and Stouffer (1978) and Johnson et al. (1982) found few significant differences in tenderness between samples from electrically stimulated and control (not-ES) sides of pork carcasses. Since pork muscle has such a rapid postmortem pH decline and does not undergo "cold shortening" like beef or lamb, this result was not unexpected. However, Johnson et al. (1982) found that pork from electrically stimulated sides of swine that had sustained prolonged periods of time without feed were more tender than pork from the control (not-ES) sides of the same pigs. The latter result suggests that electrical stimulation may have greater effect in tenderizing muscle from swine that have slow postmortem pH declines.

A possible future application of electrical stimulation is that of making feasible the "hot boning" of beef. Results of a study that TAMU scientists conducted to explore such possibility are presented in Table 4.19. "Hot boning" of beef within the first hour following slaughter has invariably

TABLE 4.19. Effect of Electrical Stimulation on Tenderness of Hot-boned Beef

Trait	Effect[a] (%)
Shear value (loin)	−16
Sensory tenderness rating (loin)	None
Shear value (top round)	−12
Sensory tenderness rating (top round)	None

Source: Seideman et al. (1979).
[a] Comparison: ES + HB (1 hr postmortem) vs control (24 hr postmortem).

been shown to decrease tenderness; in this study, when beef carcasses were electrically stimulated prior to hot boning, beef was at least as tender as that from conventionally chilled beef. Others who have studied efficacy of ES for facilitating hot boning include Bendall (1980), Berry and Stiffler (1981), Contreras and Harrison (1981), Contreras et al. (1981), Cross and Tennent (1981), Gilbert and Davey (1976), Gilbert et al. (1977), Griffin et al. (1981), Raccach and Henrickson (1978), Taylor et al. (1980), and Walker et al. (1977).

In summary, TAMU studies have shown that meat from electrically stimulated carcasses is, on the average, 18 to 30% (depending on species, grade and feeding history of animals) more tender than that from untreated (not-ES) carcasses. The procedure can be performed on either sides or unsplit carcasses in any of a number of locations on the kill floor. The increase in tenderness achieved via electrical stimulation is complemented by use of blade tenderization and by cooler aging. Other researchers who have reported that tenderness of meat is improved by electrical stimulation include Carse (1973), Chrystall and Hagyard (1976), Gilbert and Davey (1976), Gilbert et al. (1977), Grusby et al. (1976), Locker (1976), Sorinmade et al. (1978), Cross (1979), Cross et al. (1979), Judge et al. (1980), Bouton et al. (1980A,B), Ruderus and Bergquist (1980), Chrystall and Devine (1980), George et al. (1980), Ruderus (1980), Davis et al. (1981), Elgasim et al. (1981), and Sonaiya and Stouffer (1982).

Almost all of the electrical stimulation research has been reported during the decade of the 1970s. Research on electrical stimulation has been conducted worldwide in countries such as Australia (Shaw and Walker 1977; Clarke et al. 1980; Bouton et al. 1980A,B; Tume 1980; Taylor and Marshall 1980), France (Houlier et al. 1980), Great Britain (Bendall 1976; Bendall et al. 1976; George et al. 1980), New Zealand (Carse 1973; Davey et al. 1976; Chrystall and Devine 1978; Chrystall and Hagyard 1976; Chrystall and Devine 1980), Sweden (Ruderus 1980; Ruderus and Bergquist 1980), and the United States (Grusby et al. 1976; Cross 1979; Cross et al. 1979; McKeith et al. 1979, 1980A,B, 1981A,B, 1982; Riley et al. 1980A,B, 1981; Savell et al. 1976, 1977, 1978A,B,C, 1979, 1981, 1982; Smith et al. 1977, 1979A,B; Will et al. 1980). Probably no single area of meat research has been as widely studied in the last eight years as electrical stimulation. While most researchers outside the United States have concentrated on its effect on tenderness, scientists in the United States have studied the numerous other benefits associated with use of electrical stimulation. Such benefits are enumerated and discussed below.

TABLE 4.20. Flavor Improvement in Electrically Stimulated Beef

Description of animals	Increase
29 forage-fed steers	+10.9
8 grain-fed steers	+13.7
5 grain-fed steers	+11.5
10 grain-fed heifers	+5.6
Average net increase	+10.4

Source: Savell and Smith (1979).

TABLE 4.21. Effect of Electrical Stimulation on Flavor Desirability of Beef, Pork, Lamb, and Goat Meat

Species	Number of carcasses or samples	Percentage improvement
Beef	349	6
Pork	90	2
Lamb	175	0
Goat	118	6

Source: Smith et al. (1980).

FLAVOR

Sensory panel evaluations of steaks from electrically stimulated and control sides of beef have revealed significant increases in flavor desirability scores (Table 4.20). Data in Table 4.21 reveal that the effect of ES on flavor desirability of beef, pork, lamb, and goat meat is not large. Although the mechanisms by which flavor improvement can be achieved have not been thoroughly studied, it has been postulated that because electrical stimulation appears to accelerate the aging process in a manner that increases tenderness, chemical compounds that may be responsible for the "aged" meat flavor possibly are produced by this process. Calkins et al. (1982) attributed the difference in flavor between steaks from electrically stimulated and control sides to differences in the concentration of creatine phosphate, adenine nucleotides, and their derivatives caused by electrical stimulation. Therefore, an advantage for electrical stimulation over other methods of tenderization is that the flavor of the meat is also improved, thus helping to ensure a more palatable product.

LEAN COLOR

When beef carcasses are ribbed and evaluated at 18–24 hr after death, the lean color of electrically stimulated sides, when compared with their control sides, is significantly improved (Table 4.22). Electrical stimulation apparently increases the rate of postmortem glycolysis, thereby ensuring that this process will be more nearly complete and that the lean will have a brighter, more youthful color when beef carcasses are ribbed within 18 hr after death. Data in Table 4.23 reveal that ES improves lean color of beef, veal, and goat, but not that of pork.

It has been found that slight cases of "dark-cutting" beef have been alleviated by the use of electrical stimulation. However, in a test (Dutson *et al.* 1982) in which selected sides from cattle that were handled in such a manner as to produce dark-cutting beef were stimulated, electrical stimulation could not improve lean color.

TABLE 4.22. Improvement in Lean Maturity and Lean Color Scores by Electrical Stimulation

Description of animals	Lean maturity	Lean color
30 grain-fed heifers	−30	+16
8 grain-fed steers	−18	—
30 grain-fed heifers	−17	+11
12 grain-fed steers	+5	+24
50 grain-fed heifers	−31	+21
109 grain-fed cattle	−6	+6
57 grain-fed cattle	−4	+4
Average net improvement	−13.3	+10.2

Source: Savell and Smith (1979).

TABLE 4.23. Effect of Electrical Stimulation on Color and Lean Maturity Scores for Beef, Veal, Pork, Lamb, and Goat

Species	Trait	Number of carcasses or samples	Percentage improvement
Beef	Lean maturity score	1261	23
Beef	Lean color score	1261	14
Veal	Lean maturity score	40	3
Veal	Lean color score	80	12
Pork	Lean color score	90	−7
Lamb	Lean maturity score	151	4
Lamb	Lean color score	632	36
Goat	Lean maturity score	96	16

Source: Smith *et al.* (1980).

HEAT RING PREVENTION

A problem that occurs quite commonly in the meat industry is "heat ring" development in beef that has been chilled too rapidly. "Heat ring" appears to be caused by the differential chill rate within the ribeye muscle which results in differing rates of color and rigor development from the exterior to the interior portions of the muscle. This condition is especially prevalent in carcasses that have limited subcutaneous fat cover over the ribeye. Carcasses with severe heat ring are not eligible for federal meat grading since the grade factors are difficult to evaluate.

Heat ring is really a misnomer for this condition since it appears that it is probably due to severe cold temperatures slowing the color and rigor development processes along the outside portion of the ribeye. Actually, the proper term for this condition should be "cold ring" since the portion of the ribeye that is dark in color, coarse in texture, and sunken beneath the plane of the ribeye is significantly colder at early hours of chilling than the inside portion of the ribeye. Nevertheless, the term heat ring is in the vernacular of the trade and is very commonly used in the packing industry of the United States.

TABLE 4.24. Reduction in Heat Ring by Electrical Stimulation

Description of animals	Percentage heat ring reduction by electrical stimulation
30 grain-fed heifers	−37
8 grain-fed steers	−100
30 grain-fed heifers	−53
12 grain-fed steers	−43
50 grain-fed heifers	−54
109 grain-fed cattle	−20
57 grain-fed cattle	−30
Average reduction	−30

Source: Savell and Smith (1979).

Research conducted at Texas A&M University has shown that heat ring can be significantly reduced or alleviated by the use of electrical stimulation (Table 4.24). Studies on paired sides of carcasses have revealed that, with the use of electrical stimulation, heat ring formation in carcasses ribbed 18 hr after death was substantially reduced in electrically stimulated sides as compared with control sides. Electrical stimulation decreases incidence of heat ring in beef carcasses, improves lean firmness in beef and veal but not in pork, and improves lean texture of veal; however, ES increases muscle separation in pork (Table 4.25).

TABLE 4.25. Effect of Electrical Stimulation on Lean Firmness, Lean Texture and Incidence of Heat Ring in Beef, Veal, and/or Pork

Species	Trait	Number of carcasses or samples	Percentage improvement
Beef	Heat ring score	1177	23
Beef	Lean firmness score	458	4
Veal	Lean texture score	40	28
Veal	Lean firmness score	40	36
Pork	Lean firmness score	90	−11
Pork	Muscle separation score	90	−5

Source: Smith et al. (1980).

MARBLING AND QUALITY GRADE

Research has shown that marbling scores can be slightly increased by the use of electrical stimulation when carcasses are ribbed and presented for grading after 24 hr of chilling (Table 4.26). It is common knowledge in the meat industry that sides of beef that are left to chill longer than their opposite sides usually will exhibit slightly higher marbling scores when ribbed and evaluated. Also, the percentages of carcasses that grade U.S. Choice on Monday after 2 or more days of chill (called "weekend" cattle) are the highest of any day of the week. This is apparently the process that is occurring in the earlier development of a maximum marbling score in electrically stimulated carcasses. Electrical stimulation causes the lean to be firmer, finer textured, and brighter colored and therefore probably causes faster "setting up" of the fat in the ribeye. Differences in marbling between electrically stimulated and nonstimulated sides are minimal or nonexistent after additional chilling of greater than 48 hr.

Questions have arisen that concern may be registered by the USDA Meat Grading Branch regarding improvements in quality-indicating factors in electrically stimulated beef. It should be noted that the magnitude

TABLE 4.26. Increase in Marbling Score by Electrical Stimulation

Description of animals	Marbling increase
30 grain-fed heifers	+15
8 grain-fed steers	+16
30 grain-fed heifers	+5
12 grain-fed steers	+30
50 grain-fed heifers	+14
109 grain-fed cattle	+20
57 grain-fed cattle	+18
Average marbling units	+16.9

Source: Savell and Smith (1979).

144 G. C. SMITH

of the improvement in palatability is much greater than the improvement in the factors used in predicting palatability. Therefore, it is our opinion that regardless of the slight change in grade, the eating quality of the meat is far superior to what it would have been without electrical stimulation.

The most important reason that electrical stimulation is being used by packers in the United States is that it improves the quality-indicating characteristics of the lean (Savell *et al.* 1980). As a result of the accelerated postmortem glycolysis and hastened rigor mortis caused by electrical stimulation, the longissimus muscle surface that is exposed upon ribbing of the carcass at 24 hr postmortem will be brighter and more youthful in color, the marbling will be "set up" earlier and easier to evaluate, and the condition known as "heat ring" will either be greatly reduced or completely alleviated (Savell *et al.* 1978B,C, 1979; McKeith *et al.* 1980A,B; 1981B). Because of the improvement in these factors, USDA quality grade is usually improved, especially for those carcasses on the U.S. Choice-U.S. Good grade line. Such changes are pictorially revealed in Fig. 4.4. McKeith *et al.* (1982) reported that electrical stimulation can make improvements in "house grades" of veal that can result in significant increases in value of veal carcasses. Such changes in appearance for veal are pictured in Fig. 4.5. Because such grade lines are of great economic importance,

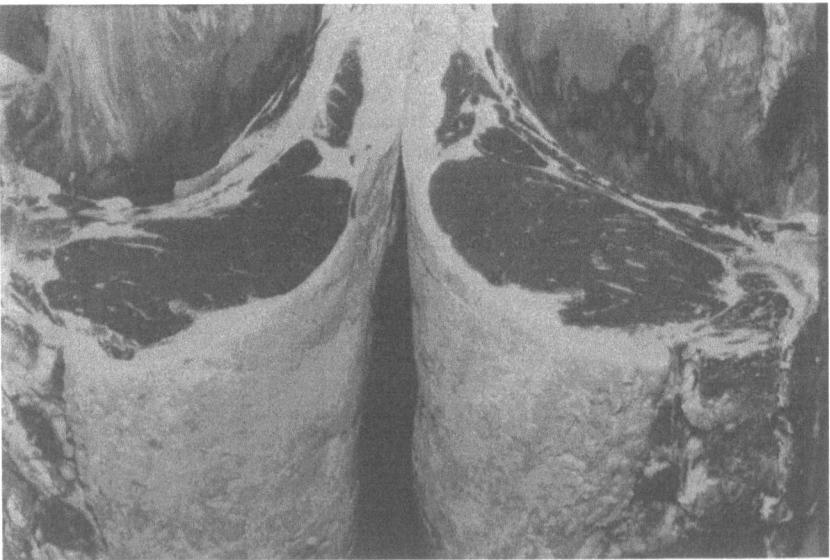

FIG. 4.4. Beef ribeye muscles from both sides of the same carcass. The side on the right was electrically stimulated; the side on the left was not.

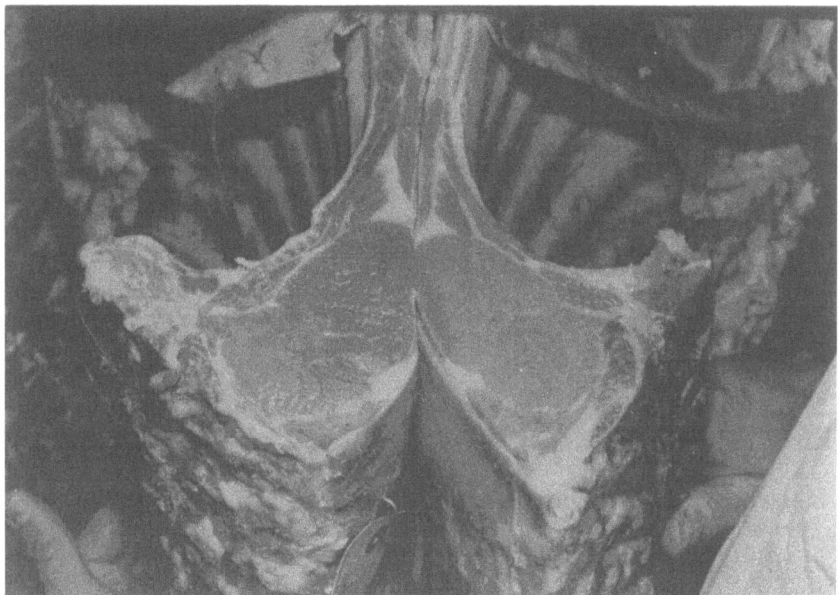

FIG. 4.5. Veal ribeye muscles from both sides of the same carcass. The side on the right was electrically stimulated; the side on the left was not.

movement of carcasses from the U.S. Good to U.S. Choice grade is financially advantageous to the packer. Electrical stimulation does not produce a higher-than-justified grade; rather, electrical stimulation causes the grade factors to reach their maximum development at an earlier time (Calkins et al. 1980). Although maximizing the quality grades of beef carcasses would best be done by chilling for 48 or 72 hr, most beef packers do not have the available space nor do they want to invest the capital for increased inventory necessary to allow for this; thus, electrical stimulation plays an important role for the packer who wishes to achieve the highest possible grade at the earliest possible time (Table 4.27).

TABLE 4.27. Effect of Electrical Stimulation on Marbling Score in Beef and on USDA Quality Grade in Beef and Lamb

Species	Trait	Number of carcasses or samples	Percentage improvement
Beef	Marbling score	1251	11
Beef	USDA quality grade	1086	8
Lamb	USDA quality grade	510	17

Source: Smith et al. (1980).

TABLE 4.28. Comparison of Retail Caselife of Hamburger Meat from Electrically Stimulated and Nonstimulated Carcasses

Trait	Day of display	Electrically stimulated	Nonstimulated
Muscle color	1	Acceptable	Acceptable
	2	Acceptable	Acceptable
	3	Unacceptable	Unacceptable
Surface discoloration	1	Acceptable	Acceptable
	2	Unacceptable	Unacceptable
	3	Unacceptable	Unacceptable
Overall appearance	1	Acceptable	Acceptable
	2	Unacceptable	Unacceptable
	3	Unacceptable	Unacceptable

Source: Derived from data of Hall *et al.* (1980).

RETAIL CASELIFE

As the first electrically stimulated beef was being sold at the retail counter, some concern was registered that retail caselife might possibly be shortened by the use of this process. In an effort to study this possible problem, Texas A&M University scientists conducted retail caselife studies using both hamburger meat and steaks (Tables 4.28 and 4.29). Caselife of hamburger meat was not affected by electrical stimulation; hamburger meat made from electrically stimulated carcasses had the same pattern of discoloration as hamburger meat made from nonstimulated carcasses (Table 4.28).

Round steaks from electrically stimulated and nonstimulated carcasses were stored under identical display conditions for 4 days (Table 4.29). In that portion of the experiment, steaks that were from electrically stimulated carcasses remained brighter for one extra day, surface discoloration

TABLE 4.29. Comparison of Retail Caselife of Steaks from Electrically Stimulated and Nonstimulated Carcasses

Trait	Day of display	Electrically stimulated	Nonstimulated
Muscle color	1	Acceptable	Acceptable
	2	Acceptable	Acceptable
	3	Acceptable	Acceptable
	4	Acceptable	Unacceptable
Surface discoloration	1	Acceptable	Acceptable
	2	Acceptable	Acceptable
	3	Marginal	Unacceptable
	4	Unacceptable	Unacceptable
Overall appearance	1	Acceptable	Acceptable
	2	Acceptable	Acceptable
	3	Marginal	Unacceptable
	4	Unacceptable	Unacceptable

Source: Derived from data of Hall *et al.* (1980).

was less severe at day 3, and the overall appearance was "marginal" at day 3 instead of "unacceptable" at day 3 as was the case for the steaks from nonstimulated carcasses. The latter results indicate that there may be an extra one-half to 1 day of retail caselife for steaks that are from carcasses that have been electrically stimulated, despite earlier concerns to the contrary.

Improved retail appearance of cuts from electrically stimulated carcasses is of advantage to the meat retailer. Both lamb chops (Riley *et al.* 1980A,B) and beef round steaks (Hall *et al.* 1980) from electrically stimulated carcasses had brighter color, less surface discoloration, and more desirable overall apperance over a 4-day retail display period than chops or steaks from their nonstimulated counterparts. This phenomenon is not well understood, but the improved appearance of the chops or steaks from electrically stimulated carcasses does not appear to be related to effects on bacteria (Gill 1980; Mrigadat *et al.* 1980; Butler *et al.* 1981), although one study (Raccach and Henrickson 1978) suggested that there were significant differences in bacterial numbers due to electrical stimulation. Rather, it has been reported (Tang and Henrickson 1980) that meat from electrically stimulated carcasses had a higher percentage of oxymyoglobin—the pigment responsible for the bright cherry-red color of beef. Regardless of the reason, retail cuts from electrically stimulated carcasses will be slightly brighter and will have a slightly longer caselife than will those retail cuts from nonstimulated carcasses.

PROCESSING PROPERTIES

Since there could be potential problems associated with the processing of meat from electrically stimulated carcasses, several studies have been conducted to determine the impact of electrical stimulation on processing properties of meat. Studies by Ockerman and Dowiercial (1980), Whiting *et al.* (1981), Swasdee *et al.* (1982), and Terrell *et al.* (1981, 1982A,B,C) reveal no (or very few) differences in processing properties that would discourage meat processors from using trimmings or other meats from electrically stimulated carcasses in sausage products. Furthermore, studies have been conducted to determine the impact of electrical stimulation on the weight loss, shrink, or purge of primals or subprimals in the transport–distribution sequence. In studies by Riley *et al.* (1980A, 1982) involving the transportation and distribution of vacuum packaged subprimal cuts, no substantial differences were observed in the characteristics of meat due to electrical stimulation. Johnson *et al.* (1982) found no differences in weight loss between wholesale cuts from electrically stimulated and nonstimulated pork sides. Dutson *et al.* (1980A) found no significant difference between samples from electrically stimulated and control sides in protein efficiency ratio (PER), thereby indicating that the protein quality of electrically stimulated beef is not affected by such treatment.

There appear to be only minor problems with characteristics associated with water-holding capacity of meat even though the prerigor conditions associated with electrical stimulation would suggest that there might possibly be major problems because of the severe pH decline while muscle temperatures are still high.

LOW- VS HIGH-VOLTAGE ELECTRICAL STIMULATION

Concern has arisen in some circles regarding the relative effectiveness of low-voltage and high-voltage electrical stimulation. Based on a limited number of studies, research has shown that low-voltage electrical stimulation is similar to high-voltage electrical stimulation in terms of its effects on carcass quality traits, but that low-voltage electrical stimulation is not as effective in improving tenderness as is high-voltage electrical stimulation. The data base from which high-voltage electrical stimulation was developed is substantial; unfortunately, the data base from which low-voltage electrical stimulation was developed is rather limited (Savell 1982). Further research in the area of low-voltage electrical stimulation,

TABLE 4.30. Means for Muscle pH, Quality Characteristics, Shear Force Values, and Purge Loss Percentages by Electrical Stimulation Treatment[a,b]

	Treatment		
Characteristic	Nonstimulated	High-voltage electrical stimulation	Low-voltage electrical stimulation
pH, 1 hr[d]	6.5[a]	5.9[b]	5.8[c]
Lean color score[e]	3.7[a]	2.9[b]	2.8[b]
Marbling score[f]	10.1[a]	10.3[b]	9.6[a]
USDA quality grade[g]	14.9[a]	14.9[a]	14.6[a]
Heat ring score[h]	2.0[a]	1.5[b]	1.3[b]
Shear force values, kg/1.27 cm[i]	5.5[a]	4.2[b]	4.9[c]
Purge loss, %[j]	2.7[a]	3.1[a]	3.0[a]

Source: West (1982).
[a] The number of sides or carcasses was 21 per treatment group.
[b] Means within a row with different letters are different ($P < 0.05$).
[c] Treatments were: nonstimulated control sides; high-voltage = sides stimulated with 500 V using 30 pulses of 1 sec with 0.5 sec intervals at 30 min postmortem; and low-voltage = carcasses stimulated with 45 V peak for 36 continuous seconds at 3 min postmortem.
[d] Measured in longissimus muscle with combination electrode probe.
[e] Scored at 20 to 24 hr postmortem using 7-point scale wherein 2 = very light cherry red and 3 = light cherry red.
[f] Scored on 27-point scale wherein 9 = slight-plus and 10 = small-minus.
[g] Scored on 21-point scale where 14 = average USDA Good and 15 = high USDA Good.
[h] Scored on 4-point scale wherein 1 = no heat ring and 2 = slight heat ring.
[i] Loin steaks broiled to 70°C internal temperature.
[j] Purge loss of top rounds during 14 days of vacuum-packaged storage at 0° to 2°C.

especially as it compares with high-voltage electrical stimulation, should be conducted in order to determine if low-voltage electrical stimulation is truly efficacious (Savell 1982).

West (1982) reported that there are some studies that show equal tenderization by low and high voltage, but that high voltage has more effect on quality characteristics of the lean. In a recently completed study involving carcasses from 42 steers that had been on feed for different times, West (1982) compared high-voltage (500 V for 30 1-sec pulses) and low-voltage (45 V peak voltage for 36 continuous seconds) electrical stimulation treatments. Quality characteristics, shear force values of broiled loin steaks, and purge loss of top round cuts stored 14 days in vacuum packaging were determined (Table 4.30). Longissimus muscle pH was lower ($P < 0.05$) at 1 hr postmortem, lean color was brighter, and heat ring scores were lower for both high- and low-voltage treatments than for the control (not-ES); no differences ($P > 0.05$) were detected between the stimulation treatments for lean color or heat ring score. Marbling increased with use of high-voltage, but not with use of low-voltage, stimulation. Shear force values were lowest ($P < 0.05$) for the steaks from the carcasses stimulated with high voltage, followed by those steaks from the carcasses stimulated with low voltage, and then the nonstimulated carcasses (Table 4.30). Purge loss was not affected ($P > 0.05$) by electrical stimulation treatment.

IMPLEMENTATION BY THE PACKING INDUSTRY

Research on electrical stimulation created an impetus for the incorporation of the necessary ES equipment into the slaughter–dressing sequence of commercial beef slaughter plants. Because space and time are important factors limiting options of particular beef slaughterers with regard to places on the kill floor at which ES equipment can be installed, it was essential that research identify the effects of electrical stimulation on the quality-indicating and palatability characteristics of carcasses stimulated (1) for different periods of time, (2) at different stages in the slaughter–dressing sequence, and (3) with different voltages.

Improvements in tenderness have been observed when electrical stimulation was performed on sides (Savell *et al.* 1977, 1978A,B,C, 1979; Davey *et al.* 1976; Gilbert and Davey 1976; Walker *et al.* 1977; McKeith *et al.* 1979) and on intact carcasses (Walker *et al.* 1977; McKeith *et al.* 1979, 1980B). Observation of the ES process reveals that carcass response to ES is much more violent when applied to sides (after splitting) than when applied to intact (unsplit) carcasses. Since Savell *et al.* (1978A) had proposed that these violent contractions might cause increased tenderness via structural damage, it was essential that it be determined whether tenderness enhancement could be achieved without having to split carcasses prior to ES treatment. Savell *et al.* (1978B, 1979) and Calkins *et al.* (1980)

had also observed improvements in quality-indicating characteristics (color, marbling, etc.), but all of their studies dealt with results from ES of sides. In the final TAMU study of such parameters (McKeith et al. 1981B), electrical stimulation improved palatability and improved quality-indicating characteristics for beef irrespective of the form (sides or intact carcasses) or of the stage in the slaughter–dressing sequence at which it was performed.

Savell et al. (1978B) found no apparent advantages in quality or palatability in response to use of many impulses as opposed to few impulses for electrical stimulation. McKeith et al. (1980B) reported that duration of ES (1 min vs 2 min) was inconsequential in assuring that beef from mature cow carcasses had been tenderized. Since a part of the tenderization effect of ES had been attributed to rate of pH decline in preventing "cold shortening" (Bendall et al. 1976), and to rupturing lysosomal membranes and freeing lysosomal enzymes (Smith et al. 1977; Dutson et al. 1980B), and since rate of pH decline may be associated with the increased firmness and brighter color of ES meat, it could be assumed that 2 min of ES might be more effective than 1 min in achieving pH-related palatability or quality enhancement. In the McKeith et al. (1981B) study, lean maturity scores were improved slightly when electrical stimulation was performed using 550 V for 2 min; however, palatability characteristics were not affected by duration (1 min vs 2 min) of stimulation. The pH decline was checked in longissimus muscles of some carcasses in the McKeith et al. (1981B) study and no consistent difference was found in pH among carcasses stimulated for 1 vs 2 min.

Because of the comparatively greater danger to workmen on the slaughter floor of 550 V as compared with 150 V, it was essential to compare effects of these two voltages on quality-indicating and palatability characteristics of beef. Most of our ES studies had been conducted with 440–550 V, with the desired results, yet Chrystall and Devine (1978) found little difference in rate of pH decline between 320, 150, and 50 V for electrical stimulation, and Taylor and Marshall (1980) accomplished tenderization of beef carcasses with ES of 32 V. Since the objective of the McKeith et al. (1981B) study was twofold—improvement of palatability and improvement in appearance of the longissimus muscle cross-sectional surface at the 12–13th rib—the effectiveness of a high (550 V) and a low (150 V) voltage for electrical stimulation was compared. Results of the McKeith et al. (1981B) study suggested sufficient advantage for high (550 V) vs low (150 V) voltage in improving lean maturity, lean color, flavor, tenderness, and overall palatability to warrant its use for commercial implementation of electrical stimulation. Time postmortem of ES application may have been a major factor in results of the latter study; high voltage may have been necessary to achieve the desired response because of the time-lapse between stunning and ES application. If ES is performed shortly after stunning, low-voltage stimulation could produce the desired quality effects.

Data of the McKeith et al. (1981B) study indicate that electrical stimulation will improve palatability of beef muscles and that it can be performed at any of the four stages in the slaughter–dressing sequence that were investigated. Electrical stimulation for 2 min rather than for 1 min had no apparent advantage in improving palatability or appearance of beef. However, in the McKeith et al. (1981B) study, electrical stimulation with 550 V rather than with 150 V improved lean maturity, lean color, flavor, tenderness, and overall palatability. McKeith et al. (1981B) concluded that electrical stimulation can be performed at any of several stages during the slaughter–dressing sequence of commercial plants using short duration (1 min) and high voltage (550 V) stimulation to improve quality-indicating characteristics and palatability of beef.

Obviously, the major benefit of electrical stimulation is that of improving the tenderness of meat, but this improvement is not recognizable as such at the packer, wholesaler, or retailer level. Of the benefits of electrical stimulation which do accrue directly to the packer, those associated with government or private grading—improved muscle color, more rapid development ("setup") of marbling, reduced incidence of heat ring, greater firmness, and finer texture of cut muscle surfaces—are of greatest monetary significance. Electrical stimulation also increases the proportion of blood loss (as drip from the carcass) that occurs on the slaughter–dressing floor. Removal of blood on the slaughter–dressing floor (1) reduces blood drip in coolers, thus saving time/labor, (2) allows for recovery or removal of blood in plant locations (on the kill floor) where it can be most efficiently handled, and (3) reduces blood staining of the surfaces on the inside of the neck, thereby reducing microbial spoilage and/or trimming prior to cutting/fabrication of carcasses. Dutson (1981) reported that several packing plants had noted (and that research studies at TAMU had confirmed) that the time for initial-chill coolers ("hot boxes") to reach a constant temperature, after having been loaded with "hot" carcasses, was decreased markedly after installation and use (on the slaughter–dressing floor) of an electrical stimulation unit. TAMU research (Dutson 1981) demonstrated that electrically stimulated sides had an initially higher temperature than nonstimulated sides but that, due to a more rapid rate of chilling, the electrically stimulated sides reached the desired temperature up to 2 hr sooner than did the nonstimulated sides.

Dutson (1981) reported results of a packer survey that was designed to determine how many of the benefits observed by scientists in research studies were actually being realized by packers who were using electrical stimulation. A total of 38 packers were interviewed who were using either low-voltage or high-voltage, manual or automated, ES units. Response of packers using low-voltage ES units indicated that improved muscle color, faster chilling, reduced incidence of heat ring, and reduced consumer complaints were the most consistent benefits that they had realized; less "drip" in coolers, improved tenderness, and increased sales also received numerous favorable responses (Table 4.31). Responses of packers using

TABLE 4.31. Results of a Survey of Meat Packers Using Electrical Stimulation

Trait[a]	Packers using low-voltage ES units		Packers using manual high-voltage ES units		Packers using automated high-voltage ES units	
	Number	%	Number	%	Number	%
Improved quality grade	—[c]	—[c]	12	92	8	88
Improved muscle color	7	100	20	70	8	100
Faster chilling	5	100	17	71	7	71
Less heat ring	6	100	18	67	8	88
Less cooler shrinkage	5	80	18	61	8	88
Less "drip" in cooler	6	83	18	89	7	86
Improved tenderness	6	83	13	92	6	100
Fewer customer complaints	6	100	16	63	8	75
Improved shelf-life	3	67	9	56	3	100
Increased sales	6	83	18	33	7	71
Easier boning	4	75	9	67	3	67

Source: Dutson (1981).
[a] Traits are those mentioned by one or more packers during interview.
[b] Number of packers interviewed: 8 packers using low-voltage ES units; 22 packers using manual, high-voltage ES units; 8 packers using automated, high-voltage ES units.
[c] None of the 8 packers interviewed graded carcasses.

manually operated high-voltage ES units indicated that improved quality grade, less drip in the cooler, and improved muscle color received the greatest number of favorable responses, with improved tenderness, faster chilling and less heat ring, less cooler shrink, fewer customer complaints, and improved grade also receiving numerous favorable responses; less important were improved shelf-life, increased sales, and easier boning (Table 4.31). Responses by those using automated high-voltage electrical stimulation in their plants are summarized in Table 4.31. Most favorable

TABLE 4.32. Recommendations of Packers Using Electrical Stimulation (Low-voltage, Manual High-voltage, and Automated High-voltage)

Recommendation	Packers responding yes	Packers responding no
Do you recommend electrical stimulation?	36[a]	1[b]
Do you recommend your type of unit?	37	1[b]
Is your electrical stimulation unit worth the cost?	36[a]	1[b]

Source: Dutson (1981).
[a] One packer felt that since his unit had been operating for a very short period of time he had insufficient experience to respond.
[b] The negative response was from one packer who was waiting to have his unit moved to a different plant location.

responses were given for improved muscle color, improved quality grade, less heat ring, and less cooler shrink. Less drip in the cooler, improved tenderness, fewer customer complaints, increased sales, and faster chilling also received numerous favorable responses.

In general, almost all packers were very enthusiastic about electrical stimulation and the benefits of its use in their operations (Dutson, 1981). Since many different types of operations were surveyed, the same benefits were not as evident to all packers. However, each packer felt his operation was aided by electrical stimulation (with one exception—see Table 4.32). Table 4.32 summarizes responses of packers when asked if they would recommend electrical stimulation and if they felt the costs were justified. Almost without exception, they felt that electrical stimulation was useful enough to their operation to recommend its use to other packers and that all of the units were economically sound investments (Dutson 1981).

REFERENCES

BENDALL, J.R. 1976. Electrical stimulation of rabbit and lamb carcasses. J. Sci. Food Agric. 27, 819.

BENDALL, J.R. 1980. The electrical stimulation of carcasses of meat animals. In Developments in Meat Science, Vol. 1 R.A. Lawrie (Editor). Applied Science Publishers, London.

BENDALL, J.R. and RHODES, D.N. 1976. Electrical stimulation of beef carcasses and its practical application. Proc. 22nd Eur. Meet. Meat Res. Workers, London, B-2, 3.

BENDALL, J.R., KETTERIDGE, C.C. and GEORGE, A.R. 1976. The electrical stimulation of beef carcasses. J. Sci. Food Agric. 27, 1123.

BERRY, B.W. and STIFFLER, D.M. 1981. Effects of electrical stimulation, boning temperature, formulation and rate of freezing on sensory, cooking, chemical and physical properties of ground beef patties. J. Food Sci. 46, 1103.

BOUTON, P.E., FORD, A.L., HARRIS, P.V. and SHAW, F.D. 1978. Effect of low-voltage stimulation of beef carcasses on muscle tenderness and pH. J. Food Sci. 43, 1392.

BOUTON, P.E., FORD, A.L., HARRIS, P.V. and SHAW, F.D. 1980A. Electrical stimulation of beef sides. Meat Sci. 4, 145.

BOUTON, P.E., WESTE, R.R. and SHAW, F.D. 1980B. Electrical stimulation of calf carcasses: Response of various muscles to different waveforms. J. Food Sci. 45, 148.

BOWLING, R.A., SMITH, G.C., DUTSON, T.R. and CARPENTER, Z.L. 1978. Effects of prerigor conditioning treatments on lamb muscle shortening, pH and ATP. J. Food Sci. 43, 502.

BRAATHEN, O. 1979. Meat research in the Nordic Countries. Proc. Recip. Meat Conf. 32, 125.

BUTLER, J.L., SMITH, G.C., SAVELL, J.W. and VANDERZANT, C. 1981. Bacterial growth in ground beef prepared from electrically stimulated and nonstimulated muscles. Appl. Environ. Microbiol. 41, 915.

CALKINS, C.R., SAVELL, J.W., SMITH, G.C. and MURPHEY, C.E. 1980. Quality-indicating characteristics of beef muscle as affected by electrical stimulation and postmortem chilling time. J. Food Sci. 45, 1330.

CALKINS, C.R., DUTSON, T.R., SMITH, G.C. and CARPENTER, Z.L. 1982. Concentration of creatine phosphate, adenine nucleotides and their derivatives in electrically stimulated and nonstimulated beef muscle. J. Food Sci. 47, 1350.

CARSE, W.A. 1973. Meat quality and the acceleration of postmortem glycolysis by electrical stimulation. J. Food Technol. *8*, 163.

CHRYSTALL, B.B. and DEVINE, C.E. 1978. Electrical stimulation, muscle tension and glycolysis in bovine sternomandibularis. Meat Sci. *2*, 49.

CHRYSTALL, B.B. and DEVINE, C.E. 1980. Electrical stimulation developments in New Zealand. Proc. 26th Eur. Meet. Meat Res. Workers, Colorado Springs, *2*, K-6.

CHRYSTALL, B.B. and HAGYARD, C.J. 1976. Electrical stimulation and lamb tenderness. N.Z. J. Agric. Res. *19*, 7.

CLARKE, F.M., SHAW, F.D. and MORTON, D.J. 1980. Effect of electrical stimulation postmortem of bovine muscle on the binding of glycolytic enzymes. Biochem. J. *186*, 105.

CONTRERAS, S. and HARRISON, D.L. 1981. Electrical stimulation and hot boning: Color stability of ground beef in a model system. J. Food Sci. *46*, 464.

CONTRERAS, S., HARRISON, D.L., KROPF, D.H. and KASTNER, C.L. 1981. Electrical stimulation and hot boning: Cooking losses, sensory properties, and microbial counts of ground beef. J. Food Sci. *46*, 457.

CROSS, H.R. 1979. Effects of electrical stimulation of meat tissue and muscle properties—A review. J. Food Sci. *44*, 509.

CROSS, H.R. and TENNENT, I. 1980. Accelerated processing systems for USDA Choice and Good beef carcasses. J. Food Sci. *45*, 765.

CROSS, H.R. and TENNENT, I. 1981. Effect of electrical stimulation and postmortem boning time on sensory and cooking properties of ground beef. J. Food Sci. *46*, 292.

CROSS, H.R., SMITH, G.C., KOTULA, A.W. and MUSE, D.A. 1979. Effects of electrical stimulation and shrouding method on quality and palatability of beef carcasses. J. Food Sci. *44*, 1560.

DAVEY, C.L., GILBERT, K.V. and CARSE, W.A. 1976. Carcass electrical stimulation to prevent cold-shortening toughness in beef. N.Z. J. Agric. Res. *19*, 13.

DAVIS, G.W., COLE, A.B., JR., BACKUS, W.R. and MELTON, S.L. 1981. Effect of electrical stimulation on carcass quality and meat palatability of beef from forage- and grain-finished steers. J. Anim. Sci. *53*, 651.

DEATHERAGE, F.E. 1980. Electrical stimulation of slaughter animals. Meat Process. *19* (Sept.) 34–36.

DEMEYER, D. and VANDENDRIESSCHE, F. 1980. Low-voltage electrical stimulation of beef carcasses: Distribution of tenderizing effect in the carcass in relation to changes in sarcomere length. Proc. 26th Eur. Meet. Meat Res. Workers, Colorado Springs, *2*, H-2.

DUTSON, T.R. 1981. Meat quality improvements and industry benefits of electrical stimulation. Electrical Stimulation Seminar, Coventry, U.K., Meat and Livestock Commission, Feb. 1981.

DUTSON, T.R., MEINERS, C.R. and SMITH, G.C. 1980A. Protein efficiency ratio of beef tenderized by electrical stimulation. Nutr. Rep. Int. *22*, 973.

DUTSON, T.R., SMITH, G.C. and CARPENTER, Z.L. 1980B. Lysosomal enzyme distribution in electrically stimulated ovine muscle. J. Food Sci. *45*, 1097.

DUTSON, T.R., SMITH, G.C., SAVELL, J.W. and CARPENTER, Z.L. 1980C. Possible mechanisms by which electrical stimulation improves meat tenderness. Proc. 26th Eur. Meet. Meat Res. Workers, Colorado Springs, *2*, J-6.

DUTSON, T.R., SMITH, G.C., SAVELL, J.W. and CARPENTER, Z.L. 1980D. Effects of electrical stimulation on meat quality. Ann. Technol. Agric. *29* (4) 573.

DUTSON, T.R., SAVELL, J.W. and SMITH, G.C. 1982. Electrical stimulation of ante-mortem stressed beef. Meat Sci. *6*, 159.

ELGASIM, E.A., KENNICK, W.H., McGILL, L.A., ROCK, D.F. and SOELDNER, A. 1981. Effects of electrical stimulation and delayed chilling of beef carcasses on carcass and meat characteristics. J. Food Sci. *46*, 340.

FORREST, J.C. and BRISKEY, E.J. 1967. Response of striated muscle to electrical stimulation. J. Food Sci. *32*, 483.

GEORGE, A.R., BENDALL, J.R. and JONES, R.C.D. 1980. The tenderizing effect of electrical stimulation of beef carcasses. Meat Sci. *4*, 51.
GILBERT, K.V. and DAVEY, C.L. 1976. Carcass electrical stimulation and early boning of beef. N.Z. J. Agric. Res. *19*, 429.
GILBERT, K.V., DAVEY, C.L. and NEWTON, K.G. 1977. Electrical stimulation and the hot boning of beef. N.Z. J. Agric. Res. *20*, 139.
GILL, C.O. 1980. Effect of electrical stimulation on meat spoilage floras. J. Food Protect. *43*, 190.
GRIFFIN, C.L., STIFFLER, D.M., RAY, E.E. and BERRY, B.W. 1981. Effects of electrical stimulation, boning time and cooking method on beef roasts. J. Food Sci. *46*, 987.
GRUSBY, A.H., WEST, R.L., CARPENTER, J.W. and PALMER, A.Z. 1976. Effects of electrical stimulation on tenderness. J. Anim. Sci. *42*, 253. (Abstract)
HALL, L.C., SAVELL, J.W. and SMITH, G.C. 1980. Retail appearance of electrically stimulated beef. J. Food Sci. *45*, 171.
HALLUND, O. and BENDALL, J.R. 1965. The long-term effect of electrical stimulation on the postmortem fall of pH in muscles of Landrace pigs. J. Food Sci. *30*, 296.
HARSHAM A. and DEATHERAGE, F.E. 1951. Tenderization of meat. U.S. Pat. 2,544,681. Mar. 13.
HOSTETLER, R.L., DUTSON, T.R. and SMITH, G.C. 1982. Effect of electrical stimulation and steak temperature at the beginning of cooking on meat tenderness and cooking loss. J. Food Sci. *47*, 687.
HOULIER, B., VALIN, C., MONIN, G. and SALE, P. 1980. Is electrical stimulation efficiency muscle dependent? Proc. 26th Eur. Meet. Meat Res. Workers, Colorado Springs, *2*, J-5.
JEREMIAH, L.E. and MARTIN, A.H. 1980. The effects of electrical stimulation on retail acceptability and case-life of beef. Proc. 26th Eur. Meet. Meat Res. Workers, Colorado Springs, *2*, H-8.
JOHNSON, D.D., SAVELL, J.W., WEATHERSPOON, L. and SMITH, G.C. 1982. Quality, palatability and weight loss of pork as affected by electrical stimulation. Meat Sci. *7*, 43.
JUDGE, M.D., REEVES, E.S. and ABERLE, E.D. 1980. Effect of electrical stimulation on thermal shrinkage temperature of bovine muscle collagen. Proc. 26th Eur. Meet. Meat Res. Workers, Colorado Springs, *2*, J-3.
LEMISCH, L.J. 1961. Benjamin Franklin: The Autobiography and Other Writings, 1st Edition. p. 236. New American Library of World Literature, New York.
LOCKER, R.H. 1976. Meat tenderness and muscle structure. Proc. N.Z. Meat Ind. Res. Conf. *18*, 1.
LOPEZ, C.A. and HERBERT, E.W. 1975. The Private Franklin, The Man and His Family, 1st Edition. p. 44. W.W. Norton and Co., New York.
McKEITH, F.K., SAVELL, J.W., SMITH, G.C., DUTSON, T.R. and SHELTON, M. 1979. Palatability of goat meat from carcasses electrically stimulated at four different stages in the slaughter-dressing sequence. J. Anim. Sci. *49*, 972.
McKEITH, F.K., SMITH, G.C., DUTSON, T.R., SAVELL, J.W., HOSTETLER, R.L. and CARPENTER, Z.L. 1980A. Electrical stimulation of intact or split steer or cow carcasses. J. Food Protect. *43*, 795.
McKEITH, F.K., SMITH, G.C., SAVELL, J.W., DUTSON, T.R., CARPENTER, Z.L. and HAMMONS, D.R. 1980B. Electrical stimulation of mature cow carcasses. J. Anim. Sci. *50*, 694.
McKEITH, F.K., SAVELL, J.W. and SMITH, G.C. 1981A. Tenderness improvement of the major muscles of the beef carcass by electrical stimulation. J. Food Sci. *46*, 1774.
McKEITH, F.K., SMITH, G.C., SAVELL, J.W., DUTSON, T.R., CARPENTER, Z.L. and HAMMONS, D.R. 1981B. Effects of certain electrical stimulation parameters on quality and palatability of beef. J. Food Sci. *46*, 13.
McKEITH, F.K., SAVELL, J.W., MURPHEY, C.E. and SMITH, G.C. 1982.

Enhancement of lean characteristics of veal carcasses by electrical stimulation. Meat Sci. *6*, 65.

MRIGADAT, B., SMITH, G.C., DUTSON, T.R., HALL, L.C., HANNA, M.O. and VANDERZANT, C. 1980. Bacteriology of electrically stimulated and unstimulated rabbit, pork, lamb and beef carcasses. J. Food Protect. *43*, 686.

OCKERMAN, H.W. and DOWIERCIAL, R. 1980. Influence of tumbling and electrical stimulation on distribution and content of sodium nitrite and sodium chloride in bacon. J. Food Sci. *45*, 1301.

ORCUTT, M.W., DUTSON, T.R., BURNS, E.E. and SMITH, G.C. 1981. Alterations of fresh beef color resulting from electrical stimulation. Am. Soc. Anim. Sci., Proc. 73rd Annu. Meet., Raleigh, NC, 1981, 226.

ORTS, F.A., SMITH, G.C. and HOSTETLER, R.L. 1971. Texas A&M Tenderstretch. Tex. Agric. Ext. Serv. Bull. *L-1003*.

RACCACH, M. and HENRICKSON, R.L. 1978. Storage stability and bacteriological profile of refrigerated ground beef from electrically-stimulated hot-boned carcasses. J. Food Protect. *41*, 957.

RENTSCHLER, H.C. 1951. Apparatus and method for the tenderization of meat. U.S. Pat. 2,544,724. Mar. 13.

RILEY, R.R., SAVELL, J.W. and SMITH, G.C. 1980A. Storage characteristics of wholesale and retail cuts from electrically stimulated lamb carcasses. J. Food Sci. *45*, 1101.

RILEY, R.R., SAVELL, J.W., SMITH, G.C. and SHELTON, M. 1980B. Quality, appearance and tenderness of electrically stimulated lamb. J. Food Sci. *45*, 119.

RILEY, R.R., SAVELL, J.W., SMITH, G.C. and SHELTON, M. 1981. Improving appearance and palatability of meat from ram lambs by electrical stimulation. J. Anim. Sci. *52*, 522.

RILEY, R.R., SAVELL, J.W., STIFFLER, D.M., EHLERS, J.G., VANDERZANT, C. and SMITH, G.C. 1982. Evaluation of retail and palatability characteristics of electrically stimulated U.S. Choice beef after commercial transport-distribution. J. Food Protect. *45*, 733.

RILEY, R.R., SAVELL, J.W., MURPHEY, C.E., SMITH, G.C., STIFFLER, D.M. and CROSS, H.R. 1983A. Effects of electrical stimulation, subcutaneous fat thickness and masculinity traits on palatability of beef from young bull carcasses. J. Anim. Sci. *56*, 584.

RILEY, R.R., SAVELL, J.W., MURPHEY, C.E., SMITH, G.C., STIFFLER, D.M. and CROSS, H.R. 1983B. Palatability of beef from steer and young bull carcasses as influenced by electrical stimulation, subcutaneous fat thickness and marbling. J. Anim. Sci. *56*, 592.

ROUQUETTE, F.M., JR., RILEY, R.R. and SAVELL, J.W. 1983. Electrical stimulation, stocking rate and creep feed effects on carcass traits of calves slaughtered at weaning. J. Anim. Sci. *56*, 1012.

RUDERUS, H. 1980. Low voltage electrical stimulation of beef. Influence of pulse types on postmortem pH fall and meat quality. Proc. 26th Eur. Meet. Meat Res. Workers, Colorado Springs, *2*, K-3.

RUDERUS, H. and BERGQUIST, A. 1980. Industry application of low voltage electrical stimulation. Ann. Technol. Agric. *29*, 659.

RUDERUS, H. and FABIANSSON, S. 1980. Research on low voltage electrical stimulation of beef carcasses in Sweden. Ann. Technol. Agric. *29*, 581.

SALM, C.P., MILLS, E.W., REEVES, E.S., JUDGE, M.D. and ABERLE, E.D. 1981. Effect of electrical stimulation on muscle characteristics of beef cattle fed a high energy diet for varying lengths of time. J. Food Sci. *46*, 1284.

SAVELL, J.W. 1979. Update: Industry acceptance of electrical stimulation. Proc. Recip. Meat Conf. *32*, 113.

SAVELL, J.W. 1982. Electrical stimulation: An overview of the worldwide science and technology associated with its use to improve meat quality and palatability. Proc. Int. Symp. Meat Sci. Technol., Lincoln, NB, 1981, Natl. Live Stock and Meat Board, Chicago, 1–18.

SAVELL, J.W. and SMITH,G.C. 1979. Electrical stimulation—Effects on meat tenderness, muscle structure and the quality indicating characteristics of meat. Proc. Annu. Meet. Res. Dev. Assoc. Mil. Food Packag. Syst., New York, 1979, 1–14.

SAVELL, J.W., SMITH, G.C., DUTSON, T.R., CARPENTER, Z.L. and SUTER, D.A. 1976. Effect of electrical stimulation on beef palatability. J. Anim. Sci. 43, 246 (Abstract)

SAVELL, J.W., SMITH, G.C., DUTSON, T.R., CARPENTER, Z.L. and SUTER, D.A. 1977. Effect of electrical stimulation on palatability of beef, lamb and goat meat. J. Food Sci. 42, 702.

SAVELL, J.W., DUTSON, T.R., SMITH, G.C. and CARPENTER, Z.L. 1978A. Structural changes in electrically stimulated beef muscle. J. Food Sci. 43, 1606.

SAVELL, J.W., SMITH, G.C. and CARPENTER, Z.L. 1978B. Effect of electrical stimulation on quality and palatability of light-weight beef carcasses. J. Anim. Sci. 46, 1221.

SAVELL, J.W., SMITH, G.C. and CARPENTER, Z.L. 1978C. Beef quality and palatability as affected by electrical stimulation and cooler aging. J. Food Sci. 43, 1666.

SAVELL, J.W., SMITH, G.C., CARPENTER, Z.L. and PARRISH, F.C., JR. 1979. Influence of electrical stimulation on certain characteristics of heavy-weight beef carcasses. J. Food Sci. 44, 911.

SAVELL, J.W., SMITH, G.C., DUTSON, T.R. and CARPENTER, Z.L. 1980. Industry application of electrical stimulation in the United States. Proc. 26th Eur. Meet. Meat Res. Workers, Colorado Springs, 2, K-2.

SAVELL, J.W., McKEITH, F.K. and SMITH, G.C. 1981. Reducing postmortem aging time of beef with electrical stimulation. J. Food Sci. 46, 1777.

SAVELL, J.W., McKEITH, F.K., MURPHEY, C.E., SMITH, G.C. and CARPENTER, Z.L. 1982. Singular and combined effects of electrical stimulation, postmortem ageing and blade tenderisation on the palatability attributes of beef from young bulls. Meat Sci. 6, 97.

SEIDEMAN, S.C., SMITH, G.C., DUTSON, T.R. and CARPENTER, Z.L. 1979. Physical, chemical and palatability traits of electrically stimulated, hot-boned, vacuum-packaged beef. J. Food Protect. 42, 651.

SHAW, F.D. and WALKER, D.J. 1977. Effect of low voltage stimulation of beef carcasses on muscle pH. J. Food Sci. 42, 1140.

SMITH, G.C. 1979A. History of research, development and industry implementation of electrical stimulation. Proc. Symp. Electr. Stimulation for Improving Meat Qual., Corpus Christi, TX, 1979, 9–14.

SMITH, G.C. 1979B. Effect of electrical stimulation on the tenderness of meat. Proc. Symp. Electr. Stimulation for Improving Meat Qual., Corpus Christi, TX, 1979, 19–31.

SMITH, G.C., DUTSON, T.R., CARPENTER, Z.L. and HOSTETLER, R.L. 1977. Using electrical stimulation to tenderize meat. Proc. Meat Ind. Res. Conf., Am. Meat Inst. Found., Chicago, 1977, 147.

SMITH, G.C., DUTSON, T.R., CROSS, H.R. and CARPENTER, Z.L. 1979A. Electrical stimulation of hide-on and hide-off calf carcasses. J. Food Sci. 44, 335.

SMITH, G.C., JAMBERS, T.G., CARPENTER, Z.L., DUTSON, T.R., HOSTETLER, R.L. and OLIVER, W.M. 1979B. Increasing the tenderness of forage-fed beef. J. Anim. Sci. 49, 1207.

SMITH, G.C., SAVELL, J.W., DUTSON, T.R., HOSTETLER, R.L., TERRELL, R.N., MURPHEY, C.E. and CARPENTER, Z.L. 1980. Effects of electrical stimulation on beef, pork, lamb and goat meat. Proc. 26th Eur. Meet. Meat Res. Workers, Colorado Springs, 2, H-5.

SONAIYA, E.B. and STOUFFER, J.R. 1982. Mechanical tensioning of electrically stimulated carcasses for improved tenderness. J. Food Sci. 47, 1010.

SORINMADE, S.O., CROSS, H.R. and ONO, K. 1978. The effect of electrical stimulation on lysosomal enzyme activity, pH decline and beef tenderness. Proc. 24th Eur. Meet. Meat Res. Workers, Kulmbach, W. Germany, 1978.

STIFFLER, D.M., SAVELL, J.W., SMITH, G.C., DUTSON, T.R. and CARPENTER, Z.L. 1982. Electrical stimulation: Purpose, application and results. Tex. Agric. Ext. Serv. Bull. *B-1375*.

SWASDEE, R.L., TERRELL, R.N., DUTSON, T.R., CRENWELGE, D.D. and SMITH, G.C. 1982. Processing properties of pork as affected by electrical stimulation, post-slaughter chilling and muscle group. J. Food Sci. *47*, 1011.

SWATLAND, H.J. 1977. Sensitivity of prerigor beef muscle to electrical stimulation. Can. Inst. Food Sci. Technol. J. *19*, 280.

TANG, B.H. and HENRICKSON, R.L. 1980. Effect of postmortem electrical stimulation on bovine myoglobin and its derivatives. J. Food Sci. *45*, 1139.

TAYLOR, A.A., SHAW, B.G. and MacDOUGALL, D.B. 1980. Hot deboning beef with and without electrical stimulation. Meat Sci. *5*, 109.

TAYLOR, D.G. and MARSHALL, A.R. 1980. Low voltage electrical stimulation of beef carcasses. J. Food Sci. *45*, 144.

TERRELL, R.N., MING, C.G., JACOBS, J.A., SMITH, G.C. and CARPENTER, Z.L. 1981. Effects of chloride salts, acid phosphate and electrical stimulation on pH and moisture loss from beef clod muscles. J. Anim. Sci. *53*, 658.

TERRELL, R.N., CORREA, R., LEU, R. and SMITH, G.C. 1982A. Processing properties of beef semimembranosus muscles as affected by electrical stimulation and postmortem treatment. J. Food Sci. *47*, 1382.

TERRELL, R.N., JACOBS, J.A., SAVELL, J.W. and SMITH, G.C. 1982B. Processing properties of beef clod muscles as affected by electrical stimulation and post-rigor frozen storage. J. Anim. Sci. *54*, 964.

TERRELL, R.N., JACOBS, J.A., SAVELL, J.W. and SMITH, G.C. 1982C. Properties of frankfurters made from electrically stimulated beef. J. Food Sci. *47*, 344.

TUME, R.K. 1980. Effect of post-mortem electrical stimulation on ovine sarcoplasmic reticulum vesicles. Aust. J. Biol. Sci. *33*, 43.

WALKER, D.J., HARRIS, P.V. and SHAW, F.D. 1977. Accelerated processing of beef. Food Technol. Aust. *29*, 504.

WEST, R.L. 1982. Commercial application of electrical stimulation in the United States. Proc. Int. Symp. Meat Sci. Technol., Lincoln, NB, 1982, Natl. Live Stock and Meat Board, Chicago, 1–23.

WESTERVELT, R.G. and STOUFFER, J.R. 1978. Relationships among spinal cord severing, electrical stimulation and postmortem quality characteristics of the porcine carcass. J. Anim. Sci. *46*, 1206.

WHITING, R.C., STRANGE, E.D., MILLER, A.J., BENEDICT, R.C., MOZERSKY, S.M. and SWIFT, C.E. 1981. Effects of electrical stimulation on the functional properties of lamb muscle. J. Food Sci. *46*, 484.

WILL, P.A., OWNBY, C.L. and HENRICKSON, R.L. 1980. Ultrastructural postmortem changes in electrically stimulated bovine muscle. J. Food Sci. *45*, 21.

5

Use of Electrical Stimulation for Hot Boning of Meat

H. R. Cross[1] and S. C. Seideman[2]

History of Hot Boning Research
Electrical Stimulation and Hot Boning
Water Holding Capacity
Appearance Properties
Microbiology
Cooking Prerigor Muscle
Utilization of Hot Boning Meat for the Production of Ground Beef
Why Hasn't Industry Accepted Hot Boning?
Economic Implications of Hot Boning
Conclusions
References

With ever-increasing processing costs, the meat industry is being forced to look into new and innovative processing methods. The efficiency of marketing meat and moving it to the consumer must be increased. The meat industry must face these new challenges and demands in order to survive.

Over the past two decades, the meat industry has dramatically shifted its manner of distributing meat, from shipping beef to retail stores in carcass form to shipping beef to retail stores in the form of vacuum packaged primal and subprimal cuts (Table 5.1). This manner of distribution has introduced a concept identified as the "boxed beef" distribution system. This system entails the process of prefabricating carcasses into primal or subprimal cuts and vacuum packaging them at locations that are near the areas of livestock production. Processing meat at a centralized location in areas that are in the general proximity to areas of consumption is a concept that is termed "centralized breaking-point". At present, 65–70% of beef distributed to retail stores is distributed in the form of vacuum packaged primal cuts. Advantages attributed to the centralized breaking-point system include more efficient use of labor and meat by-products,

[1]Meats and Muscle Biology Section, Department of Animal Science, Texas A&M University, College Station, TX.
[2]Meats Research Unit, Agricultural Research Service, United States Department of Agriculture, Roman L. Hruska U.S. Meat Animal Research Center, P.O. Box 166, Clay Center, NE.

TABLE 5.1. Scenario of Carcass Processing and Distribution Systems

Past	Carcass → retail store
Present	Carcass → centralized fabrication → vacuum packaging → retail store
Future	Carcass → electrical stimulation → hot boning → vacuum packaging → retail store

reduced tonnages for shipment, greater flexibility in marketing, increased control of inventory, and simplification of retail operations. Even though the centralized breaking-point concept offers all these advantages, increased energy and transportation costs will eventually shift all processing back toward the point of slaughter. In the future, it is likely that only the edible product, minus bone and excess fat, will leave the slaughter plant.

Hot boning is a relatively new process of carcass fabrication that involves the removal of lean meat and fat from bone prior to chilling. Hot boning has also been described as hot processing, anterigor excision, prerigor excision, accelerated processing, high temperature processing, prechill processing, hot cutting and processing, processing prior to rigor mortis, and rapid processing (Kastner 1977).

Some potential advantages and potential disadvantages of hot boning are presented in Table 5.2. Most of these advantages and disadvantages are concerned with the economics and practical application of hot boning in today's industry. The economics and practical applications of hot processing are discussed elsewhere in this chapter and in detail in Chapter 8. In general terms, the economics presently favor hot boning; however, several problems in practical applications have hindered its adoption by the meat industry.

Numerous studies have shown that the flavor, juiciness, visual color, and cooking loss characteristics of hot-boned meat are similar to those of conventionally processed meat with the only quality attribute variation being that of tenderness. Three of the most commonly researched methods

TABLE 5.2. Potential Advantages and Disadvantages of Hot Boning

Advantages	Disadvantages
1. Reduced cooler/storage space	1. Unable to quality grade
2. Reduced energy (refrigeration) input	2. Greater hygiene and temperature control required
3. Increased product turnover	3. Unconventional shape of cuts
4. Improved sanitation and shelf-life	4. Difficult to incorporate into conventional plants
5. Less drip in the vacuum bag	
6. Less staining of fat in the vacuum bag	5. No systems developed for rapid chilling of large volumes of cuts
7. Reduction in labor, material, and equipment costs	6. Reduced product quality
8. Improved processing properties	

Source: Adapted from Cross and Tennent (1980).

of hot boning are: (1) hot boning after conditioning or chilling the carcass for a specified time postslaughter; (2) hot boning, vacuum packaging, and holding the primal cuts at an elevated temperature for a specified period of time; and (3) electrically stimulating the carcass followed by hot boning at various times poststimulation. Each method can produce differences in the ultimate quality of the meat. The objectives of this chapter are to discuss the history of hot boning, the storage parameters and sensory properties of hot-boned meat, the present industry status and economic implications of hot boning in the United States, and the use of postmortem electrical stimulation in conjunction with hot boning.

HISTORY OF HOT BONING RESEARCH

The removal of meat from the carcass of an animal soon after slaughter is not new. The first humans to eat meat almost certainly would have torn the flesh from the carcass soon after it was killed. Even today, in underdeveloped countries and in large areas of South America, Asia, and the Middle East, people still practice hot boning. Why has hot boning gained international interest during the past decade? The answer lies in economics. Renewed interest has been fostered by the economic advantages, including savings in energy, space, labor, materials, and product weight loss as well as improved functional properties. The advantages of hot boning are many, but before this system can be adopted by industry, industry must be assured of being able to maintain a safe and high quality product. In addition, many technical questions have not been answered and will not be until hot boning is commercially applied under a variety of conditions.

The history of hot boning research can be traced back to research reports by Lowe and Stewart (1946), Ramsbottom and Strandine (1949), and Paul *et al.* (1952), who found that meat cooked prerigor was more tender than meat cooked after rigor mortis. Weidemann *et al.* (1967) and Cia and Marsh (1976) found similar results and reported that the immediate cooking of prerigor muscle eliminated the occurrence of rigor mortis and its detrimental effects on tenderness.

Researchers such as Lowe and Stewart (1946), Ramsbottom and Strandine (1949), Locker (1960), and Herring *et al.* (1965, 1967) performed much of the muscle biology research that ultimately led to more applied hot boning research. They generally found that muscles excised soon after slaughter and permitted to contract freely were less tender than those muscles restrained during the development of rigor mortis or excised postrigor. The extra contraction of muscles induced under these conditions is referred to as "cold shortening." There is a great amount of scientific evidence indicating that the greater the degree of shortening the tougher the meat on subsequent cooking. Hot boning requires that muscles be removed before glycolysis is complete. In this regard, research by Marsh

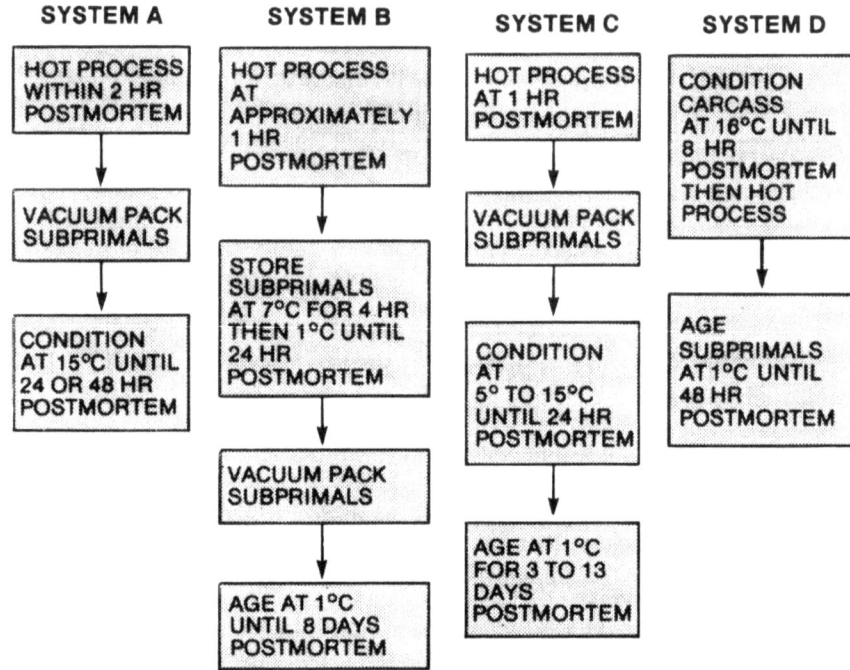

FIG. 5.1. Optimal hot-processing systems for beefsteak and roast items.
Adapted from Kastner (1983). Reprinted from Food Technology. 1983. (May):96–104. Copyright © by Institute of Food Technologists.

(1954), Marsh and Thompson (1958), Bendall (1960), Marsh and Leet (1966), Cook and Langsworth (1966), and Cassens and Newbold (1967) found that the rate of glycolysis as measured by pH decline was dependent upon the temperature of the muscle. In conventional meat processing, the negative effects of cold shortening are partially reduced by high temperature aging or slow cooling of the carcass while rigor mortis is proceeding. During this period, the muscle glycogen stores are being broken down to lactic acid. With this lactic acid production, the pH of the muscle is falling toward its ultimate value of 5.4 to 5.6. When the pH of the muscle has fallen to approximately 6.0, cold shortening will no longer occur to the extent where there is significant toughening, and hot boning can proceed.

Research on the hot boning of pork began in the mid to late 1960s. Marsh *et al.* (1972) excised pork muscles prerigor and subjected them to a temperature environment of 0°C for 24 hr. The pork muscle became significantly less tender. The relative toughening of pork due to cold shortening is much smaller when compared with beef—30% increase in pork toughness *versus* 200% increase in beef toughness (Marsh and Leet 1966; McCrae *et al.* 1971; Behnke and Fennema 1973). Because lower microbial

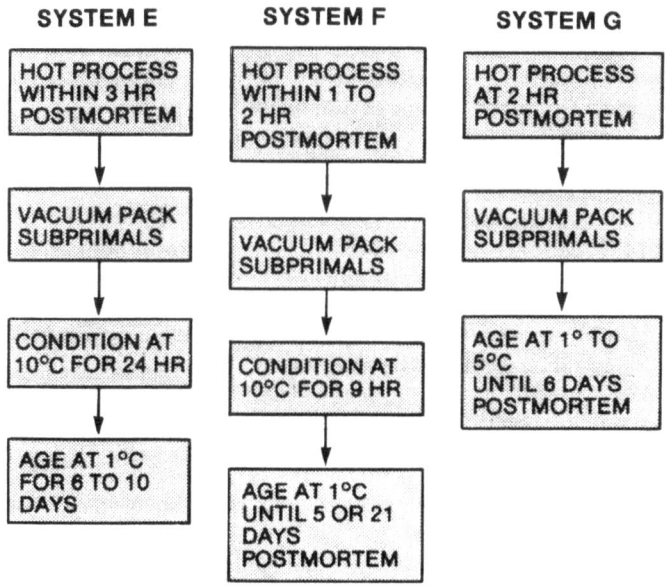

counts were found on hot-boned ham, it was hypothesized that the rapid processing of hot-boned hams offered less opportunity for postslaughter microbial contamination and growth (Barbe et al. 1966; Barbe and Henrickson 1967). Mandigo and Henrickson (1966) found hot-boned hams to be equal or superior to conventionally processed hams in yield, tenderness, juiciness, flavor, and moisture content. Trautman (1964) found that hot-boned pork had greater emulsifying capacity and more salt-soluble proteins than postrigor muscle. Because of these early studies on pork, some commercial processors are now hot boning pork.

Actual hot boning research as related to applied technology began in the early 1970s. Much of the early research on beef hot boning relied on carcass and muscle conditioning at elevated temperatures or the conventional aging of hot-boned muscles or primals to prevent or minimize any effects of cold shortening. Schmidt and Gilbert (1970) excised beef longissimus, semitendinosus, and semimembranosus muscles at 2 hr postmortem and allowed them to age for 24 or 48 hr (Fig. 5.1, System A). When compared with controls excised from opposite sides and chilled at 9°C until 24 hr postmortem, the hot-boned samples were equal or superior to controls in tenderness. Schmidt and Keman (1974) hot boned muscles from one side of six beef carcasses at 1 hr postmortem (Fig. 5.1, System B). The hot-boned muscles were stored at 7°C for 4 hr, then were placed in a 1°C cooler overnight. The muscles were vacuum packaged at 24 hr postmortem and held at 1°C for 8 days. The controls from the opposite sides were removed from the carcass at 8 days postmortem. Differences between sensory panel tenderness and shear force were not significant.

Kastner et al. (1973) boned bovine muscles at 2, 5, 8, and 48 hr postmortem and found 2 and 5 hr periods to produce slightly less tender meat than that which had been boned at 8 and 48 hr postmortem. The differences between the two treatments were not statistically significant. Will and Henrickson (1976) compared hot-boned beef boned at 3, 5, or 7 hr and then delay-chilled (16°C) for 48 hr with cold-boned (1°C) beef (Fig. 5.1, System D). They concluded that hot boning beef as early as 3 hr postmortem followed by a delayed chill resulted in satisfactory tenderness ratings. The approaches to hot boning outlined in Fig. 5.1 were designed to produce acceptable steak and roast cuts by preventing or alleviating the potential problems of cold shortening. Generally, these systems have been shown not only to ensure a product that is equal or superior in sensory traits to their control counterparts but also to produce a desirable product from an appearance and shelf-life standpoint (Cross 1980; Kastner 1981). However, these methods of hot boning may not facilitate the continuous flow of product required by the industry due to the need for carcass or muscle conditioning.

ELECTRICAL STIMULATION AND HOT BONING

Harsham and Deatherage (1951) reported that the application of electrical current to unchilled beef carcasses resulted in a more tender cooked product. Even though this research was the subject of a patent, industry elected not to pursue this approach, primarily because the negative effects of cold shortening were not recognized until the mid- to late-1950s. In addition, the efficiency of the meat industry's chillers was such that the effects of rapid temperature decline were not evident. The degree of fatness in U.S. beef prevented rapid postmortem temperature decline in muscle. The concept of electrical stimulation to reduce the effects of cold shortening was first realized in New Zealand since scientists there were seeking a means to overcome toughening problems in frozen lamb.

The New Zealand research defined the role of electrical stimulation in accelerating the onset and development of rigor mortis. A practical procedure for using electrical current to condition lamb carcasses with a subsequent reduction in the toughening during freezing was devised and reported by Chrystall and Hagyard (1976). These studies stimulated interest in other countries, especially in the United States, England, and Australia, where the aim has been to study the use of electrical stimulation of beef rather than lamb. Published data to date have shown that a wide range of applied voltages will achieve acceleration of glycolysis in beef. In England, workers use 600 to 700 V DC (Bendall et al. 1976); in New Zealand (Gilbert and Davey 1976), workers use 3600 V AC; and in the United States, workers use 200 to 600 V AC (Berry and Kotula 1982; West and Oblinger 1979).

Electrically stimulating carcasses soon after slaughter can accelerate the onset of rigor mortis, thereby eliminating or minimizing tenderness problems associated with cold shortening. Therefore, carcass or cut conditioning periods (Fig. 5.1) used to avoid potential tenderness problems associated with rapid chilling prerigor can be eliminated or reduced by using electrical stimulation. Also, postmortem electrical stimulation may enhance tenderness by other mechanisms (Dutson et al. 1980). For these reasons, electrical stimulation has been incorporated into much of the recent hot boning research.

A number of electrical stimulation/hot boning systems are outlined in Fig. 5.2. Gilbert and Davey (1976) used electrical stimulation to accelerate the onset of rigor mortis to allow early boning of beef muscles (Fig. 5.2, System A). Rigor developed in 3–4 hr in stimulated carcasses; thus, they could be boned at 5 hr postmortem as compared with 24 hr for controls. Electrically stimulated muscles had all reached a pH of less than 6.0 at 5 hr postmortem. The authors reported that "stimulated carcasses had achieved rigor in 5 hr and it should be possible to bone them without the risk of cold-shortening despite subsequent rapid chilling or freezing." Tenderness of unaged cuts transferred immediately to the freezer is the palatability characteristic most likely to be affected by processing treatment. Cuts from the stimulated sides had a moderate to high degree of tenderness. Gilbert and Davey (1976) concluded "that stimulation reduced the need for conventional chilling to achieve carcass setting, overcame cold and thaw shortening and still permitted additional tenderizing from aging." They further concluded that the quality of the cuts from electrically stimulated/hot-boned beef sides were as acceptable as unstimulated/cold-boned (24 hr) ones and were further improved by aging.

Gilbert et al. (1976) hot boned/stimulated beef muscles at 1 hr and conventionally boned at 24 hr (Fig. 5.2, System B). Except for the fillet, the unstimulated, unaged cuts were all tougher and less uniform than their stimulated counterparts. Stimulation greatly reduced vulnerability of the cuts to shortening despite very early boning and rapid freezing. Gilbert et al. (1976) concluded "that hot-boned cuts from stimulated carcasses aged before freezing attained a high and uniform degree of tenderness. The major potential of carcass stimulation followed by hot-boning lies in reducing the chilling and aging period to two days from the 10–20 days often used commercially." Pierce (1977) found that hot-boned beef that had been previously electrically stimulated was significantly more tender than unstimulated, hot-boned beef.

Cross and Tennent (1980) compared the effects of electrical stimulation

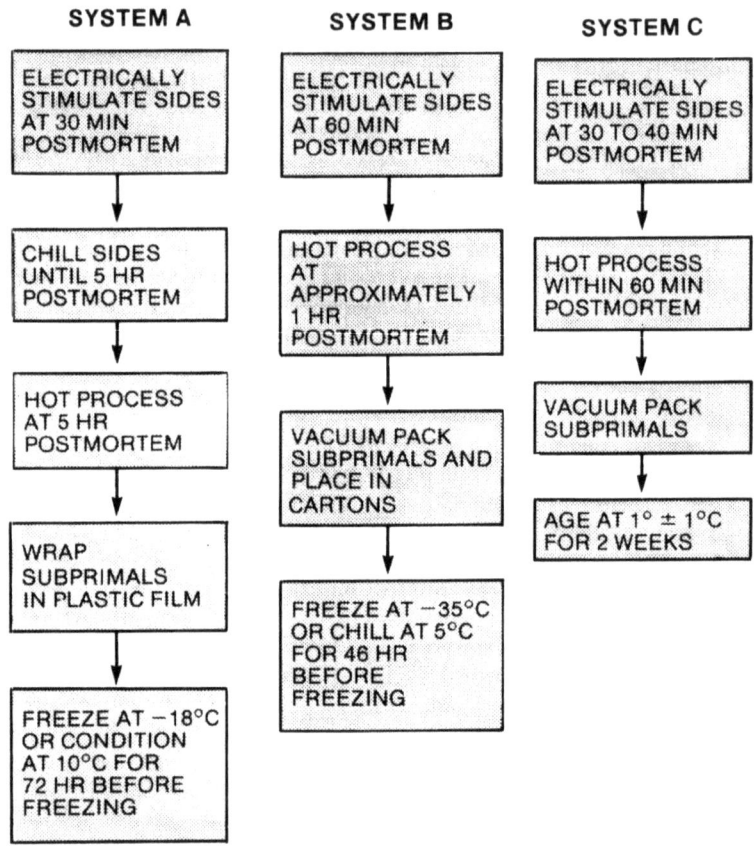

FIG. 5.2. Optimal electrical stimulation and hot-processing systems for beefsteak and roast items.
Adapted from Kastner (1983). Reprinted from Food Technology. 1983. (May):96–104. Copyright © by Institute of Food Technologists.

on USDA Choice and Good beef carcasses boned at 1, 4, and 48 hr postmortem (Table 5.3). Electrically stimulated carcasses were more tender than nonstimulated carcasses at all postmortem excision times. They also found that electrical stimulation tended to offset the negative effects of early boning time on tenderness and that, with electrical stimulation, muscles can be frozen after 24 hr. Numerous additional studies have been reported that vary the electrical stimulation treatment and boning times (Table 5.4). The interrelationships among stimulation method, current

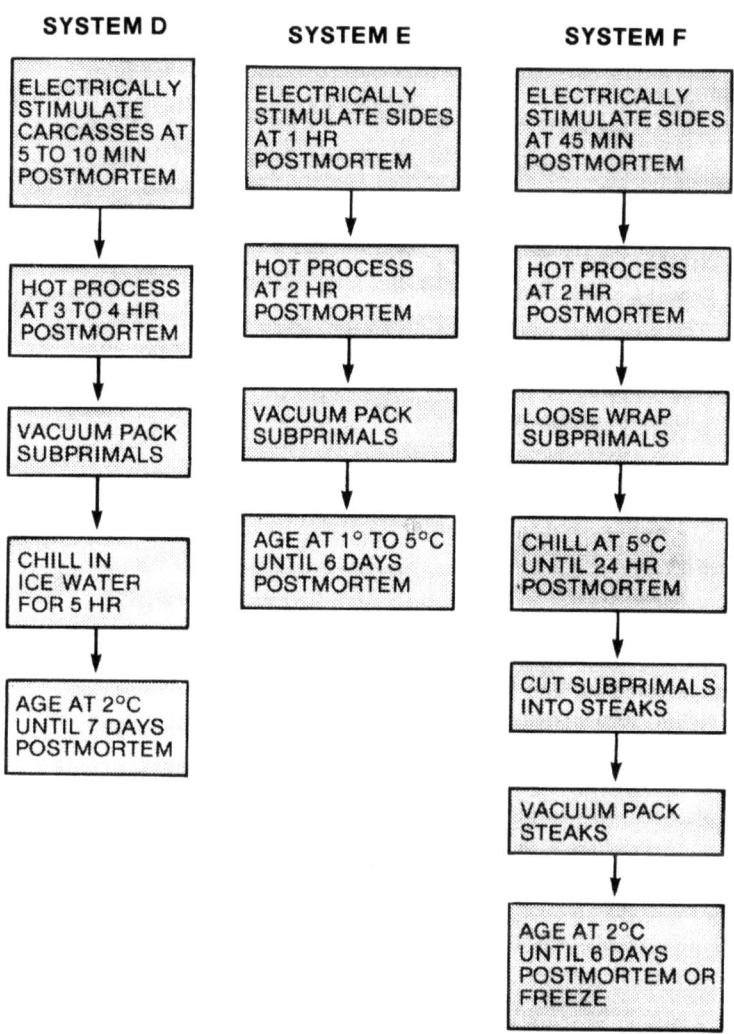

distribution, time postmortem for current application, muscle excision time poststimulation, and rate of chill are complicated and have not been thoroughly studied. Considerable work is needed in this area before optimal systems can be developed.

It appears from the literature that some carcasses can be hot boned within 1 hr postmortem without affecting tenderness whereas others cannot. Electrical stimulation may be a useful tool in allowing these in the latter group to be hot boned at 1 hr also.

TABLE 5.3. Effect of Postmortem Boning Time and Electrical Stimulation on Palatability and Shear Force of the Beef Longissimus

	Postmortem boning time (hr)[a]					
	1		4		48	
Trait	ES[b]	NS[b]	ES[b]	NS[b]	ES[b]	NS[b]
Tenderness[c]	5.6[a]	5.4[ab]	5.5[a]	5.1[b]	6.0[a]	5.6[a]
Connective tissue[c]	6.5[bc]	6.8[ab]	6.6[abc]	6.3[c]	6.8[ab]	7.0[a]
Juiciness[c]	5.3[a]	5.2[a]	5.1[a]	5.2[a]	5.2[a]	5.5[a]
Flavor intensity[c]	5.1[a]	5.0[a]	4.9[a]	4.8[a]	5.0[a]	5.1[a]
Shear force (kg)	7.3[a]	7.9[a]	6.5[ab]	6.8[ab]	5.1[c]	5.6[bc]

Source: Adapted from Cross and Tennent (1980).
[a] Means in the same row with different letters are significantly different ($P < 0.05$).
[b] ES = electrically stimulated. NS = nonstimulated.
[c] Tenderness: 8 = extremely tender and 1 = extremely tough. Connective tissue: 8 = none and 1 = abundant. Juiciness: 8 = extremely juice and 1 = extremely dry. Flavor intensity: 8 = extremely intense and 1 = extremely bland.

WATER HOLDING CAPACITY

Carcass shrinkage values are usually about 2% in the first 24 hr. Falk (1974) reported that hot-boned beef had a lower shrinkage value at 3, 5, and 7 hr holding periods as compared with 48 hr control sides. Taylor *et al.* (1980) also reported that hot boning reduced overall evaporative losses by more than 1%.

The capacity of primal cuts to retain their unbound water during storage has been measured by practical manifestations of this property such as purge (fluid) loss during vacuum-packaged storage, loss during retail display, and thaw and cooking losses.

Table 5.5 outlines comparisons of water holding capacity between hot and cold boning treatments without electrical stimulation. Kastner *et al.* (1973) conditioned muscles at 16°C for 2, 5, and 8 hr prior to boning. Hot-boned cuts removed after 8 hr at 16°C had a lower water holding capacity (WHC) than did comparable controls, but the differences were not large enough to be of practical importance. Follett *et al.* (1974) evaluated different conditioning temperatures for cuts removed at 1 hr postmortem. Differences in percentages for purge loss indicated that cuts conditioned at 5°, 10°, or 15°C for 24 hr lost less weight during storage than did cold-boned cuts. During retail display, cuts conditioned at 15°C had higher weight losses than cold-boned cuts. Cooking losses were greater for cuts conditioned at 5° and 10°C as compared with the controls.

Several studies that utilized electrical stimulation prior to hot boning are summarized in Table 5.6. Cross and Tennent (1979A) studied the effect of electrical stimulation and hot boning at 1 hr postmortem on purge loss differences of 10 primal cuts. Overall, the hot-boned treatment resulted in

TABLE 5.4. Variations in Electrical Stimulation (ES) and Boning Conditions

Authors	Product	Voltage	Current	Pulses (duration)	ES application time (PM)	Boning time (PM)
Berry and Kotula (1982)	Beef	250–400	1.5	40 (1 s)	60 min	2 hr
Butler et al. (1981)	Beef	500	5.0	16 (1.8 s)	Immediately	2 hr
Gill (1980)	Lamb	1130	1.8	858 (10 min)	905/30 min	30 min
Kotula and Emswiler-Rose (1981)	Beef	250–400	1.5	12 (10 s)	60 min	Immediately
Mrigadat et al. (1980)	Beef, lamb, pork	550	5.0	16 (1.8 s)	30–45 min	Immediately
	Rabbit	220		20 (2)	20 min	Immediately
Raccach and Henrickson (1978)	Beef	300	1.6–1.8	400 (0.5 min)	15/30 min	1.5 hr
Stern (1980)	Lamb	250–400	1.5	30 (1 s)	60 min	Immediately
West and Oblinger (1979)	Beef	500	1.5–2.0	60 (1 s)	30 min	45 min

Source: Adapted from Oblinger (1983). Reprinted from Food Technology. 1983. 37:86. Copyright © by Institute of Food Technologists.

TABLE 5.5. Difference Comparison of Water Holding Capacity Measures Between Cold- and Hot-boned Cuts as Related to Carcass or Cut Conditioning Treatment

Reference	Treatments[a]	Muscle or cut[b]	Storage time	Measures (difference, %)[b]		
				Purge	Retail	Cooking
Kastner et al. (1973)	CB—48 hr, 2°–3°C[c]	4 muscles	48 hr			(13.8 to 18.2%)[d]
	HB—2 hr, at 16°C, then 2°–3°C					0.4
	HB—5 hr at 16°C, then 2°–3°C					1.8
	HB—8 hr at 16°C, then 2°–3°C					−8.6
Follett et al. (1974)	CB—36 hr at 2°–3°C	Top round	7 days			(20.0 to 28.0%)[d]
	HB—1 hr, conditioned at:					
	15°C for 24 hr, then 0°–1°C			50.0	−25.0	0.3
	10°C for 24 hr, then 0°–1°C			80.0	5.0	−7.2
	5°C for 24 hr, then 0°–1°C			75.0	20.0	−2.9
	−5°C for 12 hr, then 0°–1°C			−40.0	5.0	3.9
Taylor et al. (1980)	CB—15°C for 7 hr, then 0°–1°C, 48 hr	15 primals	5 days	(0.11%)[d]		
	HB—1–2 hr, 9 hr at 10°C, then 1°C			0		
	ES (50 min) + HB at 1–2 hr, −1°C for 24 hr			36.4		
	CB		21 days	(0.37%)[d]		
	HB			56.8		
	ES + HB			29.7		

Source: Adapted from West (1983). Reprinted from Food Technology. 1983. 37:57. Copyright © by Institute of Food Technologists.
[a] CB = cold-boned. HB = hot-boned. ES = electrical stimulation.
[b] Difference values calculated as follows: Difference (%) = (CB value − HB value ÷ CB value) × 100.
[c] Paired control sides were conditioned at appropriate times prior to chilling.
[d] Actual values for cold-boned controls.

TABLE 5.6. Difference Comparison of Water Holding Capacity Measures Between Cold- and Hot-boned Cuts as Related to Electrical Stimulation and Chilling Treatments

Reference	Treatments[a]	Muscle or cut	Storage time (days)	Measures (difference, %)[b]		
				Purge	Cooking	Juiciness[c]
Cross and Tennent (1979A)	CB—48 hr, 2°–3°C ES + HB—1 hr, 2°–3°C	10 primals	20	(1.02%)[d] 71.6		
Seideman et al. (1979)	CB—24 hr, 1°–3°C ES (30–40 min) + HB (45–60 min) 1°–3°C	Loin	14	(0.69%)[d] 17.9	(28.12%)[d] −3.1	HB < CB
	CB ES + HB	Eye of round	14	(1.97%)[d] 97.5	(29.82%)[d] −21.0	ND
Cross and Tennent (1980)[e]	NES + CB (48 hr) ES + CB (48 hr) NES + HB (1 hr) ES + HB (1 hr) NES + HB (4 hr) ES + HB (4 hr)	Loin	—	(1.7%)[d] −29.4 0 5.9 −47.1 17.6	(34.3%)[d] −2.0 3.2 7.9 −2.3 0.9	ND
Berry and Kotula (1982)	NES + CB (48 hr, 2°C) ES + CB (48 hr, 2°C) ES + HB (2 hr, 2°C) NES + HB (2 hr, 2°C)	Loin	7 and 14	(0.33%)[d] −45.5 −236.4 −93.9		
	NES + CB ES + CB ES + HB NES + HB	Eye of round	7 and 14	(0.64%)[d] −81.3 −71.9 −3.1		

Source: Adapted from West (1983). Reprinted from Food Technology. 1983. 37:57. Copyright © by Institute of Food Technologists.
[a] CB = cold-boned. ES = electrical stimulation. HB = hot-boned. NES = nonstimulated.
[b] Difference values calculated as follows: Difference (%) = (CB value − HB value ÷ CB value) × 100.
[c] ND = no difference at probability level used in reference.
[d] Actual value of CB control.
[e] Used three storage treatments involving (1) immediate freezing; (2) freezing after 24 hr chill; and (3) chilling for 20 days then freezing.

71.6% less purge than did cold boning at 48 hr. Seideman et al. (1979) reported comparable findings for eye of round cuts but detected no differences in the purge loss of loin cuts. The eye of round cuts that were hot boned had a higher cooking loss than cold-boned cuts, suggesting a compensatory loss. In contrast, Berry and Kotula (1982) presented results suggesting that both electrical stimulation and hot boning lowered the ability of loin and eye of round cuts to hold moisture during vacuum-packaged storage.

Cross and Tennent (1980) evaluated various combinations of hot boning, electrical stimulation, cold boning, and storage methods on purge and cooking losses. When the combined effects of boning times and stimulation treatments were compared, stimulation of carcasses prior to cold boning was found to cause higher purge and cooking losses than did the control (nonstimulated) group. Boning at 1 hr resulted in loins with losses similar to the controls. Boning at 4 hr resulted in purge and cooking losses greater than the control group.

The majority of the literature indicates that if a treatment is used to promote a rapid pH decline prior to boning, water holding capacity (WHC) of hot-boned cuts is similar to that of cold-boned cuts. The losses at the various stages in the product flow may be compensatory. Advantages may be evident at initial stages (purge losses) but not later (cooking losses). Detrimental losses of WHC in hot-boned cuts do not appear to be a problem unless chilling of the cuts is too slow. One could expect more uniformity of WHC characteristics within hot-boned muscles since chilling is more uniform (Tarrant 1977; Tarrant and Mothersill 1977).

APPEARANCE PROPERTIES

Hot-boned beef cuts packaged in oxygen-permeable film have often been reported to be darker in color than cold-boned cuts, but the color was not considered unacceptable (Kastner et al. 1973; Kastner and Russell 1975; Hunt et al. 1980). Electrical stimulation coupled with hot boning tended to minimize the color differences between hot- and cold-boned cuts. Color uniformity of hot-boned muscle has been observed to be superior to cold-boned controls. This was due to the uniform pH decline of the muscle (electrical stimulation and/or uniform rate of temperature decline).

Cross and Tennent (1979A) reported that after 20 days of vacuum-packaged storage at 2°–3°C, hot-boned (1 hr postmortem) and conventionally processed (48 hr postmortem) primal cuts did not significantly differ in lean color; however, hot-boned cuts had significantly whiter fat. Conventionally processed cuts were rated more normal in shape and had greater weight losses (as purge) during storage as compared with hot-boned cuts (Cross and Tennent 1979A).

Buchter (1980) reported that hot-boned beef aged 1 week in an 80% O_2 + 20% CO_2 controlled atmosphere was slightly superior in retaining color and sensory traits as compred with conventionally processed beef aged in the same atmosphere.

MICROBIOLOGY

As is the case with any form of new technology, it is critical to examine all aspects of the process as well as the product that results from such a system. Few researchers have studied the microbiological aspects of hot boning. Whether one is dealing with conventional slaughter and chilling operations or innovative systems, such as hot boning, it is important that product integrity be maintained throughout the process. A vital portion of this integrity deals with the microbiology of the processing system from the natural microflora of the live animal to the microflora that develop during processing, storage, and distribution.

A major consideration of hot boning is the significant increase in exposed surface area available for cross-contamination as compared with conventional cold-boned meat. Hot-boned meat is meat still close to body temperature, with an initial microflora that reflects its environment. After packaging, the differences in handling begin to affect the microflora. With conventional cold boning, the heterogeneous population begins to change quickly as the carcass is chilled. There are combined effects of reduced temperature and surface desiccation. These conditions tend to favor the development of psychrotrophic microorganisms. With hot-boned meat, there is no comparable chilling period; hot meat is placed directly into bags and/or boxes within 3 to 4 hr postmortem. The hot meat is placed in a cooler or freezer; thus, there is no 24–48 hr of cold-temperature selection and desiccation.

In evaluating the effects of hot boning systems on the microflora, one must consider the effects of electrical stimulation, the packaging system, size of cut, and the temperature profile. Kotula (1981) reviewed the available research on the effects of electrical stimulation and growth of microorganisms and concluded that there was little or no influence of electrical stimulation on resident microflora.

Falk and Henrickson (1974) compared hot-boned ground beef with conventionally processed ground beef and found hot-boned ground beef to have slightly higher bacterial counts than conventionally processed beef but concluded the differences were not large enough for hot boning to be detrimental to shelf-life. Schmidt and Gilbert (1970) reported that conventionally processed wholesale cuts had a surface bacteria count of less than $10^3/cm^2$, whereas hot-boned wholesale cuts had a surface bacteria count of $10^2/cm^2$ to $10^5/cm^2$. They concluded that acceptable meat (as evaluated by a sensory panel) having satisfactory microbiological standards could be produced when primal cuts were hot boned, vacuum packaged, and conditioned for 48 hr at 15°C.

Fung et al. (1980) reported that mesophilic and psychrotrophic bacterial counts of hot-boned and conventionally processed beef were low (log $0.2/cm^2$) at 0 time, but after 14 days of vacuum packaged storage at 2.2°C, hot-boned cuts had higher microbial counts than conventionally processed cuts. Mesophilic and psychrotropic counts of hot-boned cuts were log $5.26/cm^2$ and log $5.15/cm^2$, respectively, and log $4.64/cm^2$ and log $4.43/cm^2$, respectively, for conventionally processed cuts (Fung et al. 1980). In the study by Fung et al. (1980), hot-boned cuts were vacuum packaged and boxed prior to chilling. This resulted in a slower chilling rate for hot-boned cuts which could have contributed to higher microbial loads and subsequently increased growth of bacteria in cold storage (Fung et al. 1980). Emswiler and Kotula (1979) reported that aerobic plate counts (APC) of ground beef made from hot-boned beef (2 hr postmortem) were either significantly lower or not significantly different from APC of ground beef made from conventionally processed beef. No significant differences in Most Probable Numbers (MPN) of coliforms and *Escherichia coli* were found between ground beef made from hot-boned and conventionally processed beef (Emswiler and Kotula 1979). They concluded that bacterial quality of ground beef made from hot-boned carcasses does not limit and might enhance the feasibility of boning carcasses before chilling.

Temperature Control

A major barrier to the adoption of hot boning is the uncertainty regarding the cooling procedures necessary to maintain the microbiological integrity of hot-boned meat. Herbert and Smith (1980) reported on the refrigeration requirements to meet the microbial demands of hot-boned meat. These workers sought to define temperature and microbiological parameters that would enable processors to chill and freeze beef while avoiding excessive bacterial growth. Herbert and Smith (1980) based much of their work on the observation of Meynall (1958) that rapid cooling to below 8°C of blended meat samples on which bacteria are growing results in a substantial decline in numbers. This work led Herbert and Smith (1980) to recommend that hot-boned meat be cooled to 8°C or below within 4 hr of boning when the initial temperature of the boned meat is 40°C. Their recommendation was 6 hr when initial temperature was 30°C and 9.5 hr for 20°C.

It can be concluded, based on the presently available data, that the practice of hot boning with or without electrical stimulation does not alter the microbiological quality of the resultant products. The major concern to processors and merchandisers is the temperature profile or history of products that originate as hot-boned meat.

COOKING PRERIGOR MUSCLE

The precooking of hot-boned muscle can result in considerable energy savings, particularly if little or no heating is required before serving (Berry et al. 1980). Cooked prerigor meat has been found to be more tender than cooked postrigor meat (Ramsbottom and Strandine 1949; Paul et al. 1952; Pearson 1971). Weidemann et al. (1967) and Cia and Marsh (1976) found similar results and reported that the immediate cooking of prerigor muscle eliminated the occurrence of rigor mortis and its detrimental effects on tenderness.

Cia and Marsh (1976) cooked sternomandibularis muscle at various stages of rigor mortis and concluded that although prerigor muscles shortened considerably more than postrigor muscle, cooking losses for prerigor muscle were lower and tenderness ratings were higher, particularly if cooked within 3 hr of slaughter.

Weidemann et al. (1967) found that after broiling and oven roasting, prerigor muscle became more tender product and postulated that the production of supercontraction clots disrupted the protein filaments and produced the resulting tenderization. Streitel et al. (1977) found microwave cookery to tenderize prerigor beef by as much as 50% when compared with the microwave cookery of postrigor beef. The speed of heat application of microwave cookery could, perhaps, produce even better tenderization results than conventional methods of cookery (Streitel et al. 1977). They also observed the clots of coagulated proteins in cooked prerigor beef as observed by Weidemann et al. (1967) and suggested that these clots were an indication of a disruption of the muscle's internal structure.

Berry et al. (1980) reported that cooked prerigor semimembranosus and semitendinosus roasts when served as cubes had significantly higher shear force values, higher amounts of sensory panel detectable connective tissue, and lower tenderness and juiciness scores than postrigor cooked roasts. However, no significant differences in sensory characteristics were found between prerigor and postrigor semitendinosus roasts when they were evaluated in a thinly sliced form (Berry et al. 1980).

Ray et al. (1981A,B) compared hot-boned (1 hr postmortem) semitendinosus (ST) and semimembranosus (SM) muscles with their conventionally processed (7 days postmortem) counterparts. Prerigor-cooked roasts from SM and ST muscles exhibited greater shortening (27 vs 18%) than those of postrigor muscles, whereas the width of the roasts from the prerigor ST was greater than roasts from postrigor muscles (+2 vs −7%). Roasts from prerigor muscles were deeper in width (23 vs 6.5%) than those from postrigor muscles, suggesting cooked roasts from prerigor muscles were than their counterparts. Prerigor roasts from ST and SM muscles had higher cooking yields, 84 vs 78% and 86 vs 79%, respectively, than postrigor roasts. Meat from postrigor-cooked roasts was significantly more tender than prerigor meat (3.6 vs 7.4 kg/1.27 cm). Prerigor roasts required significantly less (22%) cooking time (93.2 vs 119.9 min/kg) than chilled postrigor roasts (Ray et al. 1981A,B).

UTILIZATION OF HOT BONING MEAT FOR THE PRODUCTION OF GROUND BEEF

Ground beef is a very important commodity in the United States. If advantages in functional properties, particularly water holding capacity (WHC) of prerigor meat, could be maintained by hot boning, tremendous savings could accrue. However, trimmings for the production of ground beef may have originated from carcasses from which primal cuts were hot boned and, therefore, may have undergone treatments to prevent cold shortening.

Table 5.7 illustrates the effects of these postmortem treatments on cooking loss and juiciness ratings. Most studies used carbon dioxide (CO_2) in some form to chill the hot trimmings and stored the finished product in the frozen state.

Jacobs and Sebranek (1980) compared ground beef patties made from hot-boned beef with those from conventionally processed beef. They concluded that ground beef patties made from hot-boned beef had a higher pH value, sustained less cooking loss, and were preferred by a consumer panel ($n > 100$) for tenderness, juiciness, and overall acceptability as compared with ground beef patties made from conventionally processed meat. Cross et al. (1979) concluded that ground beef made from hot-boned beef was

TABLE 5.7. Comparison of Cooking Loss and Juiciness Values of Ground Beef Prepared from Cold- and Hot-boned Meat

			Comparison[b]	
Reference	Treatments[a]	Storage	Cooking loss	Juiciness
Lester (1979)	CB (48 hr); HB (45 min); ES + HB; HTC (5 hr at 16°C) + HB; CO_2	Fresh	ND	ND
Cross and Tennent (1979B)	CB (48 hr); HB (3 hr), CO_2	Frozen (−10°C)	HB < CB	HB > CB
Jacobs and Sebranek (1980)	CB; HB (1–3 hr), CO_2	Frozen (CO_2)	HB < CB	HB > CB
Cross and Tennent (1980)	CB (24 hr), ES, HB (1 and 3 hr); CO_2	Frozen (−10°C)	HB < CB	HB > CB
Contreras et al. (1981)	CB (48 hr), ES + HB (2 hr), 3°C for 24 hr	Frozen (−26°C)	HB > CB	HB < CB

Source: Adapted from West (1983). Reprinted from Food Technology. 1983. 37:57. Copyright © by Institute of Food Technologists.
[a] CB = cold-boned. HB = hot-boned. ES = electrical stimulation. HTC = conditioning prior to boning. CO_2 = use of CO_2 snow for initial chilling.
[b] ND = no difference.

superior to ground beef made from conventionally processed beef in palatability, cooking properties, and shelf-life. Cross et al. (1979) reported that ground beef patties made from hot-boned beef were significantly more tender and juicy and lost less water during cooking than patties prepared from chilled beef. In addition, patties made from hot-boned beef had significantly less change in configuration during cooking (diameter change was less in patties made from hot-boned beef) than patties made from chilled beef (Cross et al. 1979).

Thus, preparation of ground beef from hot-boned beef appears to offer many advantages with few problems. For the maintenance of the prerigor advantages in hot-boned ground beef, rapid fabrication and freezing of the product appear necessary. Hot boning, with or without electrical stimulation, does not appear to cause detrimental changes in the physical or sensory properties of ground beef.

WHY HASN'T INDUSTRY ACCEPTED HOT BONING?

Although hot boning may have numerous economic advantages and produce meat of equal or superior quality, several problems exist that prohibit the utilization of hot boning. A decade or so ago, hot boning was found to produce beef that was less tender than conventionally processed beef, but the advent of electrical stimulation and postmortem high temperature conditioning virtually eliminated any problems in this regard. However, some commercial processors have indicated that the hot-boned primal cuts, when vacuum packaged hot, undergo a distortion in shape; Cross and Tennent (1979B), however, did not find this to be a problem.

One large problem with hot boning centers around the chilling of hot-boned vacuum-packaged cuts. If several hot-boned, vacuum-packaged primal cuts are boxed, the temperature within the box may be too high for too long a period of time. This high temperature may lead to the proliferation of spoilage bacteria or, worse yet, food poisoning microorganisms such as *Clostridium botulinum* and *Staphylococcus* sp., among others. Very little research is available on chilling methods for large volumes of boxed, hot-boned primal cuts; however, this potential problem acts as a disincentive for industry acceptance of hot boning. Some small meat processors have tried vacuum packaging of hot-boned meat and placed the vacuum packages on shelves for a period prior to boxing. This practice was considered to be very laborious, used a substantial amount of cooler space, and offered no great advantage because of energy (refrigeration) input.

Another somewhat related problem is that most conventional meat processing plants within the United States would have difficulty in introducing hot boning into their existing plants due to their original design. Major renovations in plant design would be necessary to situate boning lines nearer to the abattoir, and major changes would be necessary in refrigeration systems to accommodate boxes of hot-boned meat.

Another problem with hot boning that has limited its industrial acceptance is the inability to grade the unchilled carcass. The lack of a mechanism to quality- and yield-grade beef carcasses is perhaps the greatest single factor in preventing the U.S. industry from moving toward hot boning. The U.S. livestock and meat industry relies heavily on USDA grades as a marketing tool as do many other countries. Many feel that they cannot market their product effectively without grades.

Another problem that must be overcome in connection with hot boning concerns the dark-cutting (DFD) condition in some carcasses. If an animal prior to slaughter has been stressed sufficiently to deplete its muscle glycogen, the meat is likely to be dark and coarse textured. The lack postmortem of a sufficient quantity of muscle glycogen will result in a relatively high ultimate pH (6.0 or higher). This high pH will allow increased microbial growth, and thus reduced shelf-life. Thus, it is critical that DFD carcasses be identified prior to vacuum packaging so that the primal cuts can be marketed separately. This identification is not a problem in conventionally chilled carcasses, but potentially DFD prerigor muscle at 1–3 hr postmortem is difficult to segregate from normal muscle. Research is needed in this area to develop a means to identify these DFD carcasses at the time of hot boning.

Although the hot boning of beef has numerous industry problems regarding its acceptance, such has not been the case with pork. Pork carcasses are generally not quality graded in the United States, so that is not a problem. Hot-boned pork has been found to have exceptional emulsifying, binding, and water holding properties. Since a large proportion of pork is used in processed meat items, numerous pork processing plants utilize hot boning. One such example is breakfast sausage. Hogs can be hot boned, and the meat ground, formulated, stuffed into tubes, and rapidly chilled in a propylene glycol or supercooled solution, thereby preventing any microbial proliferation.

ECONOMIC IMPLICATIONS OF HOT BONING

Rosoff (1975) claimed that the meat industry accounts for 9% of the energy used by the entire food industry. Unger (1975) estimated that food and kindred products ranked sixth in energy use and first in labor use among all industries. The U.S. Department of Commerce (U.S. Dep. Commer. 1977) reported that within the food and kindred product group, meat packing and processing was the fifth highest user of energy. A Kansas State University study on the economics of hot boning (Erickson et al. 1980) reported that the high use of resources in meat processing basically reflects the large quantity of products involved, meat's highly perishable nature, and the comparatively long transport distances between production and consumption areas.

Kastner (1981) reported that when compared with conventional processing practices, it has been estimated that hot boning could: (1) require 40–50% less refrigeration input; (2) result in a 50–55% reduction in cooler space; (3) eliminate the need for shrouding, neckpinning, scribing, and the operations needed to support these functions; (4) reduce labor used in fabrication operations by as much as 25%; (5) decrease cooler shrinkage up to 2%; and (6) reduce product in-plant residence time. Therefore, significant savings in energy, yield, materials and supplies, labor, and interest on fixed capital and inventory may be accrued due to hot boning (Kastner 1977; Dvorak 1979; Nason 1979; Cross and Tennent 1980; Erickson et al. 1980).

Because hot boning requires the chilling of only edible meat and not excess fat and bone, a distinct savings in cooling energy should result. Erickson et al. (1980) compared conventional processing (72 hr postmortem) with hot boning (within 8 hr postmortem) and with hot boning preceded by electrical stimulation. The meat that was hot boned reduced energy usage by 32%, and hot boning coupled with electrical stimulation reduced energy usage by 42%. Henrickson and McQuiston (1977) reported that the chilling of a 270 kg carcass would require 31,500 BTUs of energy transfer to reduce it from 40° to 0°C. The edible portion of the same carcass (420 lb) would require only 22,050 BTUs to lower the same edible product to a temperature of 0°C, which is nearly a 30% reduction in energy requirement. In addition to the reduced energy requirement, hot-boned meat can move more rapidly through the packing plant's inventory. Hot boning lends itself to boning on the rail. Brasington and Hammons (1971) indicated that on-the-rail boning resulted in a higher yield of meat than did normal table cutting.

Historically, beef has been distributed in the carcass form. Due to recent changes in methods of distribution, 65 to 70% of the beef in the United States is currently distributed in the form of vacuum-packaged primal cuts. This change in the method of distribution has resulted in a decrease in transportation costs due to a reduction in space requirements and the removal of excess fat and bone prior to shipment. Henrickson et al. (1974) reported that there could be a 30 to 35% reduction in the amount of required chilling space if beef is hot boned and chilled rather than handled in the conventional manner. Henrickson (1975) reported that hot boning could reduce refrigeration costs by 78% by the removal of excess bone and fat. Henrickson and Ferguson (1977) claimed that there could be a 65% savings in transportation space if carcasses were hot boned before shipping rather than shipped as carcasses. In this regard, truckers could haul much larger quantities of product and reduce the number of trips. Electrical stimulation has a cost of operation figure of approximately 3¢ per carcass. This figure does not include the wages for the operator, cost of the stimulator, sanitation, and the space required for stimulation.

CONCLUSIONS

Hot boning yields a quality product under a variety of processing conditions, and it offers a number of processing advantages. Even so, not all the questions about hot boning have been answered. More complete evaluations of integrated systems that incorporate the presently available knowledge are needed. Combined efforts between research and commercial application personnel are needed to determine which system is best suited for today and the future.

REFERENCES

BARBE, D.D. and HENRICKSON, R.L. 1967. Bacteriology of rapid cured ham. J. Food Sci. *21*, 9.
BARBE, D.D., MANDIGO, R.W. and HENRICKSON, R.L. 1966. Bacterial flora associated with rapid-processed ham. J. Food Sci. *31*, 998.
BEHNKE, J.R. and FENNEMA, O. 1978. Quality changes in prerigor beef muscle at $-3°C$. J. Food Sci. *38*, 539.
BENDALL, J.R. 1960. Post-mortem changes in muscle. *In* The Structure and Function of Muscle III. pp. 227–272. G.H. Bourne (Editor). Academic Press, New York.
BENDALL, J.R., KETTERIDGE, C.C. and GEORGE, A.R. 1976. The electrical stimulation of beef carcasses. J. Sci. Food Agric. *27*, 1123.
BERRY, B.W. and KOTULA, A.W. 1982. Effects of electrical stimulation, temperature of boning and storage time on bacterial counts and shelflife characteristics of beef cuts. J. Food Sci. *47*, 852.
BERRY, B.W., RAY, E.E. and STIFFLER, D.M. 1980. Effects of electrical stimulation and hot boning on sensory and physical characteristics of prerigor cooked beef roasts. Proc. Annu. Meet. Eur. Meat Res. Workers, Colorado Springs, CO, *26*, I-7.
BRASINGTON, C.F. and HAMMONS, D.R. 1971. Boning carcass beef on the rail. U.S. Dep. Agric. *ARS 52-63*.
BUCHTER, L. 1980. Hot-boned and traditionally-chilled beef as raw material for the production of controlled atmospheres retail packs-comparison of sensory and microbiological properties. Proc. Annu. Meet. Eur. Meat Res. Workers, Colorado Springs, CO, *26*, I-14.
BUTLER, J.L., SMITH, G.C., SAVELL, J.W. and VANDERZANT, C. 1981. Bacterial growth in ground beef prepared from electrically stimulated and nonstimulated muscles. Appl. Environ. Microbiol. *41*, 915.
CASSENS, R.G. and NEWBOLD, R.P. 1967. Effect of temperature on the time course of rigor mortis in ox muscle. J. Food Sci. *32*, 269.
CHRYSTALL, B.B. and HAGYARD, C.L. 1976. Electrical stimulation and lamb tenderness. N.Z. J. Agric. Res. *19*, 7.
CIA, G. and MARSH, B.B. 1976. Properties of beef cooked before rigor onset. J. Food Sci. *41*, 1259.
CONTRERAS, S., HARRISON, D.L., KROPF, D.H. and KASTNER, C.L. 1981. Electrical stimulation and hot-boning: Cooking losses, sensory properties, and microbial counts of ground beef. J. Food Sci. *46*, 457.
COOK, C.F. and LANGSWORTH, R.F. 1966. The effect of preslaughter environmental temperature and post-mortem temperature upon some characteristics of ovine muscle. Shortening and pH. J. Food Sci. *31*, 497.
CROSS H.R. 1980. Optimal systems for rapid processing of beef. Proc. Annu. Meet. Eur. Meat Res. Workers, Colorado Springs, CO, *26*, H-4.

CROSS, H.R. and TENNENT, I. 1979A. The effect of electrical stimulation and postmortem boning time on sensory and cookery properties of ground beef. J. Food Sci. 46, 292.
CROSS, H.R. and TENNENT, I. 1979B. Storage properties of primal cuts of hot- and cold-boned beef. J. Food Qual. 4, 289.
CROSS, H.R. and TENNENT, I. 1980. Accelerated processing systems for USDA Choice and Good beef carcasses. J. Food Sci. 45, 765.
CROSS, H.R., BERRY, B.W. and MUSE, D. 1979. Sensory and cooking properties of ground beef prepared from hot and chilled beef carcasses. J. Food Sci. 44, 1432.
CUTHBERTSON, A. 1977. Hot boning of beef carcasses. Institute of Meat Bull., Meat and Livestock Commission, Great Britain.
DUTSON, T.R., SMITH, G.C., SAVELL, J.W. and CARPENTER, Z.L. 1980. Possible mechanisms by which electrical stimulation improves meat tenderness. Proc. Annu. Meet. Eur. Meat Res. Workers, Colorado Springs, CO, 26, II.
DVORAK, N. 1979. On the rail hot beef boning—An idea whose time has come. Proc. 21st Meat Sci. Inst., 1979, Univ. Georgia, Athens.
EMSWILER, B.S. and KOTULA, A.W. 1979. Bacteriological quality of ground beef prepared from hot and chilled beef carcasses. J. Food Prot. 42, 561.
ERICKSON, D.B., McCOY, J.H., RILEY, J.B., CHUNG, D.S., NASON, P.G., ALLEN, D.M., DIKEMAN, M.E., FUNG, D.Y.C., HUNT, M.C., KASTNER, C.L. and KROPF, D.H. 1980. Economic feasibility of hot processing beef carcasses. Kans. Agric. Exp. Stn. Bull. 639.
FALK, S.N. 1974. Feasibility of "hot" processing the bovine carcass. Ph.D. Dissertation. Oklahoma State Univ., Stillwater.
FALK, S.N. and HENRICKSON, R.L. 1974. Feasibility of hot boning the bovine carcass. Okla. Agric. Exp. St. MP-92, 145.
FOLLETT, M.J., NORMAN, G.A. and RATCLIFF, P.W. 1974. The ante-rigor excision and air cooling of beef semimembranosus muscles at temperatures between −5C and +15C. J. Food Technol. 9, 509.
FUNG, D.Y.C., KASTNER, C.L., HUNT, M.C., DIKEMAN, M.E. and KROPF, D.H. 1980. Mesophilic and psychrotrophic populations on hot-boned and conventionally processed beef. J. Food Prot. 43, 547.
GILBERT, K.V. and DAVEY, C.L. 1976. Carcass electrical stimulation and early boning of beef. N.Z. J. Agric. Res. 19, 139.
GILBERT, K.V., DAVEY, C.L. and NEWTON, K.G. 1976. Electrical stimulation and hot-boning of beef. N.Z. J. Agric. Res. 20.
GILL, C.O. 1980. Effect of electrical stimulation on meat spoilage floras. J. Food Prot. 43, 957.
HARSHAM, A. and DEATHERAGE, F. 1951. Tenderization of meat. U.S. Pat. 2,544,681.
HENRICKSON, R.L. 1975. Hot boning. Proc. Meat Ind. Res. Conf., 25.
HENRICKSON, R.L. and FERGUSON, E.J. 1977. Energy conservation in the meat industry. Energy Res. Dev. Admin., Washington, DC, Contract EY 76-S-05-5097, Prog. Rep. ORO-5097-4.
HENRICKSON, R.L. and McQUISTON, F.C. 1977. A study of hot beef boning for energy conservation. Presented before Am. Soc. Heating, Refrig. Air Conditioning Eng., Feb. 16, 1977, Chicago.
HENRICKSON, R.L., FALK, S.N. and MORRISON, R.D. 1974. Beef quality resulting from muscle boning the unchilled carcass. Proc. 4th Int. Congr. Food Sci. Technol. 4, 124.
HERBERT, L.S. and SMITH, M.G. 1980. Hot-boning of beef: Refrigeration requirements to meet microbiological demands. CSIRO Food Res. Q., 65.
HERRING, H.K., CASSENS, R.G. and BRISKEY, E.J. 1965. Sarcomere length of free and restrained bovine muscles at low temperature as related to tenderness. J. Sci. Food Agric. 16, 379.
HERRING, H.K., CASSENS, R.G., SUESS, G.G., BRUNGARDT, V.H. and BRISKEY,

E.J. 1967. Tenderness and associated characteristics of stretched and contracted bovine muscles. J. Food Sci. *32,* 317.

HUNT, M.C., KENDALL, J.L.A., DIKEMAN, M.E., KASTNER, C.L. and KROPF, D.H. 1980. Ground beef from electrically stimulated and hot-boned carcasses. 13th Annu. Meet. Midwest. Sect. Am. Soc. Anim. Sci. Abstr. *30,* 70.

JACOBS, D.K. and SEBRANEK, J.G. 1980. Use of prerigor beef for frozen ground beef patties. J. Food Sci. *45,* 648.

KASTNER, C.L. 1977. Hot processing: Update on potential energy and related economics. Proc. Meat Ind. Res. Conf. Am. Meat Inst. Found., Chicago, 43.

KASTNER, C.L. 1981. Hot boning of beef carcasses. Proc. Natl. Beef Grading Conf. Iowa State Univ., Ames, Coop. Ext. Serv. *CE-1633,* 104.

KASTNER, C.L. 1983. Optimal hot-processing systems for beef. Food Technol. (May) 96–104.

KASTNER, C.L. and RUSSELL, T.S. 1975. Characteristics of conventionally and hot-boned muscle excised at various conditioning periods. J. Food Sci. *40,* 747.

KASTNER, C.L., HENRICKSON, R.L. and MORRISON, R.D. 1973. Characteristics of hot-boned bovine muscle. J. Anim. Sci. *36,* 484.

KOTULA, A.W. 1981. Microbiology of hot-boned and electrostimulated meat. J. Food Prot. *44,* 544.

KOTULA, A.W. and EMSWILER-ROSE, B.S. 1981. Bacteriological quality of hot-boned primal cuts from electrically stimulated beef carcasses. J. Food Sci. *46,* 471.

LESTER, T.I. 1979. Comparison of ground beef quality prepared from hot and cold chilled beef. M.S. Thesis. Univ. of Florida, Gainesville.

LOCKER, R.H. 1960. Degree of muscular contraction as a factor in tenderness of beef. Food Res. *25,* 304.

LOWE, B.A. and STEWART, G.F. 1946. The cutting of the breast muscle of poultry soon after killing and its effect on tenderness after subsequent storage and cooking. Adv. Food Res. *1,* 232.

MANDIGO, R.W. and HENRICKSON, R.L. 1966. Influence of hot-processing pork carcasses on cured ham. Food Technol. *20,* 186.

MARSH, B.B. 1954. Rigor mortis in beef. J. Sci. Food Agric. *2,* 70.

MARSH, B.B. and LEET, N.G. 1966. Studies in meat tenderness. III. The effect of cold shortening on tenderness. J. Food Sci. *31,* 450.

MARSH, B.B. and THOMPSON, J.F. 1958. Rigor mortis and thaw rigor in lamb. J. Sci. Food Agric. *7,* 417.

MARSH, B., CASSENS, R.G., KAUFFMAN, R.G. and BRISKEY, E.J. 1972. Hot boning and pork tenderness. J. Food Sci. *37,* 179.

McCRAE, S.E., SECCOMBE, C.B., MARSH, B.B. and CARSE, W.A. 1971. Studies in meat tenderness. 9. The tenderness of various lamb muscles in relation to their skeletal restraint and delay before freezing. J. Food Sci. *36,* 566.

MEYNALL, G.G. 1958. The effect of sudden chilling on *Escherichia coli.* J. Gen. Microbiol. *19,* 380.

MRIGADAT, B., SMITH, G.C., DUTSON, T.R., HALL, C.C., HANNA, M.O. and VANDERZANT, C. 1980. Bacteriology of electrically stimulated and unstimulated rabbit, pork, lamb, and beef carcasses. J. Food Prot. *43,* 686.

NASON, P.G. 1979. Energy comparison of hot and cold beef processing. M.S. Thesis. Kansas State Univ., Manhattan.

OBLINGER, J.L. 1983. Microbiology of hot-boned beef. Food Technol. *37,* 86.

PAUL, P.C., BRATZLER, L.J., FARWELL, E.D. and KNIGHT, K. 1952. Studies on tenderness of beef. I. Rate of heat penetration. Food Res. *17,* 504.

PEARSON, A.M. 1971. Muscle function and postmortem changes. *In* The Science of Meat and Meat Products. J.F. Price and B.S. Schweigert (Editors). W.H. Freeman and Co., San Francisco.

PIERCE, B.N. 1977. The effect of electrical stimulation and hot boning on beef tenderness. M.S. Thesis. Oklahoma State Univ., Stillwater.

RACCACH, M. and HENRICKSON, R.L. 1978. Storage stability and bacteriological profile of refrigerated beef from electrically stimulated hot-boned carcasses. J. Food Prot. *41*, 957.

RAMSBOTTOM, J.M. and STRANDINE, E.J. 1949. Initial physical and chemical changes in beef as related to tenderness. J. Anim. Sci. *8*, 398.

RAY, E.E., STIFFLER, D.M. and BERRY, B.W. 1981A. Effects of hot-boning and cooking method upon physical changes, cooking time and losses, and tenderness of beef roasts. J. Food Sci. *45*, 769.

RAY, E.E., STIFFLER, D.M. and BERRY, B.W. 1981B. Effects of electrical stimulation and hot-boning on physical changes, cooking time and losses, and tenderness of beef roasts. J. Food Sci. *47*, 210.

ROSOFF, H.D. 1975. Here's status report of AMI energy task force. Natl. Provisioner Nov. 15, 1975, 91.

SCHMIDT, G.R. and GILBERT, K.V. 1970. The effect of muscle excision before the onset of rigor mortis on the palatability of beef. J. Food Technol. *5*, 331.

SCHMIDT, G.R. and KEMAN, S. 1974. Hot boning and vacuum packaging of eight major bovine muscles. J. Food Sci. *39*, 140.

SEIDEMAN, S.C., SMITH, G.C., DUTSON, T.R. and CARPENTER, Z.L. 1979. Physical, chemical and palatability traits of electrically stimulated, hot-boned, vacuum-packaged beef. J. Food Prot. *42*, 8.

STERN, N.J. 1980. Effect of boning, electrical stimulation, and medicated diet on the microbiological quality of lamb cuts. J. Food Sci. *45*, 1749.

STREITEL, R.H., OCKERMAN, H.W. and CAHILL, V.R. 1977. Maintenance of beef tenderness by inhibition of rigor mortis. J. Food Sci. *42*, 583.

TARRANT, P.V. 1977. The effect of hot-boning on glycolysis in beef muscle. J. Sci. Food Agric. *28*, 927.

TARRANT, P.V. and MOTHERSILL, C.J. 1977. Glycolysis and associated changes in beef carcasses. J. Sci. Food Agric. *28*, 739.

TAYLOR, A.A., SHAW, B.G. and McDOUGALL, D.B. 1980. Hot deboning beef with and without electrical stimulation. Proc. Annu. Meet. Eur. Meat Res. Workers, Colorado Springs, CO, *26*, I-3.

TRAUTMAN, J.C. 1964. The influence of delayed chilling on beef tenderness. M.S. Thesis. Oklahoma State Univ., Stillwater.

UNGER, S.G. 1975. Energy conservation goals for the food industry in the 1980's. Presented before Am. Soc. Heat., Refrig., Air Conditioning Eng., 1975, Chicago.

U.S. DEP. COMMER. 1977. Annual Survey of Manufacturers, 1976. U.S. Bureau of Census, Washington, DC.

WEIDEMANN, J.R., KAESS, G. and CARRUTHERS, L.E. 1967. Histology of pre-rigor and post-rigor ox muscle before and after cooking and its relation to tenderness. J. Food Sci. *32*, 7.

WEST, R.L. 1983. Functional properties of hot-boned meat. Food Technol. *37*, 57.

WEST, R.L. and OBLINGER, J.L. 1979. Futuristic beef processing systems. Final Rep., W.R. Grace & Co., Cryovac Div., Duncan, SC.

WILL, P.A. and HENRICKSON, R.L. 1976. The influence of delayed chilling and hot boning on tenderness of bovine muscle. J. Food Sci. *41*, 1102.

6

Scientific Basis for Electrical Stimulation

A. M. Pearson[1] and T. R. Dutson[1]

Structure of Muscle
Energy Changes in Prerigor Muscle
Basic Causes of Cold Shortening
Other Gains from Electrical Stimulation
Summary
References

The development of electrical stimulation as a means of meat tenderization dates back to the basic work of Harsham and Deatherage (1951) and the patents assigned by them to the Kroger Company. Although the process was reported to tenderize beef carcasses, it was not adopted by the meat industry. The reasons for its failure to be accepted are probably related to the fact that cold shortening of prerigor muscle was not yet recognized, and thus means of preventing its development were not deemed necessary. Furthermore, the importance of meat tenderness had not yet been fully emphasized by consumer studies. The role of meat tenderness in consumer satiety was fully realized only when large-scale consumer studies emphasized its importance. Development of large supermarkets and the ready availability of precut and packaged steaks in self-service meat display cases did not occur until the late 1940s and early 1950s; consequently, the resulting lack of meat tenderness was only beginning to be recognized at this time so no emphasis was placed on the importance of the discovery that electrical stimulation improved meat tenderness.

In recent years, research has clearly demonstrated that electrical stimulation not only improves meat tenderness but also increases the desirability of lean meat color, enhances the appearance of marbling, prevents development of the two-toned color of the ribeye known as "heat ring" in the meat trade, enhances the flavor, and improves USDA carcass grades. Savell (1979) has summarized the beneficial effects of electrical stimulation and discussed its acceptance by the meat industry. Additional details on the advantages of electrical stimulation are also discussed elsewhere in this book, especially in Chapters 4, 5, and 7.

[1]Department of Food Science and Human Nutrition, Michigan State University, East Lansing, MI.

The present discussion will center upon the mechanism by which electrical stimulation improves the quality characteristics of meat. However, first it will be necessary to review some facts on the basic structure of muscle and some of the energy changes that occur during muscle contraction and development of rigor mortis. Finally, the specific mechanisms involved in the improvement of the physical properties of meat from the action of electrical stimulation will be reviewed as far as they are understood.

STRUCTURE OF MUSCLE

The degree of structural detail observed in skeletal muscle is determined by the level of magnification, varying from the gross morphology of muscle as seen by the naked eye to the ultrastructural details revealed by the electron microscope. In order to fully understand the structural components of muscle and their involvement in the improvement by electrical stimulation, the gross structure of muscle will be described first followed by a description of some of the ultrastructural elements involved. Thus, the gross structure of muscle as observed with the naked eye will be reviewed, then that shown with light microscopy, and finally details in the ultrastructure as observed by transmission and scanning electron microscopy will be covered.

Gross Structure

The bodies of meat-producing animals are composed of about 300 anatomically distinct structural units that differ greatly in size, shape, and appearance (Sisson and Grossman 1953). Muscle is supported and surrounded by connective tissues, which give structural integrity to the muscle tissue and connect it to the skeleton in order to coordinate body movements in the intact living animal. On examination of muscle tissue in cross-section, there is a fairly thick envelope of connective tissue surrounding each individual muscle (Fig. 6.1). This thick layer of connective tissue around an entire muscle is known as the epimysium, whereas the smaller layer of connective tissue dividing the muscle into bundles or fasciculi is called the perimysium. A still finer connective tissue layer extends from the perimysium into the muscle fiber bundles and encircles each muscle fiber. This thin layer of connective tissue consists of reticular fibers carrying the blood capillaries and is called the endomysium. Although it is in close association with the muscle cell membrane or sarcolemma, it is a distinctly separate structure. The organizational relationship of the epimysium, perimysium, and endomysium is shown in Fig. 6.1.

The connective tissue layer is composed of two proteins, collagen and elastin. Both of these proteins have their own characteristic amino acid composition and can also be identified by their different staining characteristics. Collagen is by far the more abundant of these two proteins, being found not only in the fascia surrounding the epimysium, perimysium, and endomysium but also in the epithelium, skin, tendon, cartilage, teeth, and bone. It constitutes some 25% of the total protein content of the animal body, being the most abundant protein in mammalian tissue (Stryer 1981).

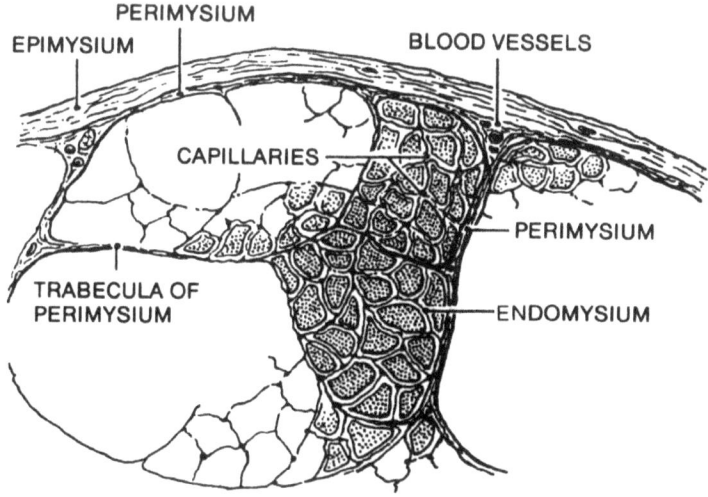

FIG. 6.1. Diagram of muscle in cross-section showing the arrangement of connective tissue into epimysium, perimysium, and endomysium and their relationship to the muscle fibers and fasciculi and the position of the blood vessels.
From Ham (1969).

Microscopic Appearance

Properly fixed and stained skeletal muscle shows both longitudinal and cross striations under the light microscope, the appearance of which has been reviewed by Birkner and Auerbach (1960). The longitudinal striations are due to the presence of long thin fibrils, approximately 2 to 3 μm in diameter, arranged parallel to each other. The cross striations are due to alternating bands along the length of each fibril. The bands that stain dark with iron hematoxylin are called A-bands or discs. Areas along the fibril do not take the stain and occur at regular intervals. These areas are known as I-bands. The A-bands and I-bands of adjacent myofibrils generally lie next to the same type of bands in the adjoining myofibrils, which makes the composite muscle fiber appear to be cross-striated or banded.

Each I-band also contains a narrow stainable band in its center, which is called the Z-line. These structures are shown in Fig. 6.2, which clearly labels each of the bands and the functional unit of the myofibril, called the sarcomere.

Sarcomeres occur as repeating units along the entire length of the muscle fiber (Fig. 6.2) and vary in the width of the I-band as a consequence of muscle contraction. During muscle contraction, the I-band becomes narrower, and upon relaxation it again becomes wider. Thus, the adjacent Z-lines (sarcomeres) in the same muscle fiber are closer together during muscle contraction and, upon relaxation, again move farther apart. In fact, sarcomere length is commonly measured as an index of the degree of muscle contraction. Locker (1960) reported that the sarcomere lengths of beef muscle varied from 2.4–3.7 μm in the relaxed state to 0.7–1.5 μm in the contracted condition. These extreme differences in sarcomere length cover a somewhat wider range than is normally found in beef muscle. Herring et al. (1965) reported sarcomere length to be influenced by the degree of tension placed on the individual muscles by their skeletal attachments. The sarcomere lengths of carcasses suspended vertically by conventional hanging by the Achilles tendon varied from 1.8 to 3.6 μm, a range of 1.8 μm. On the other hand, the sarcomere lengths for muscles from the opposite side of the same carcasses, which were laid horizontally on a flat surface, varied from 2.0 to 2.7 μm, a range of only 0.7 μm. Herring et al. (1965) demonstrated that, when muscles shorten, there is a corresponding

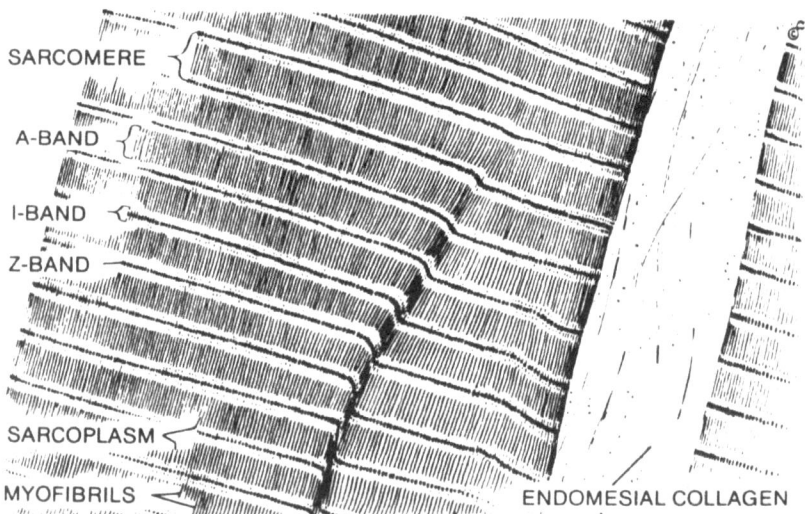

FIG. 6.2. Diagrammatic sketch showing the cross-banding of the myofibrils in the muscle fibers.
From Cassens (1971).

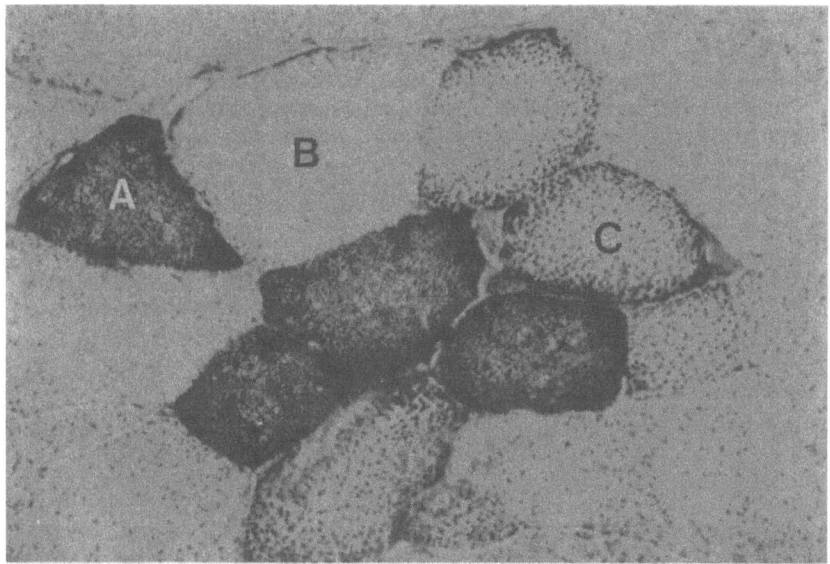

FIG. 6.3. Photomicrograph of cross-section of pig longissimus muscle showing red (A), white (B), and intermediate (C) muscle fibers stained with diphosphopyridine nucleotide tetrazolium reductase (DPNH-TR). Note how the red fibers tend to be clumped together in groups (×1605).
From Cassens (1971).

decrease in sarcomere length and an increase in fiber diameter, which is accompanied by a decrease in tenderness.

The characteristic banding pattern observed by light microscopy is a result of the ordered array of the principal myofibrillar proteins, especially of actin and myosin. As will be shown later, the I-band contains the thin or actin filaments on each side of the dark-staining Z-disc. The relatively wide A-band, which has a dark appearance on staining with hematoxylin, is composed of the thick or myosin filaments with some overlapping of the thin filaments near the junction with the I-band at each end of the thick filaments. Thus, the striated appearance of skeletal muscle is due to the ordered arrangement of the thick and thin filaments in each muscle fiber. More details on this arrangement become clear upon examination of skeletal muscle with the transmission electron microscope.

Most muscles are composed of red and white fibers with some mixed fiber types, which are a gradation between the two. The fiber types can be recognized on the basis of their staining characteristics, depending upon whether they are predominantly oxidative or glycolytic. Figure 6.3 shows a photomicrograph in which the red fibers from pig muscle are stained with diphosphopyridine nucleotide tetrazolium reductase (DPNH-TR).

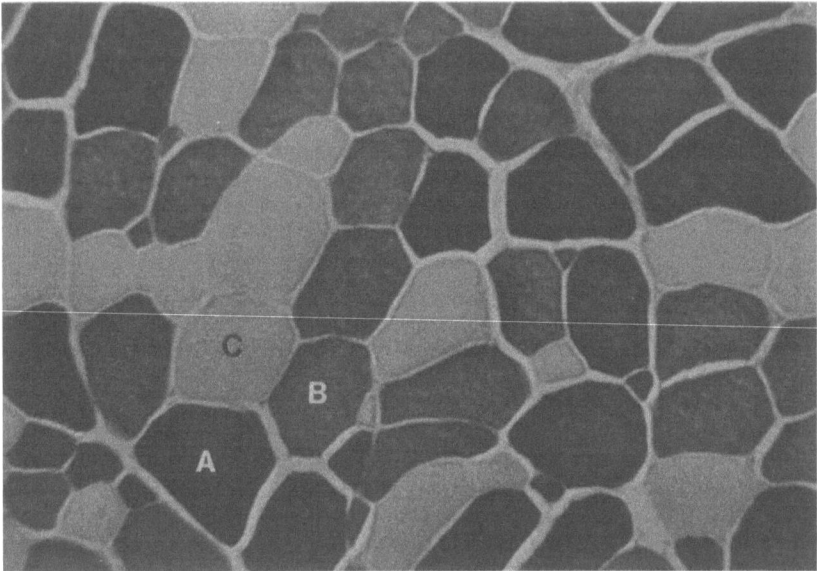

FIG. 6.4. Photomicrograph of bovine longissimus muscle in cross-section showing red (C), white (A), and intermediate (B) muscle fibers stained for alkaline-stable ATPase activity. Note the checkerboard pattern with the lack of clumping into groups of the same fiber types (×136).

The light colored or unstained fibers are presumed to be white fibers and would be expected to stain with glycolytic stains, such as phosphorylase. Under normal staining conditions, however, some of the fibers stain in varying degrees with both oxidative and glycolytic stains, and these fibers are classified as being intermediate. This classification is simplistic but will suffice for the purposes of this discussion.

Figure 6.4 is a photomicrograph showing red (light), white (dark), and intermediate (less dark) fibers from bovine muscle stained for alkaline-stable ATPase activity. The dark-staining fibers would be classified as α-white, whereas, the light colored fibers would be β-red and the less dark ones α-red fibers, using the classification system of Ashmore and Doerr (1971). This illustrates that the staining characteristics of the fibers are dependent upon the particular stain used, with the apparent darkness of the individual fiber being proportional to the extent of the reaction with the dye. Greater detail and more elaborate classification systems have been discussed elsewhere (Dubowitz and Pearse 1960; Stein and Padykula 1962; Beecher 1966; Beatty et al. 1967).

The arrangement of the fibers into groups with similar staining characteristics is found in some species, such as the pig (Fig. 6.3), where the red fibers occur in discrete clumps (Cassens and Cooper 1971). The fiber arrangement in bovine and rabbit muscle occurs in the familiar checkerboard pattern with the red and white fibers being intermixed throughout the muscles (Fig. 6.4).

Red fibers contract more slowly than white fibers and, thus, are called slow and fast, respectively. White fibers have narrower, less dense Z-lines, longer but less numerous mitochondria, and a more extensively developed sarcoplasmic reticulum than red fibers (Gauthier 1970; Dutson *et al.* 1974). These differences in structure, which can be observed by light microscopy, are important factors in the development of cold shortening. Both the mitochondria and the sarcoplasmic reticulum play important roles in calcium binding and release such that the fundamental differences in structure help to explain the differential responses between red and white muscles to cold temperatures. A more detailed explanation of differences occurring in cold shortening on the basis of structure will be given later.

Electron Microscopic Appearance

Figure 6.5 presents a diagrammatic sketch of a single skeletal muscle fiber and some associated structures. The covering of endomysial connective tissue is stripped back from one end in order to reveal the sarcolemma or muscle cell membrane underneath and to demonstrate how it encloses the muscle fiber. The nuclei are clearly shown and lie in the immediate proximity of the cell membrane, in contrast to most other cells in which the nuclei are centrally located in the cell. Another characteristic of the muscle fiber is the large number of nuclei, which may number several

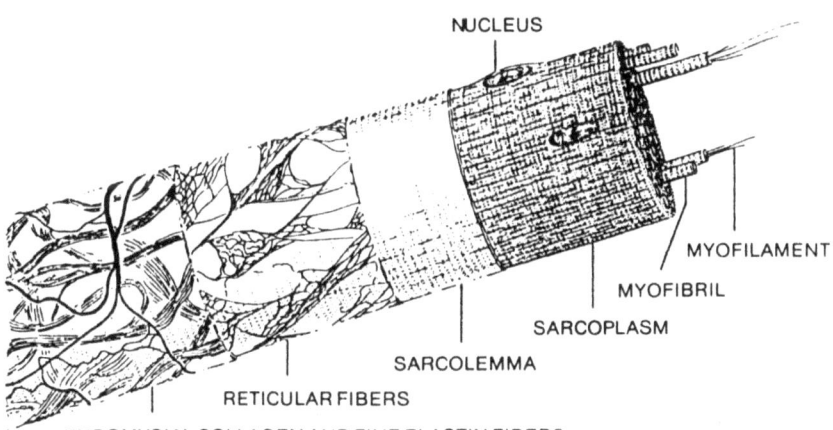

FIG. 6.5. Diagrammatic sketch of a skeletal muscle fiber showing some associated structural components.
From Cassens (1971).

hundred in an average sized myofiber (Smith and Copenhaver 1948). This is in contrast to most other cells, including smooth muscle fibers, which contain only one nucleus per cell except when in the process of mitosis or cell division.

The diagrammatic representation of the muscle fiber and its structures shown in Fig. 6.5 reveals the sarcolemma as well as the endomysium, both of which can be seen with the light microscope under favorable conditions. Each muscle fiber is composed of groups of myofibrils surrounded by the sarcoplasm as shown in Fig. 6.5. Differences in the uptake of the stains by the myofibrils and sarcoplasm, which is referred to as the fields of Cohnheim, result in a characteristic mottled appearance under the light microscope. However, differentiation of the individual myofibrils into myofilaments is possible only with electron microscopy. As illustrated, each myofibril contains several myofilaments in which the muscle proteins are arranged in a regularly ordered manner. This ordered array of the contractile proteins results in the cross-striated pattern that is characteristic of skeletal muscle as shown in Fig. 6.2.

Figure 6.6 presents a transmission electron micrograph of skeletal muscle showing its structural organization in the relaxed condition. The dis-

FIG. 6.6. Electron micrograph showing portions of seven myofibrils sectioned longitudinally from noncontracted rabbit psoas muscle. The dark A-bands are bisected by the lighter H-bands, which contain the thin dark M-bands. The I-bands are bisected by the thick dark Z-lines, with a single sarcomere extending to the next Z-line in the same myofibril (×5184).
Courtesy of Dr. H. E. Huxley.

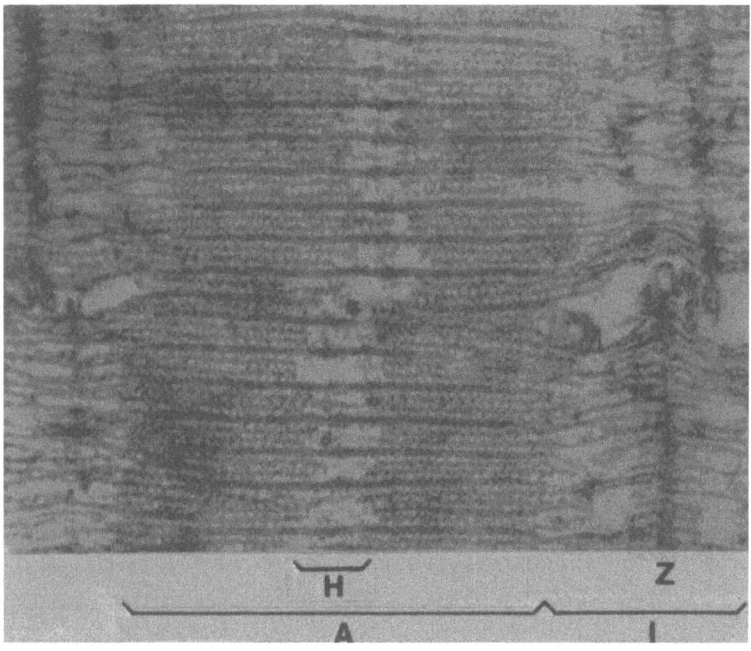

FIG. 6.7. Electron micrograph of relaxed rabbit psoas muscle in longitudinal section. The thick myosin filaments extend throughout the length of the A-band, whereas the thin actin filaments are located in the I-band and extend into the M-band to the point where the H-band is located. Two actin filaments are visible between each two myosin filaments. Cross-bridges between the actin and myosin filaments are clearly discernible (×59,500).
Courtesy of Dr. H. E. Huxley.

tance between two adjacent Z-lines in the same myofibril represents a sarcomere and is the basic structural repeating unit involved in muscle contraction. The A-band is clearly discernible and is composed of the thick or myosin filaments with some overlapping with the thin or actin filaments. The lighter H-band in the middle of each A-band is also easily identified and is characterized by the dark line in the center which is called the M-line. The I-band, shown on each side of and including the Z-line, contains the thin filaments that are composed of F-actin and various regulatory proteins. In muscle contraction, the thin filaments slide past the thick filaments as explained by Hanson and Huxley (1955). This results in the thin filaments sliding into the A-band, thus narrowing the I-bands. Under extreme contraction, the I-band almost completely disappears as the thin filaments slide between the thick filaments into the A-band. As a result of the sliding action, the distance between Z-lines shortens. Thus, a single sarcomere becomes much shorter.

Figure 6.7 presents an electron micrograph showing a further enlargement of a longitudinal skeletal muscle section. More detail is shown of the muscle structure. The thin actin filaments not only can be seen in the I-band but extend into the A-band as far as the H-band. Thus, the lighter appearance of the H-band is due in part to the fact that only the thick filaments are present in this area. The darker appearance in the center of the H-zone is the result of the darker staining M-line, which appears to be associated with two or more specialized connecting proteins in this area (Porzio *et al.* 1979). The less dense staining area immediately adjacent to the M-line is related to the absence of cross-bridges on the thick filaments in this area and is called the pseudo H-zone. Examination of Fig. 6.7 also reveals the presence of two actin filaments between every two myosin filaments.

Figure 6.8 is an electron micrograph of skeletal muscle in cross-section, showing the hexagonal array of six thin filaments around each thick filament at the point of overlap in the A-band on either side of the H-zone. In cross-section at the H-band, only the thick filaments can be seen as there is no overlap of thick and thin filaments in this area. Similarly, only the actin filaments are present in the I-band, since thick filaments do not extend into the I-band. The cross-bridges between actin and myosin are found only at the point of overlap in the A-bands. As already indicated, the

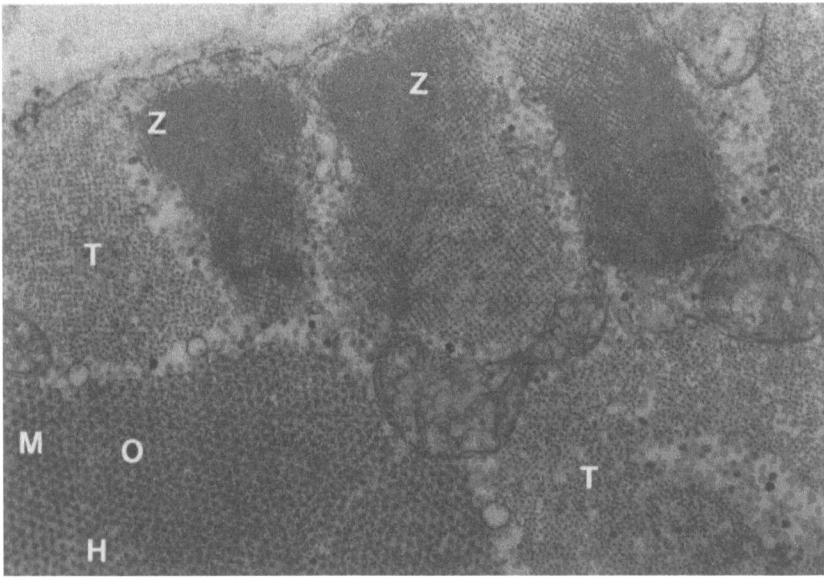

FIG. 6.8. Cross-section of bovine longissimus myofibrils showing the Z-line (Z), thin filaments (T), M-line (M), the overlap area (O) showing both thick and thin filaments, and the H-zone (H) containing only thick filaments (×42,780).

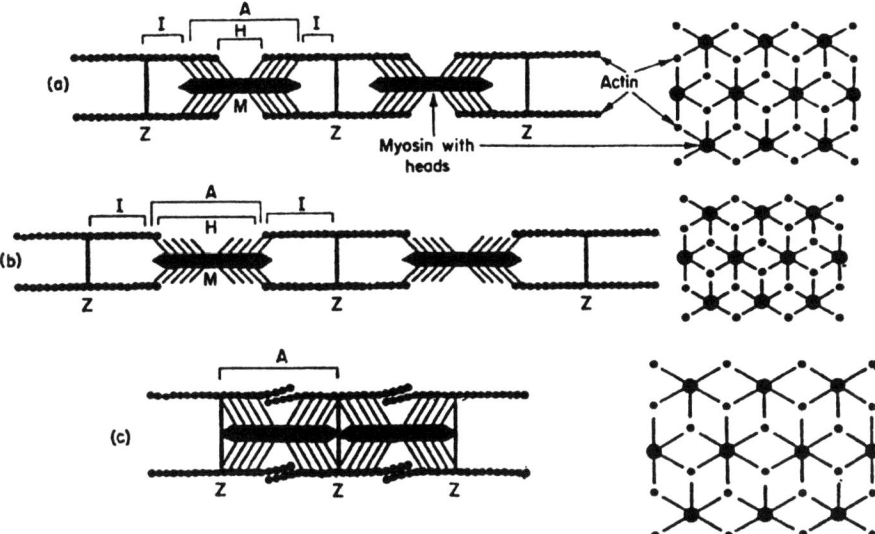

FIG. 6.9. Diagram of the fine structure of a sarcomere showing the actin and myosin filaments in longitudinal (left) and cross-sectional (right) views. Diagram (a) shows a sarcomere near its rest length, (b) represents a stretched sarcomere, and (c) shows a contracted sarcomere.
From Bendall (1969).

sliding of the actin filaments during contraction greatly narrows the area in the I-band and also decreases the area in the H-band. The net effect is to shorten the sarcomere length or the distance between Z-lines in the same sarcomere.

Figure 6.9 diagrammatically illustrates the changes that occur in the actin and myosin filament banding pattern during relaxation and contraction of skeletal muscle. The hexagonal arrangement of the six actin filaments around each myosin filament is more easily visualized when viewed in cross-section. Although the A-band is unaltered during contraction, the I-band is virtually absent on contraction. The cross-bridges on the myosin molecules are clearly evident, and during contraction a greater proportion become attached to the sites for interaction on the actin molecules. These interaction sites play an important role during contraction, serving as the points for moving the thin past the thick filaments (Huxley 1965). Although the cross-bridges are all attached in the illustration shown of contracted muscle in Fig. 6.9, only a small portion of the available cross-bridges are engaged at any one time. Furthermore, the cross-bridges can be engaged, disengaged, and reengaged during contraction, thus pulling the actin filaments past the myosin filaments.

ENERGY CHANGES IN PRERIGOR MUSCLE

The structure of muscle plays an important role in meat tenderness, with the changes in sarcomere length being central to these alterations. The sliding of the thin filaments between the thick filaments during either contraction or relaxation is an important determinant in meat toughness or tenderness, respectively. The movement of the thin filament requires energy and is responsible for the changes occurring in sarcomere length of prerigor postmortem muscle. The degree of relaxation or contraction of prerigor muscle is closely related to the status of the high energy phosphate compounds and the control mechanisms involved in their synthesis and degradation (Nauss and Davies 1966; Bendall 1969).

The cross-bridges of myosin provide the sites for interaction with actin and thus, are involved in the contraction-relaxation cycle in muscle. The energy for the movement is provided by ATP (adenosine triphosphate), which is broken down to release energy as shown:

$$\text{ATP} + \text{stimulus} \xrightarrow{\text{ATPase}} \text{energy} + \text{ADP} + P_i$$

The ADP (adenosine diphosphate) formed in this reaction, which still contains one high-energy bond, can be reconverted to ATP by combining with CP (creatine phosphate), which also contains one high-energy bond:

$$\text{ADP} + \text{CP} \xrightleftharpoons{\text{phosphocreatine kinase}} \text{ATP} + \text{creatine}$$

or else by combining with another mole of ADP:

$$2\text{ADP} \rightarrow \text{ATP} + \text{AMP (adenosine monophosphate)}$$

The ATPase involved in muscle contraction is also bound to the crossbridges on the myosin heads. Thus, the energy released by breakdown of ATP is readily available as a driving force for movement of the crossbridges so that the actin filaments are propelled along the myosin filaments to shorten the sarcomeres and cause muscle contraction. This movement occurs only in prerigor muscle and is characteristic of the movement that takes place on electrical stimulation. However, on electrical stimulation, the electrical impulse must be of sufficient voltage to simultaneously cause contraction in almost all of the major carcass muscles.

In the living animal, ATP is readily formed so that there is generally an ample supply for all metabolic processes, including movement through muscle contraction. The potential chemical energy of glucose can be converted to ATP by way of aerobic glycolysis and the tricarboxylic acid (TCA) cycle with its associated phosphorylation. These reactions are aerobic and result in complete oxidation of glucose to CO_2. The glucose is converted to pyruvate, which is oxidized by way of acetyl-coenzyme A, thus providing entrance into the tricarboxylic acid cycle to form 12 moles of ATP for each mole of acetyl-CoA utilized in the cycle (White et al. 1964).

Glycolysis, which is the breakdown of glucose and/or glycogen in the absence of oxygen to produce lactic acid, results in the synthesis of ATP. This occurs not only in living muscle under anaerobic conditions but also in prerigor muscle during the early postmortem state. Virtually all cells, including muscle fibers, are capable of partially oxidizing glucose under anaerobic conditions. This results in a net yield of two moles of ATP for each mole of glucose that is converted to lactic acid. This reaction is shown schematically:

$$\text{Glucose} + 2\text{ADP} + 2\text{P}_i \rightleftharpoons 2 \text{ lactate} + 2\text{ATP} + 2\text{H}_2\text{O}$$

Glycolysis provides a means for obtaining a limited supply of ATP under anaerobic conditions. It occurs not only in normal living muscle in times of need or stress but also in postmortem muscle until the source of glucose is depleted. Muscle glycogen is the major source of glucose in postmortem prerigor muscle. However, the amount is rather limited and is generally depleted fairly soon after death. If all of the muscle glycogen is not broken down to form glucose soon after death, the drop in pH associated with lactate formation inactivates the enzymes, phosphofructokinase and phosphorylase, thus inhibiting the breakdown of glycogen. Thus, glycolysis ceases either as a result of the lack of glycogen or else from the inhibition of glycogen breakdown as a result of the drop in pH associated with lactic acid formation. Regardless, available glucose is depleted and ATP is no longer generated by anaerobic glycolysis (Bodwell *et al.* 1965).

Although ATP, CP, and glycogen are generally present in muscle at the time of death, as already mentioned these sources of energy are not available to furnish energy after their depletion or as a result of inactivation of the enzymes that cause their hydrolysis as a consequence of the decrease in muscle pH. When ATP and the other sources of muscle energy (CP and glycogen) are depleted or the muscle pH falls below 6.0, the muscle enters rigor mortis. Thus, electrical stimulation is most effective in the early prerigor state while energy reserves are still high. Electrical stimulation continues to evoke contraction for about 2 hr postmortem, at which time muscle ceases to respond, apparently because the membranes lose their ability to transmit the electrical impulse to the contractile apparatus. This is the case even though the muscle has adequate ATP to support further contraction.

Figure 1.8 (Chapter 1) presents the relationships between pH, creatine phosphate, and acid-labile phosphorus and is explained by Newbold (1966). The drop in acid-labile phosphorus and the decline in CP are shown to be closely related to the pH fall, and are indicative of changes in ATP. CP falls rapidly following death, during which time the ATP concentration, which is reflected by acid-labile phosphorus, remains relatively high. After CP reaches a comparatively low level, however, the acid-labile phosphorus or ATP concentration declines rapidly. On the other hand, pH declines fairly steadily until it levels off at about 5.6 to 5.8. Extensibility decreases rapidly upon reaching pH 6.0 or lower and is related to the changes in the concentrations of CP and ATP.

BASIC CAUSES OF COLD SHORTENING

As already indicated, cold shortening is associated with meat toughness. Cold-induced toughening of meat was first reported by Locker and Hagyard (1963), who discovered that prerigor muscle shortened on exposure to cold temperatures. The influence of cold on meat toughness was later confirmed by other New Zealand researchers (Marsh and Leet 1966; Marsh et al. 1968), who demonstrated that cold shortening created a serious tenderness problem in lamb carcasses frozen in the prerigor state. It was shown that cold shortening did not occur after the onset of rigor, which led to development of a process called "conditioning and aging" to circumvent its development. In this process, it was recommended that lamb carcasses be held at 15°C for 15–20 hr in order to allow them to go into rigor before subjecting them to cold temperatures. During the holding period, the pH usually reached 5.8 or below so that the unrestrained muscles would no longer respond to cold by shortening.

Temperature Effects

The effect of temperature upon shortening of prerigor meat is shown in Fig. 1.6 (Chapter 1). Examination of the plot shows that maximum shortening occurs near 0°C and then reaches a minimum at 10° to 20°C as pointed out by Locker and Hagyard (1963). Shortening increases again on raising the temperature after attainment of the minimum but does not reach the level attained at low temperatures, even during cooking. This plot clearly demonstrates that cold has its greatest effect on muscle shortening at temperatures near freezing but has little influence at temperatures between 10° and 20°C.

Figure 1.1 (Chapter 1) presents a scatter diagram showing the relationship between shear force (a mechanical measurement indicative of tenderness) and percentage shortening which was first observed by Locker (1960). The diagram, which comes from Marsh and Leet (1966), demonstrates that maximum toughness occurs at about 40% shortening, whereas up to 20% shortening has little influence upon tenderness. Above 40% shortening, however, the meat again begins to improve in tenderness until at somewhere between 55 and 60% shortening, it is again as tender as it was at 20% shortening.

Marsh and Carse (1974) have presented evidence showing that at 35–40% shortening, which produces maximum toughening in muscle, the thick filaments penetrate the Z-discs and may interact with actin filaments in adjacent sarcomeres to form a continuum of myosin throughout. They concluded that this structure is responsible for the increase in toughness of cold-shortened meat. In contrast, Voyle (1969) concluded that there is an increase in the proportion of actively contracting fibers in cold-shortened muscle, which he suggested is responsible for the increased toughness. Marsh et al. (1974) have demonstrated that shortening beyond

40% produces an increasing number of contracture bands. This causes tearing of the muscle, especially in the proximity of the Z-disc, which they theorized is responsible for the increase in tenderness associated with massive muscle shortening. Massive muscle shortening, however, does not appear to occur in carcasses or cuts under practical chilling conditions, but if, in fact, it did occur, it would be desirable. Thus, the real problem is cold shortening with the related toughening of the meat.

Locker and Hagyard (1963) demonstrated that the temperature effects are reversible during the early-postmortem state. They found that pre-rigor muscle strips would shorten at 0°C, but on placing the already-contracted muscle at 37°C, shortening was reversed and the muscle again relaxed. The degree of relaxation depended upon the time it had been held at the low temperature, although it never completely relaxed to its original length.

Effects of pH

Normally, muscle pH begins to fall immediately after death and continues to decline until rigor mortis is complete. Completion of rigor mortis in beef and lamb carcasses usually requires about 16–18 hr at 0°–4°C. By this time, the major muscles of the carcass have declined to about pH 5.7 and are no longer responsive to cold-induced shortening and toughening. Higher temperatures will accelerate glycolysis and speed up the decline in pH so that rigor will set in earlier, as shown in Fig. 1.8 (Chapter 1).

Electrical stimulation, which was first shown to prevent cold shortening by Carse (1973), will greatly accelerate glycolysis and speed up the pH drop. Bendall (1976) reported that electrically-stimulated lamb carcasses went into rigor mortis earlier, reaching pH 5.7 in about 3 hr. The differences in pH between electrically-stimulated and unstimulated sides from beef carcasses are shown in Fig. 6.10. The rapid acceleration of the pH decline in the stimulated carcasses is clearly evident in comparing their pH values with those for unstimulated control sides. On the basis of pH differences, Bendall et al. (1976) concluded that electrical stimulation would permit rapid chilling or freezing of lamb carcasses within 3 hr of slaughter without any problems from cold shortening. However, unstimulated carcasses would require at least a 10-hr holding period in order to prevent cold shortening.

Working on electrical stimulation of beef carcasses, Bendall et al. (1976) noted that 50% of the ATP had disappeared at pH 6.0 and that 90% was depleted by the time the carcass pH reached 5.7. Within 1 hr of electrical stimulation, the pH of the major muscles had dropped to 6.0 and by 2½ hr had reached pH 5.7. Furthermore, the muscle temperatures of the stimulated sides were much higher, averaging 36°C compared with 27°C for similar unstimulated sides. The higher temperature of the stimulated sides is, no doubt, an important factor in acceleration of glycolysis, thereby permitting earlier chilling. In addition, the higher temperature of the

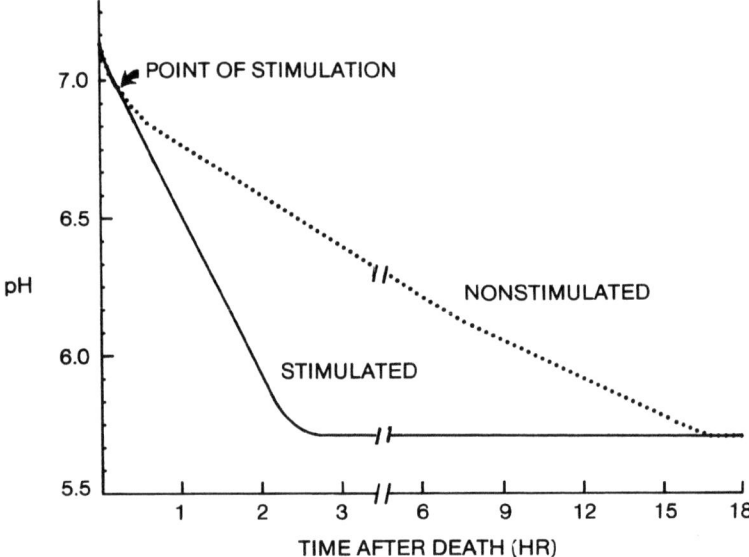

FIG. 6.10. Differences in rate of pH decline in electrically-stimulated and unstimulated control beef sides from the same carcasses.

electrically stimulated sides may contribute to faster aging of the meat, which will be discussed in more detail later.

Bendall et al. (1976) have suggested that electrical stimulation will allow rapid chilling of beef carcasses within an hour of application, which would require but little delay before rapid chilling or freezing. Other workers using different voltages and times after death before electrical stimulation have found the pH drop to be somewhat slower (Bouton et al. 1980; Chrystall et al. 1980; George et al. 1980). However, the pH drop in most studies was rapid enough to permit chilling within 2 hr of stimulation and seems to be adequate to prevent cold shortening (Fig. 6.10).

Chrystall et al. (1980) have shown that the nervous system begins to lose its responsiveness to electric current approximately 30 min following death. Thus, direct stimulation is necessary to accelerate glycolysis if stimulation is delayed, and consequently there are advantages in stimulating the carcasses soon after death. If stimulation is delayed, higher voltages are necessary to depolarize the muscle membranes so that those muscles remote from the site of stimulation receive the necessary stimulus to accelerate glycolytic changes in the muscles.

Although muscle without any glycogen reserves will go into rigor, the acceleration of glycolysis and the associated rapid pH decline do not occur. Under these conditions, muscle develops "alkaline rigor," which was first described by Claude Bernard in 1877. Dutson et al. (1982) have recently demonstrated that electrical stimulation of carcasses from stressed beef cattle does not improve muscle color, tenderness, or any of the other qualitative measurements of meat desirability. In fact, the flavor scores for the steaks from all stressed carcasses were not acceptable, which was true for both the stimulated and unstimulated sides. The results demonstrate that the pH decline associated with glycolysis is necessary to develop both flavor and tenderness in meat. Electrical stimulation hastens glycolysis with the associated pH drop that is responsible for more rapid development of these desirable attributes in meat.

Role of Relaxing Factor

Stimulation of muscle causes contraction, while withdrawal of the stimulus allows relaxation. Contraction was early recognized to be the result of a nervous impulse to the muscle or muscles involved. However, the mechanism of relaxation was not known until Marsh (1952A,B) discovered a factor causing relaxation in the low-speed centrifugates of muscle fibrils. On adding this "relaxing factor" back to the myofibrils in the presence of ATP and Mg^{2+} ions, the factor completely inhibited the ATPase activity, thereby preventing contraction. It was later shown that addition of traces of Ca^{2+} ions ($1 \times 10^{-4}\,M$) back to the fibrils reversed the process, causing contraction (Marsh 1952B). This suggested that the relaxing factor functioned by removal of Ca^{2+} from the system. Other researchers (Portzehl 1957; Ebashi 1961; Hasselbach and Makinose 1962) demonstrated that the relaxing factor came from the breakdown of the sarcoplasmic reticulum and exerted its action by removal of Ca^{2+} ions from the system. Weber and Herz (1962) then demonstrated with Ca^{2+} chelators that traces of calcium are necessary for splitting of ATP and contraction of muscle filaments.

In summary, relaxation of muscle takes place since the Ca^{2+} ions are normally pumped out of the contractile system and bound by the sarcoplasmic reticulum. Contraction is initiated by a nervous stimulus, which depolarizes the fiber membrane or sarcolemma. The sarcolemma is contiguous with the T-tubule system, which carries the impulse to the myofibrils. The impulse is then transferred from the T-tubules to the terminal cisternae of the sarcoplasmic reticulum. At this point, Ca^{2+} ions are released from the sarcoplasmic reticulum into the sarcoplasm and immediately come into contact with troponin-C. Calcium binding to troponin-C abolishes the interaction between troponin-I and actin, which allows the troponin–tropomyosin complex to rotate, thus exposing the site for interaction between actin and myosin and the site of ATPase activity on the myosin head. ATP interacts with the ATPase and is split, releasing the energy that is necessary for muscle contraction. Almost simultaneously, the myosin cross-bridges engage with actin, and the sliding of the thin filaments past the thick filaments occurs (Pearson 1984).

Factors Influencing Ca^{2+} Binding and Release. The binding and release of Ca^{2+} ions not only play an important role in muscle contraction and relaxation, but Ca^{2+} binding is also involved in rigor mortis. In the case of rigor development, the attachment of the cross-bridges of the myosin molecule to the binding site on actin is not reversible since there is no ATP available. Thus, the Ca^{2+} cannot be reaccumulated by the sarcoplasmic reticulum in the absence of ATP, and the muscles become inextensible. Electrical stimulation hastens the development of rigor mortis as already pointed out, thereby permitting chilling without development of cold shortening. In prerigor meat, however, both Ca^{2+} binding and release can occur, with the release of Ca^{2+} being responsible for cold shortening.

Mitochondria can also bind Ca^{2+} ions. Buege and Marsh (1975) have suggested that anoxic muscle mitochondria release Ca^{2+} ions at low temperatures and may be capable of initiating cold shortening. This theory was evolved on demonstrating that mitochondrial uncouplers enhanced shortening in chilled muscle under anaerobic conditions, whereas the presence of oxygen inhibited shortening. This viewpoint was supported by the fact that muscles containing a high proportion of red fibers (beef and sheep) are subject to cold shortening, but muscles high in white fibers (rabbit and to a lesser extent pig) are not subject to cold shortening. Furthermore, white muscles contain fewer mitochondria and a more extensively developed sarcoplasmic reticulum than red muscles. Therefore, white muscles are capable of recapturing and binding any excess Ca^{2+} ions spilled by the mitochondria under the influence of cold. These two factors together help to explain why red muscles are subject to cold shortening, whereas white muscles do not cold shorten (Cornforth et al. 1980).

The effects of temperature and pH upon Ca^{2+} accumulation and release by beef muscle mitochondria are presented in Fig. 6.11. Examination of the plot reveals that chilling initiates the release of Ca^{2+} ions by the mitochondrial preparations. On lowering the temperature from 37° to 0°C, the beef mitochondria released 13% of their initial Ca^{2+} load. On the other hand, raising the temperature from 0° to 37°C more than doubled the amount of Ca^{2+} accumulated by the beef muscle mitochondria. Lowering the pH from 7.3 to 5.0 resulted in an 80% decrease in Ca^{2+} binding by the mitochondria. Thus, results demonstrate that both low temperatures and low pH decrease the ability of mitochondria to bind Ca^{2+} ions and are important contributors to development of cold shortening.

The influence of pH and temperature upon Ca^{2+} accumulation and release by beef sarcoplasmic reticulum vesicles is presented in Fig. 6.12. Examination of the plot clearly demonstrates the great capacity of the sarcoplasmic reticulum vesicles from beef muscles to accumulate large amounts of Ca^{2+} ions at 37°C. On the other hand, holding the vesicles at 0°C greatly limited their ability to accumulate Ca^{2+} ions. Raising the temperature from 0° to 37°C markedly increased the ability of the sarcoplasmic reticulum to accumulate Ca^{2+}. Lowering the pH from 7.3 to 5.0 virtually abolished the binding of Ca^{2+} ions by the sarcoplasmic reticular

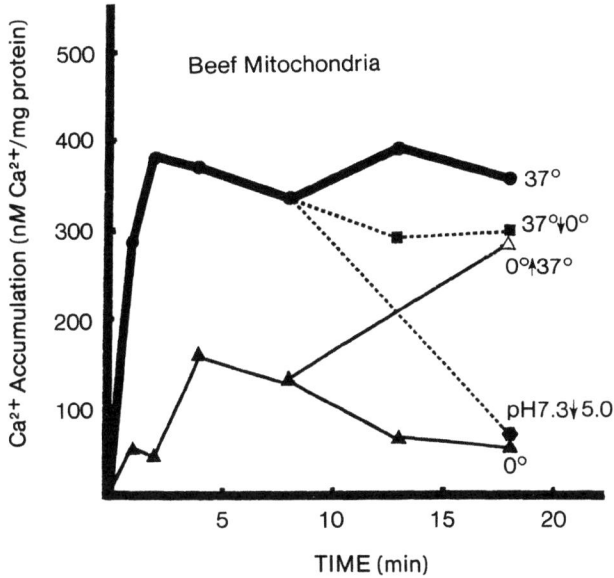

FIG. 6.11. Effects of temperature (0° vs 37°C) and pH (7.3 and 7.3 → 5.0) upon calcium accumulation and release by beef muscle mitochondria.
From Cornforth et al. (1980).

vesicles. Results clearly show that both low temperature and low pH markedly decrease the Ca^{2+} binding by beef muscle sarcoplasmic reticulum. This indicated that cold temperatures and the drop of pH in postmortem muscle influence Ca^{2+} binding by the sarcoplasmic reticulum and play an important role in cold shortening.

Mechanism of Cold Shortening

Cornforth *et al.* (1980) demonstrated that both sarcoplasmic reticulum vesicles and mitochondrial preparations isolated from red (beef) and white (rabbit) muscles respond in essentially the same manner and to the same degree to temperature and pH changes. Intact red muscles, however, cold shorten, whereas white muscles do not shorten on exposure to cold temperatures. Apparently, the cause of cold shortening is related to the relative concentrations of mitochondrial and sarcoplasmic membranes in red and white muscles.

The fundamental differences in red and white muscles are due to a high concentration of mitochondria in the red muscles. Under the influence of cold temperatures, the mitochondria spill large amounts of Ca^{2+} into the

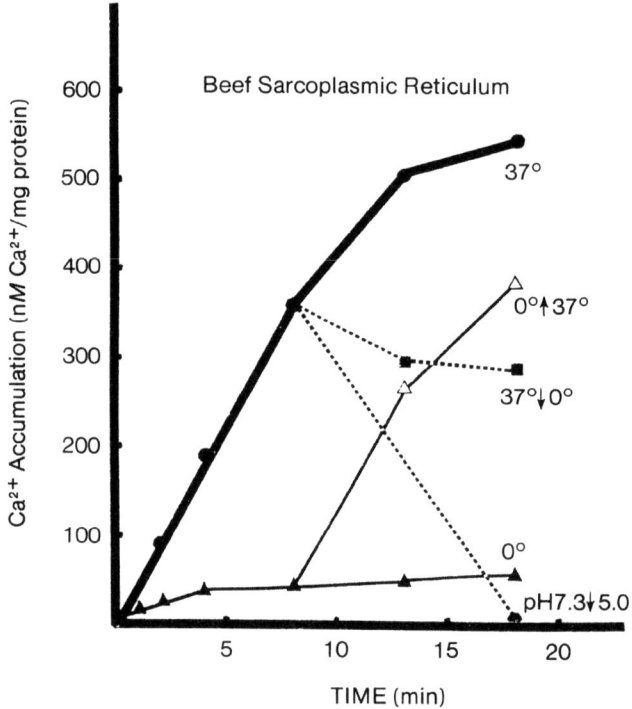

FIG. 6.12. Influence of pH (7.3 → 5.0) and temperature (37° → 0°C) upon calcium accumulation and release by beef sarcoplasmic reticulum vesicles.
From Cornforth et al. (1980).

intracellular spaces. Since red muscles also contain fewer sarcoplasmic reticular membranes, they become readily overloaded with Ca^{2+}. The excess amounts of Ca^{2+} cannot be sequestered, and the free Ca^{2+} causes shortening in the red muscles. In white muscles, there are considerably fewer mitochondria so that fewer Ca^{2+} ions are released. Those Ca^{2+} ions released are also more readily recaptured by the more abundant sarcoplasmic reticular membranes present in white muscle. Thus, white muscles are not subject to cold shortening, but red muscles readily cold shorten.

Further support for this viewpoint is obtained from the work of Weiner and Pearson (1966), who demonstrated that EDTA (a Ca^{2+} ion chelator) prevents cold shortening in prerigor muscle. It has also been shown that injection of Ca^{2+} ions into rabbit and sheep muscle will cause muscle shortening, even in the absence of cold temperatures (Weiner and Pearson

1969; Pearson et al. 1973). Muscles treated with Ca^{2+} ions in this manner also are tougher, as indicated by greater shear force values, thus further supporting the role of Ca^{2+} ions in shortening and the accompanying toughening of meat. These studies suggest that cold temperatures result in the release of a sufficient quantity of Ca^{2+} from the sarcoplasmic reticular membranes to initiate cold shortening in prerigor red muscle, a theory first proposed by Pearson et al. (1973) and later supported by Davey and Gilbert (1974).

Role of ATP

Another difference is that white muscles have more ATP available than red muscles, which further aids in preventing cold shortening by providing energy for the reaccumulation of Ca^{2+} ions by the sarcoplasmic reticulum (more in white muscles). ATP plays a role in Ca^{2+} accumulation by the mitochondria, also.

Cornforth et al. (1980) have discussed three possible mechanisms by which ATP may support mitochondrial accumulation of Ca^{2+} ions. First, ATP may be required for retention of Ca^{2+} ions accumulated during mitochondrial respiration. Lehninger et al. (1967) pointed out that two Ca^{2+} ions are accumulated per pair of electrons passing each energy-conserving site in the electron transport chain during mitochondrial respiration in the presence of Ca^{2+} ions. As a result of electron transport-driven accumulation of Ca^{2+}, no ATP is synthesized, i.e., Ca^{2+} ions uncouple mitochondrial synthesis. However, Ca^{2+} ions accumulated by this mechanism are not retained if the intramitochondrial ATP is depleted (Kimura and Rasmussen 1977). Secondly, Ca^{2+} accumulation may be supported by actual ATP hydrolysis, even in the absence of mitochondrial respiration (Brand and Lehninger 1975).

A third possible mechanism by which ATP may enhance mitochondrial accumulation is by simply providing a source of inorganic phosphate. Mitochondrial ATPase, as well as contaminating ATPase from other sources, will almost certainly produce increasing quantities of inorganic phosphate in the medium with time. Mitochondrial Ca^{2+} retention in the presence of a respiratory substrate is greatly increased if inorganic phosphate is also present in the medium (Lehninger et al. 1967). Thus, mechanisms are present in skeletal muscle whereby ATP plays an important role in both mitochondrial respiration and in the reaccumulation of Ca^{2+} ions by the sarcoplasmic reticulum, both of which will ameliorate cold shortening in white muscles.

Summary of Events

The series of events leading to cold shortening of prerigor red muscle have been theorized as following a systematic sequence, according to Cornforth *et al.* (1980). Chilling serves as the direct stimulus, which causes the already-saturated mitochondria to spill excess Ca^{2+}. The extra Ca^{2+} overloads the sarcoplasmic reticulum membranes that already have a decreased ability to bind Ca^{2+} ions as a result of the lower temperature and simultaneous fall in muscle pH. The excess free Ca^{2+} ions then interact with troponin-C and abolish the interaction between troponin-I and actin. This allows the troponin-tropomyosin complex to rotate, which exposes the site for interaction between actin and myosin on the myosin head. ATP is split, releasing the energy for muscle contraction, thus permitting actin and myosin to engage each other to produce shortening. Since the sarcoplasmic reticulum is already saturated and the calcium pump is not active, the interaction becomes fixed. This leaves the muscle permanently shortened and causes the toughness that is characteristic of cold shortening.

Cold shortening can be prevented by holding at temperatures above 10°C until the muscle goes into rigor mortis, which will require holding at this temperature for about 10 to 12 hr. It can also be circumvented by electrical stimulation, which speeds up glycolysis and hastens the onset of rigor. Electrical stimulation of beef and lamb carcasses will allow for rapid chilling or freezing within 2 hr of application without any problems from cold shortening.

OTHER GAINS FROM ELECTRICAL STIMULATION

So far, the discussion has focused upon the role of electrical stimulation in preventing cold shortening. However, Savell (1979) has indicated that other improvements occur in the physical and chemical characteristics of meat during electrical stimulation. These include better meat color as well as improvement in a number of palatability traits. However, the mechanisms for improvement are less well understood. Some of these gains and the basis for their improvement will be discussed in greater detail.

Improvement in Lean Meat Color

Savell (1979) concluded that the color of the lean in electrically stimulated beef is significantly improved, with the lean meat having a brighter, more attractive appearance. He suggested that the improvement in color is related to the effects of more complete glycolysis at the time of ribbing, which is normally done at about 18–24 hr postmortem. Regardless of the reason for the improvement in lean color, electrically stimulated beef carcasses were on the average rated as showing 13% improvement in youthfulness and 10% improvement in brightness of lean color scores (Savell 1979).

It was suggested by Savell (1979) that electrical stimulation improved the color of carcasses that tended to be slightly dark. However, Dutson *et al.* (1982) found that electrical stimulation was not effective in improving the color of carcasses that cut dark. This indicates that glycolysis is necessary in order to obtain good muscle color, which supports earlier research showing a close relationship between the potential reducing sugar content and the development of bright color in beef carcasses (Hall *et al.* 1944). Thus, the influence of electrical stimulation seems to be related to the increased rate of glycolysis with a faster decline in pH.

It seems probable that the brighter, more youthful appearance of the lean in electrically stimulated beef results from a lower pH and its effects in increasing reflectance. The lower pH would cause more denaturation of the muscle proteins and would result in more free water at the surface of the meat. The greater content of free water at the meat surface in turn increases reflectance and would account for the more youthful, brighter color that tends to develop in electrically stimulated beef. The differences in the amount of denaturation of the proteins between electrically stimulated and unstimulated meat are very likely small and difficult to measure, but their effects on color are visually discernible.

Although not fully understood, another aspect of meat color improved by electrical stimulation is a 30% decrease in the incidence of heat ring (Savell *et al.* 1978A). In this condition, there is a dark-appearing partial ring-like area near the lumbar-dorsal fascia, which extends inward toward the twelfth rib for 1–2 cm, giving the exposed ribeye muscle a two-toned appearance.

The decreased incidence of heat ring as a consequence of electrical stimulation is probably due to the increased rate of glycolysis and improvement in reflectance as a result of a lower pH. However, the cause of heat ring is believed to be complicated by the fact that rapid chilling immediately postmortem may decrease the activity of the glycolytic enzymes near the outer surface of the ribeye (Tarrant and Mothersill 1977). Electrical stimulation would accelerate glycolysis throughout the muscle and allow development of acidity before the meat is exposed to the low temperature, thus circumventing the development of heat ring. This has recently been verified by Orcutt *et al.* (1982).

Improvement in Tenderness

Savell (1979) has reported about 20% improvement in sensory tenderness scores and approximately 21% lower shear force values for electrically stimulated as compared with unstimulated beef sides. The major improvement in tenderness of electrically stimulated meat was originally due to prevention of cold shortening (Carse 1973), which has already been discussed. However, electrical stimulation also appears to improve tenderness above that which can be accounted for by inhibition of cold shortening. Support of this viewpoint is found in a report by Dutson *et al.* (1980B), who concluded that electrical stimulation generally improves tenderness even though no differences in sarcomere length may be evident between stimulated and unstimulated control muscles. Evidence suggests that electrical stimulation may also benefit tenderness by causing the rapid release of lysosomal enzymes and/or by physical disruption of the electrically stimulated muscle fibers (Dutson et al. 1977). Since prevention of cold shortening has already been reviewed, only the evidence for lysosomal enzyme release and physical disruption will be discussed further.

Dutson et al. (1980A) demonstrated that electrical stimulation increased the total, free, and specific activities of β-glucuronidase and cathepsin-C, both of which are lysosomal enzymes. This suggests that electrical stimulation causes disruption of the lysosomal membranes, thus releasing the lysosomal enzymes into the intracellular and intercellular compartments. Lysosomal enzymes are known to have the ability to degrade the myofibrillar proteins under the high temperature and low pH conditions prevailing in postmortem muscle (Schwartz and Bird 1977; Dutson and Yates 1978). Sorinmade et al. (1982) observed empty lysosomal vesicles and evidence of some proteolysis of electrically stimulated beef muscle, which also supports the concept that release of the lysosomal enzymes plays an important role in tenderization of meat following application of electrical stimulation. The disruption of the lysosomal membranes with the release of the lysosomal enzymes appears to be responsible, at least in part, for the increased tenderness of electrically stimulated meat.

The disruption of the muscle fibers may also contribute to the increased tenderness of electrically stimulated meat. Savell et al. (1978B) first reported that electrical stimulation caused physical disruption as shown by the presence of contracture bands accompanied by some disarrangement of the normal orderly array of the proteins within the muscle filaments. They suggested that stretching of myofilaments resulted from the pressure exerted on either side of the contracture bands, thus lowering the resistance to either chewing or mechanical shearing. Sorinmade et al. (1982) have since presented ultrastructural evidence that electrical stimulation causes contracture bands with superstretching of the myofibrils, resulting in the absence or presence of poorly defined A-bands, I-bands, and Z-lines. This further substantiates the fact that physical disruption is another mechanism whereby tenderization occurs as a consequence of electrical stimulation. Nevertheless, the voltage, the duration of stimulation, and the method of application are all important factors that may influence the tenderness changes occurring as a result of electrical stimulation and the mechanism by which these changes occur.

FIG. 6.13. Photomicrographs of beef longissimus muscle in longitudinal section showing the effects of electrical stimulation. A—Control muscle from unstimulated side. B—Muscle from the opposite side of the same carcass that was electrically stimulated. Note the dark, dense-appearing contracture bands in B and their absence in A ($\times 1962$).

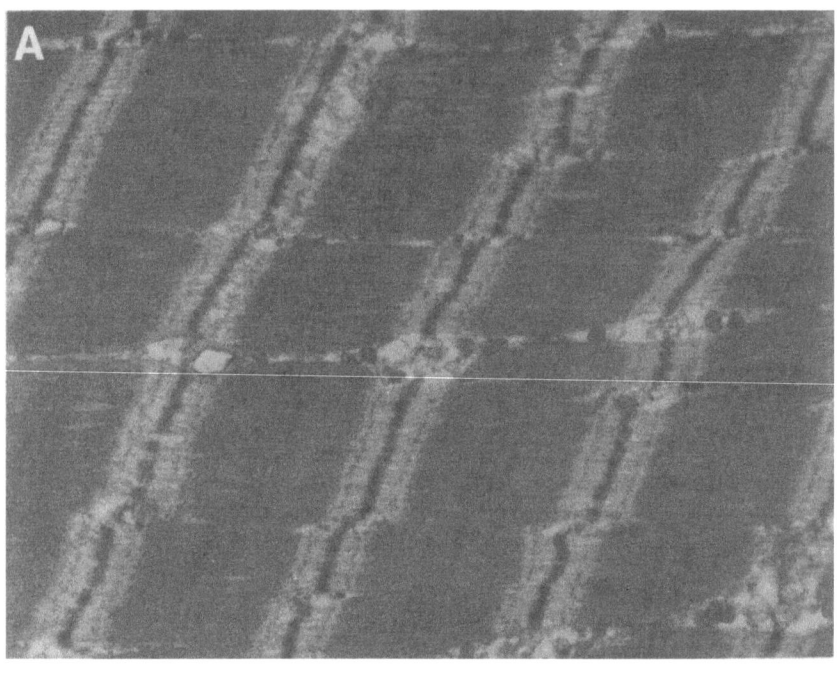

6 SCIENTIFIC BASIS FOR ELECTRICAL STIMULATION 211

Figure 6.13 presents a photomicrograph showing beef muscle before and after electrical stimulation. The dark-appearing contracture bands resulting from electrical stimulation are clearly evident on comparing Fig. 6.13A (before stimulation) with Fig. 6.13B (after stimulation). Figure 6.14A shows an electron micrograph of relaxed beef muscle. Examination of the micrograph reveals that the I-bands are readily discernible on both sides of the Z-line, which is characteristic of relaxed muscle. On the other hand, Fig. 6.14B presents an electron micrograph of the same muscle after cold shortening. It shows that the I-band has virtually disappeared, which is typical of the effect of cold on muscle from species having primarily red muscle (bovine and ovine).

Figure 6.15 presents electron micrographs of unstimulated (Fig. 6.15A) and electrically stimulated (Fig. 6.15B) beef muscle. Examination of Fig. 6.15A shows that the muscle is in the relaxed state similar to that shown in Fig. 6.14A. However, Fig. 6.15B shows a contracture band in the right-hand portion of the micrograph. The myofibrils exhibit active contraction and the contracted areas are more or less amorphous, with the ultrastructural elements being difficult to recognize. Where the sarcomeres can be differentiated, they show marked contraction and are much shorter than the uncontracted muscle (Fig. 6.15A). Cia and Marsh (1976) concluded that rapid cooking of prerigor beef results in development of contracture bands that tear and disrupt the muscle, thereby resulting in improvement of tenderness. If electrical stimulation induces contracture bands that are indicative of physical disruption of the muscle fibers, this would lend further support to the concept that electrical stimulation of prerigor meat improves tenderness through disruption of the myofibrillar elements.

Marsh et al. (1981) have suggested that the early postmortem aging period at high temperatures and high pH may play an important function in development of meat tenderness. Since application of electrical stimulation causes a rise in muscle temperature and since chilling of stimulated carcasses is usually delayed for 2–3 hr postmortem in actual practice (Marsh et al. 1981), a combination of these two factors may also contribute to the tenderness of electrically stimulated carcasses.

Regardless of the individual effects of electrical stimulation on meat tenderness, the most dramatic is the improvement in tenderness from preventing cold shortening, often resulting in a 2- or 3-fold improvement in tenderness (Marsh et al. 1968). The increased tenderness induced from the release of lysosomal enzymes, from physical disruption of the myofibers, and from the high temperature and lowered pH immediately follow-

---

FIG. 6.14. Electron micrographs showing a comparison of relaxed and cold-shortened beef longissimus muscle in longitudinal section. A—Relaxed restrained muscle showing the wide I-band extending on both sides of the Z-line. B—Cold-shortened muscle from the opposite side of the same carcass held for 24 hr at 0°C. Note the virtual absence of the I-band in B (×16,800).

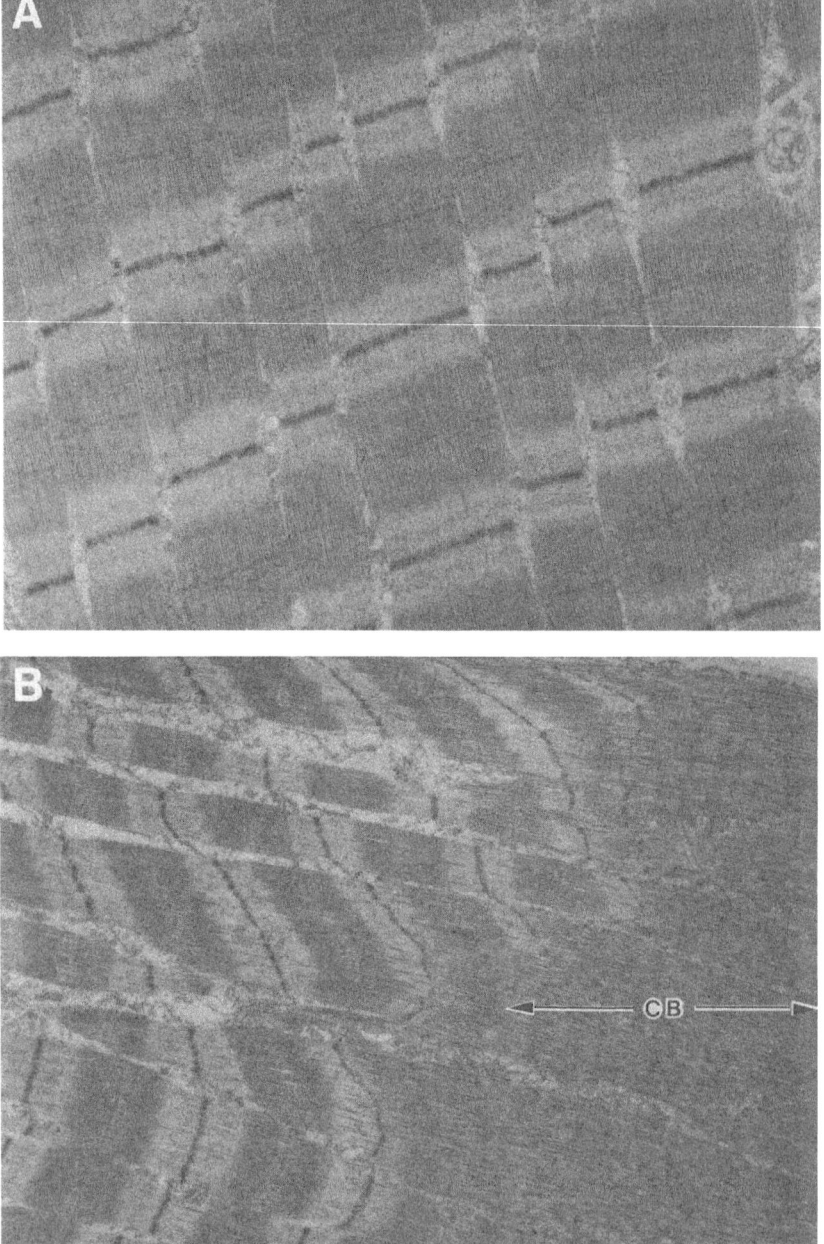

ing death, although it may be of a lesser magnitude than that obtained from preventing cold shortening, is probably important in improvement of the meat tenderness in electrically stimulated beef. No doubt, some or all of these factors are involved in the improvement of tenderness, above and beyond prevention of cold shortening.

Effects of Electrical Stimulation on Subsequent Aging

Savell (1979) concluded that electrically stimulated beef carcasses require less aging than similar unstimulated carcasses in order to reach the same level of tenderness. He summarized data showing that aging of stimulated beef loins for 14 days resulted in only 15% improvement in tenderness as compared with 26% improvement for similarly aged loins from unstimulated controls. These results along with the fact that loins from electrically stimulated carcasses were about 10% more tender than similar loins from unstimulated controls aged for 21 days indicated that aging can be substantially reduced by electrical stimulation.

The basis for the improved tenderness of the electrically stimulated carcasses is probably related to the release of lysosomal enzymes and/or to the structural changes already discussed. It may well be that the lysosomal enzymes released by electrical stimulation act much faster in the meat as a consequence of the accompanying structural damage. The added effects of the high temperature and low pH aging of prerigor beef as explained by Dutson et al. (1977) may also be a factor in accounting for the shorter aging period required to achieve the same degree of tenderness in electrically stimulated meat. Although more research will be necessary to sort out the importance of different mechanisms in tenderization, there is little doubt that less aging is necessary to obtain the same degree of tenderness in electrically stimulated beef.

Improvement in Meat Flavor

Electrical stimulation has been reported to result in a 10% improvement in flavor (Savell 1979). The exact mechanism of flavor improvement is not known, although it may be related to acceleration of the aging process and development of flavor enhancers (Calkins et al. 1982). Acceleration of glycolysis may be involved in the flavor improvement since low pH meat has a more desirable flavor than high pH meat (Dutson et al. 1982). Glycolysis will also result in the breakdown of ATP to ADP and could increase

---

FIG. 6.15. Electron micrographs of beef longissimus muscle showing a comparison between unstimulated and electrically stimulated muscle in longitudinal section. A—Unstimulated control (×10,400). B—Electrically stimulated muscle from the opposite side of the same carcass (×8000). CB—Contracture band.

the proportion of IMP (inosine monophosphate), which is an important meat flavor enhancer (Batzer et al. 1962). However, there is little evidence that indigenous levels of IMP are low enough in the red meats to be a quantitatively limiting factor in meat flavor development (Dannert and Pearson 1977).

Improvement in Marbling and Carcass Grades

Marbling is an important factor in beef carcass grading. All carcass grading in the United States now requires ribbing in order to determine the degree of marbling or the apparent amount of intramuscular fat. Savell (1979) pointed out that longer periods of chilling tend to increase marbling scores, although grading is normally done following a 24 hr chill period. Observations within the industry suggest that although longer chilling improves marbling scores, the same degree of enhancement of marbling scores can be achieved within a few hours by electrical stimulation (Savell 1979).

Improvement of marbling scores appears to be due to the faster rate of glycolysis in electrically stimulated carcasses. This seems to cause the lean tissues to be firmer, to have a finer texture, and to be brighter in color, resulting in solidifying of the fat within the ribeye (Savell 1979). The difference in marbling between electrically stimulated and unstimulated sides is either minimal or disappears upon chilling for 48 hr or more (Calkins et al. 1980).

The end result of the increased marbling scores in electrically stimulated beef and the corresponding improvement in carcass grading were covered in Chapter 4.

Improvement in Retail Caselife

Savell (1979) and Hall et al. (1980) have reported that electrical stimulation resulted in one-half to 1 day longer retail caselife for round steaks but was not different for hamburger produced from electrically stimulated postrigor meat. The longer shelf-life for the steaks from electrically stimulated carcasses was quite likely associated with the brighter color and the increased rate of glycolysis since low pH values tend to inhibit microbial growth (Lechowich 1971). The failure of hamburger meat made from electrically stimulated carcasses to have a longer caselife in these studies is probably related to the fact that the meat from both the electrically stimulated and nonstimulated muscles had already achieved its low ultimate pH in the carcasses before grinding. It should be remembered, however, that grinding per se will also speed up glycolysis (Newbold and Lee 1965), although less rapidly than electrical stimulation. Thus, hamburger prepared from electrically stimulated prerigor meat generally has a longer caselife than hamburger made from nonstimulated prerigor meat, the faster fall in pH apparently being responsible.

SUMMARY

The structure of muscle is reviewed along with the changes that occur during muscle contraction and development of rigor mortis. The shortening that develops during exposure of prerigor postmortem meat to cold is similar to that occurring during muscle contraction. During cold shortening, however, the energy sources are no longer available for relaxation and the muscle is locked into the shortened state. The shortening is responsible for the toughness of cold-shortened meat, which may be 2- to 3-fold more resistant to chewing and/or shearing than similar muscle in which cold shortening has been avoided.

The failure of the muscle to relax after shortening is related to the fact that red muscles, which predominate in sheep and cattle, have a larger number of mitochondria and a less well developed sarcoplasmic reticulum. Normally, both the mitochondria and the sarcoplasmic reticulum bind Ca^{2+} ions, which play an important role in contraction. Upon exposure of prerigor muscle to cold, however, the mitochondria become overloaded with Ca^{2+} ions, which are then spilled into the intracellular spaces. Since the poorly developed sarcoplasmic reticulum is unable to recapture the free Ca^{2+} ions due to being already overloaded with the Ca^{2+} ions released from the mitochondria as a result of exposure to cold, shortening occurs and the meat becomes tough. Thus, both cold temperatures and low pH contribute to the inability of the mitochondria and sarcoplasmic reticulum membranes to bind Ca^{2+} ions and inhibit shortening.

Although cold shortening is responsible for most of the toughening in meat exposed to cold temperatures, electrical stimulation improves tenderness above and beyond that anticipated from preventing cold shortening. Evidence suggests that electrical stimulation causes disruption of the muscle by releasing the lysosomal enzymes and by causing stretching and tearing of the myofibers, both of which appear to contribute to increased tenderness beyond prevention of cold shortening. Electrical stimulation also results in improvement in meat color, prevention of heat ring, increasing marbling, and enhancement of flavor, all of which appear to be associated with the effects of electrical stimulation in accelerating glycolysis.

REFERENCES

ASHMORE, C.R. and DOERR, L. 1971. Comparative aspects of muscle fiber types in different species. Exp. Neurol. *31*, 431.

BATZER, O.F., SANTORO, A.T. and LANDMANN, W.A. 1962. Identification of beef flavor precursors. J. Agric. Food Chem. *10*, 94.

BEATTY, C.H., BASINGER, G.M. and BOCEK, R.M. 1967. Differentiation of red and white fibers in muscle from fetal, neonatal and infant *Rhesus* monkeys. J. Histochem. Cytochem. *15*, 93.

BEECHER, G.R. 1966. Biochemical characteristics of red and white striated muscle. Ph.D. Thesis. Univ. of Wisconsin, Madison.

BENDALL, J.R. 1969. Muscles, Molecules and Movement. Heinemann Educational Books, London.

BENDALL, J.R. 1976. Electrical stimulation of rabbit and lamb carcasses. J. Sci. Food Agric. 27, 819.

BENDALL, J.R., KETTERIDGE, C.C. and GEORGE, A.R. 1976. The electrical stimulation of beef carcasses. J. Sci. Food Agric. 27, 1123.

BERNARD, C. 1877. Lessons upon diabetes and glycogen formation in animals. In Lectures on Experimental Physiology. Bailliere et Fils, Paris. (French)

BIRKNER, M.L. and AUERBACH, E. 1960. Microscopic structure of animal tissues. In The Science of Meat and Meat Products. pp. 10–55. American Meat Institute Foundation (Editor). W.H. Freeman & Co., San Francisco.

BODWELL, C.E., PEARSON, A.M. and SPOONER, M.E. 1965. Post-mortem changes in muscle. I. Chemical changes in beef. J. Food Sci. 30, 766.

BOUTON, P.E., FORD, A.L., HARRIS, P.V. and SHAW, F.D. 1980. Electrical stimulation of beef sides. Meat Sci. 4, 145.

BRAND, M.D. and LEHNINGER, A.L. 1975. Superstoichiometric Ca^{2+} uptake supported by hydrolysis of endogenous ATP in rat liver mitochondria. J. Biol. Chem. 250, 7958.

BUEGE, D.R. and MARSH, B.B. 1975. Mitochondrial calcium and post-mortem muscle shortening. Biochem. Biophys. Res. Commun. 65, 478.

CALKINS, C.R., SAVELL, J.W., SMITH, G.C. and MURPHEY, C.E. 1980. Quality-indicating characteristics of beef as affected by electrical stimulation and postmortem chilling time. J. Food Sci. 45, 1330.

CALKINS, C.R., DUTSON, T.R., SMITH, G.C. and CARPENTER, Z.L. 1982. Concentration of creatine phosphate, adenine nucleotides and their derivatives in electrically stimulated and non-stimulated beef. J. Food Sci. 47, 1350.

CARSE, W.A. 1973. Meat quality and the acceleration of postmortem glycolysis by electrical stimulation. J. Food Technol. 8, 163.

CASSENS, R.G. 1971. Microscopic structure of animal tissues. In The Science of Meat and Meat Products, 2nd Edition. pp. 11–75. J.F. Price and B.S. Schweigert (Editors). W.H. Freeman & Co, San Francisco.

CASSENS, R.G. and COOPER, C.C. 1971. Red and white muscle. Adv. Food Res. 19, 1.

CHRYSTALL, B.B., DEVINE, C.E. and DAVEY, C.L. 1980. Studies in electrical stimulation: Post-mortem decline in nervous response in lambs. Meat Sci. 4, 69.

CIA, G. and MARSH, B.B. 1976. Properties of beef cooked before the onset of rigor. J. Food Sci. 41, 1259.

CORNFORTH, D.P., PEARSON, A.M. and MERKEL, R.A. 1980. Relationship of mitochondria and sarcoplasmic reticulum to cold shortening. Meat Sci. 4, 103.

DANNERT, R.D. and PEARSON, A.M. 1977. Concentration of inosine 5'-monophosphate in meat. J. Food Sci. 32, 49.

DAVEY, C.L. and GILBERT, K.V. 1974. The mechanisms of cold induced shortening in beef muscle. J. Food Technol. 9, 51.

DUBOWITZ, V. and PEARSE, A.G.E. 1960. Reciprocal relationship of phosphorylase and oxidative enzymes in skeletal muscle. Nature 185, 701.

DUTSON, T.R. and YATES, L.D. 1978. Molecular and ultrastructural alterations in bovine muscle caused by high temperature and low pH incubation. Proc. 24th Eur. Meat Res. Worker's Conf. 24, E6.

DUTSON, T.R., PEARSON, A.M. and MERKEL, R.A. 1974. Ultrastructural postmortem changes in normal and low quality porcine muscle fibers. J. Food Sci. 39, 32.

DUTSON, T.R., YATES, L.D., SMITH, G.C., CARPENTER, Z.L. and HOSTETLER, R.L. 1977. Rigor onset before chilling. Proc. Recip. Meat Conf. 30, 79.

DUTSON, T.R., SMITH, G.C. and CARPENTER, Z.L. 1980A. Lysosomal enzyme distribution in electrically stimulated ovine muscle. J. Food Sci. 45, 1097.

DUTSON, T.R., SMITH, G.C., SAVELL, J.W. and CARPENTER, Z.L. 1980B. Possible mechanisms by which electrical stimulation improves meat tenderness. Proc. Eur. Meet. Meat Res. Workers 26 (II) 84.

DUTSON, T.R., SAVELL, J.W. and SMITH, G.C. 1982. Electrical stimulation of ante-mortem stressed beef. Meat Sci. 6, 159.

EBASHI, S. 1961. The role of relaxing factor in the contraction relaxation cycle of muscle. Progr. Theor. Phys. (Kyoto) 17 (Suppl.) 35.

GAUTHIER, G.R. 1970. The ultrastructure of three fiber types in mammalian skeletal muscle. In The Physiology and Biochemistry of Muscle as a Food, Vol. 2. P. 103. E.J. Briskey, R.G. Cassens and B.B. Marsh (Editors). Univ. of Wisconsin Press, Madison.

GEORGE, A.R., BENDALL, J.R. and Jones, R.C.D. 1980. The tenderising effect of electrical stimulation of beef carcasses. Meat Sci. 4, 51.

HALL, J., LATSCHAR, E.E. and MACKINTOSH, D.L. 1944. Characteristics of darkcutting beef. Survey and preliminary investigation. Kans. Agric. Exp. Stn. Bull. 58.

HALL, L.C., SAVELL, J.W. and SMITH, G.C. 1980. Retail appearance of electrically stimulated beef. J. Food Sci. 45, 171.

HAM, A.W. 1969. Histology, 6th Edition. J.B. Lippincott Co., Philadelphia.

HANSON, J. and HUXLEY, H.E. 1955. The structural basis of contraction in striated muscle. Symp. Soc. Exp. Biol. 9, 228.

HARSHAM, A. and DEATHERAGE, F. 1951. Tenderization of meat. U.S. Pat. 2,544,681.

HASSELBACH, W. and MAKINOSE, M. 1962. ATPase and active transport. Biochem. Biophys. Res. Commun. 7, 132.

HERRING, H.K., CASSENS, R.G. and BRISKEY, E.J. 1965. Further studies on bovine muscle tenderness as influenced by length and fiber diameter. J. Food Sci. 30, 1049.

HUXLEY, H.E. 1965. The mechanism of muscular contraction. Sci. Am. 213, 18.

KIMURA, S. and RASMUSSEN, H. 1977. Adrenal glucocorticoids, adenine nucleotide translocation, and mitochondrial calcium accumulation. J. Biol. Chem. 252, 1217.

LECHOWICH, R.V. 1971. Microbiology of meat. In The Science of Meat and Meat Products, 2nd Edition. pp. 151–184. J.F. Price and B.S. Schweigert (Editors). W.H. Freeman & Co., San Francisco.

LEHNINGER, A.L., CARAFOLI, E. and ROSSI, C.S. 1967. Energy-linked ion movements in mitochondrial systems. Adv. Enzymol. 29, 259.

LOCKER, R.H. 1960. Degree of muscular contraction as a factor in tenderness of beef. Food Res. 25, 304.

LOCKER, R.H. and HAGYARD, C.J. 1963. A cold shortening effect in beef muscles. J. Sci. Food Agric. 14, 787.

MARSH, B.B. 1952A. The effects of ATP on the fiber volume of a muscle homogenate. Biochim. Biophys. Acta 9, 247.

MARSH, B.B. 1952B. Observations on rigor mortis in whale muscle. Biochim. Biophys. Acta 9, 127.

MARSH, B.B. and CARSE, W.A. 1974. Meat tenderness and the sliding filament hypothesis. J. Food Technol. 9, 129.

MARSH, B.B. and LEET, N.G. 1966. Studies in meat tenderness. III. The effects of cold shortening on tenderness. J. Food Sci. 31, 450.

MARSH, B.B., WOODHAMS, P.R. and LEET, N.G. 1968. Studies on meat tenderness. V. Effects of carcass cooling and freezing before completion of rigor mortis. J. Food Sci. 33, 12.

MARSH, B.B., LEET, N.G. and DICKSON, M.R. 1974. The ultrastructure and tenderness of highly cold shortened muscle. J. Food Technol. 9, 141.

MARSH, B.B., LOCKNER, J.V., TAKAHASHI, G. and KRAGNESS, D.D. 1981. Effects of early post-mortem pH and temperature on beef tenderness. Meat Sci. 5, 479.

NAUSS, K.M. and DAVIES, R.E. 1966. Changes in phosphate compounds during the development and maintenance of rigor mortis. J. Biol. Chem. *241*, 2918.
NEWBOLD, R.P. 1966. Changes associated with rigor mortis. *In* The Physiology and Biochemistry of Muscle as a Food. E.J. Briskey, R.G. Cassens and J. C. Trautman (Editors). Univ. of Wisconsin Press, Madison. pp. 213–224.
NEWBOLD, R.P. and LEE, C.A. 1965. Postmortem glycolysis in skeletal muscle. The extent of glycolysis in diluted preparations of mammalian muscle. Biochem. J. *97*, 1.
ORCUTT, M.W., DUTSON, T.R., CORNFORTH, D. and DUTSON, P.J. 1982. Analysis of the heat ring phenomenon in Holstein-Friesian steer carcasses. Proc. Am.-Can. Soc. Anim. Sci. Abstr., 44.
PEARSON, A.M. 1984. Muscle function and post-mortem change. *In* The Science of Meat and Meat Products, 3rd Edition. J.F. Price and B.S. Schweigert (Editors). Food and Nutrition Press, Westport, CT. (In press)
PEARSON, A.M., CARSE, W.A., DAVEY, C.L., LOCKER, R.H. and HAGYARD, C.J. 1973. Influence of epinephrine and calcium upon glycolysis, tenderness and shortening of sheep muscle. J. Food Sci. *38*, 1124.
PORTZEHL, H. 1957. The binding of the relaxing factor of Marsh on muscle. Biochim. Biophys. Acta *26*, 373. (German)
PORZIO, M.A. PEARSON, A.M. and CORNFORTH, D.P. 1979. M-line protein: Presence of two non-equivalent high molecular weight compounds. Meat Sci. *3*, 31.
SAVELL, J.W. 1979. Industry acceptance of electrical stimulation. Proc. Recip. Meat Conf. *3*, 31.
SAVELL, J.W., SMITH, G.C. and CARPENTER, Z.L. 1978A. Effect of electrical stimulation on quality and palatability of light-weight beef carcasses. J. Anim. Sci. *46*, 1221.
SAVELL, J.W., DUTSON, T.R., SMITH, G.C. and CARPENTER, Z.L. 1978B. Structural changes in electrically stimulated beef muscle. J. Food Sci. *43*, 1606.
SCHWARTZ, W.N. and BIRD, J.W.C. 1977. Degradation of myofibrillar proteins by cathepsin D. Biochem. J. *167*, 811B.
SISSON, S. and GROSSMAN, J.D. 1953. The Anatomy of Domestic Animals, 4th Edition. W.B. Saunders Co., Philadelphia.
SMITH, P.E. and COPENHAVER, W.M. 1948. Bailey's Textbook of Histology, 10th Edition. Williams and Wilkins Co., Baltimore.
SORINMADE, S.O., CROSS, H.R., ONO, K. and WERGIN, W.P. 1982. Mechanisms of ultrastructural changes in electrically stimulated beef longissimus muscle. Meat Sci. *6*, 71.
STEIN, J.M. and PADYKULA, H.A. 1962. Histochemical classification of individual skeletal muscle fibers of the rat. Am. J. Anat. *110*, 103.
STRYER, L. 1981. Biochemistry, 2nd Edition. W.H. Freeman & Co., San Francisco.
TARRANT, P.V. and MOTHERSILL, C. 1977. Glycolysis and associated changes in beef carcasses. J. Sci. Food Agric. *28*, 739.
VOYLE, C.A. 1969. Some observations on the histology of cold-shortened muscle. J. Food Technol. *4*, 275.
WEBER, A. and HERZ, R. 1962. Requirement for calcium in the synaeresis of myofibrils. Biochem. Biophys. Res. Commun. *6*, 364.
WEINER, P.D. and PEARSON, A.M. 1966. Inhibition of rigor mortis by ethylenediamine tetraacetic acid. Proc. Soc. Exp. Biol. Med. *123*, 185.
WEINER, P.D. and PEARSON, A.M. 1969. Calcium chelators influence some physical and chemical properties of rabbit and pig muscle. J. Food Sci. *34*, 592.
WHITE, A., HANDLER, P. and SMITH, E.L. 1964. Principles of Biochemistry, 3rd Edition. McGraw-Hill, New York.

7

Industrial Applications of Electrical Stimulation[1]

J. W. Savell[2]

Industry Interest in Electrical Stimulation
Electrical Stimulation Equipment
Installations of Electrical Stimulators Within the
 Slaughter–Dressing Sequence
Electrical Parameters
Safety, Installation, and Sanitation
Trade Names and Promotion of Electrically Stimulated Beef
Industry Adoption of Electrical Stimulation
Summary
Acknowledgments
References

The industrial application of electrical stimulation in slaughter plants in the United States and throughout the world has been phenomenal. According to Savell et al. (1980), some of the benefits of using electrical stimulation realized by meat packers include improved tenderness of meat, brighter lean color, more rapid setup of marbling after ribbing, less "heat-ring" development and reduced postmortem aging requirements for assurance of tender meat (color photographs delineating the improvement in lean color and prevention of heat-ring by electrical stimulation are included in a bulletin by Stiffler et al. 1982). These benefits along with the low relative costs at which electrical stimulation can be incorporated into slaughter plants have influenced both large and small slaughter facilities to adopt this procedure. In many slaughter plants, electrical stimulation has become an integral part of the process of converting live animals to meat and meat products.

With the interest shown by the U.S. meat industry regarding the incorporation of electrical stimulation into the slaughter–dressing sequence of beef packing plants, several companies began to manufacture commer-

---

[1]Technical Article *18044* from the Texas Agricultural Experiment Station. Reference to commercial products or trade names is made with the understanding that no discrimination is intended and no endorsement by the Texas Agricultural Experiment Station is implied.
[2]Meats and Muscle Biology Section, Department of Animal Science, Texas Agricultural Experiment Station, Texas A&M University, College Station, TX.

cially-available electrical stimulators. The degree of complexity of commercial electrical stimulators manufactured by these companies is varied; stimulators are made for slaughterers with capacities of 10 head/day up to 325 head/hr. Furthermore, because not all slaughter plants are alike, the layout and installation of most electrical stimulators vary from plant to plant. Problems arising in the installation of electrical stimulators in plants with differing slaughter capacities, plant layouts, and space limitations have generally been solved by the combined efforts of slaughterers and equipment manufacturers and, with few exceptions, these problems have not been a hindrance for most potential users of electrical stimulation.

This chapter will deal with the industrial application of electrical stimulation. Because electrical stimulation is most widespread in New Zealand and the United States—as compared with other countries of the world—and because determining the industrial applications of electrical stimulation in many foreign countries would be a difficult task, the discussions involved within this chapter will deal with the industrial application of electrical stimulation in the United States in which the writer has experience. The chapter will be divided into major sections in order to better address the topics associated with the equipment, installation, and operation involved with electrical stimulation.

INDUSTRY INTEREST IN ELECTRICAL STIMULATION

The U.S. beef slaughtering industry became interested in electrical stimulation when it began to read and hear reports—especially those from the Texas Agricultural Experiment Station—regarding the effects of electrical stimulation on the quality and palatability of beef. Although the beef slaughtering industry was interested in improving the palatability of its beef, the fact that electrical stimulation could be used to improve certain quality-related factors—lean color and marbling, especially—of beef was very appealing. The effects of electrical stimulation on the quality-indicating characteristics of beef (Savell 1979) suggest that the use of electrical stimulation for the improvement of these factors could be very beneficial—in financial terms—to beef slaughterers. Because USDA beef quality standards are so dependent on the degree of marbling and the lean color, firmness, and texture of the ribeye muscle and because the best evaluations of the ribeye muscle are made when complete pH decline and full rigor mortis have occurred (both pH decline and rigor mortis are accelerated by electrical stimulation), electrical stimulation of beef carcasses results in these factors' being at their optimum at an early (19--24 hr) instead of at a later time (48 hr postmortem). Electrical stimulation results in more beef carcasses being eligible for the U.S. Choice grade during the first day postmortem and sharply reduces the number of carcasses that must be held two days or more waiting for these quality-indicating characteristics—especially marbling—to develop fully. Electrical stimulation does not result in beef carcasses receiving higher-than-justified grades; rather, chilling for up to 48 hr before ribbing and presenting carcasses for grading has been found to be a more desirable method for optimizing quality grades, but this method is not as efficient for most beef slaughterers (Savell et al. 1980).

In 1977, as researchers from the Texas Agricultural Experiment Station were conducting field studies on electrical stimulation (using the B and D "Electro-Sting" Hog Stunner® used in their early experiments) in beef slaughtering plants, the results—especially the lean color, marbling, and heat-ring characteristics—were so dramatic that personnel from these plants began to inquire as to where they could purchase such equipment. In response to the growing demand for more information regarding electrical stimulation and the need to develop a prototype electrical stimulator, researchers from the Agricultural Research Service, USDA, College Station, TX, along with the LeFiell Company, San Francisco, CA, developed the first prototype unit, which could generate different voltages, pulse durations, and pulse intervals. With this unit, a single electrode was inserted into the neck of the carcass and electricity was administered automatically. In April 1978, the Texas Agricultural Experiment Station and Agricultural Research Service researchers conducted field studies at H & H Meat Products evaluating different electrical parameters using this prototype unit. The response in quality-indicating characteristics to electrical stimulation was so evident that H & H Meat Products personnel persuaded the group of researchers to leave the prototype unit at their plant for further studies. With the subsequent purchase and installation of an electrical stimulator after several months of using the prototype unit, H & H Meat Products became the first beef slaughterer in the United States to use electrical stimulation, and the process has become a vital part of this company's production scheme.

Based on their earlier work with the prototype electrical stimulator and the growing interest in electrical stimulation, the LeFiell Company became the first developer of commercial electrical stimulators. The first installation of a commercially manufactured electrical stimulator was in the Sam Kane Beef Processors plant in Corpus Christi, TX, where, in November 1978, the first "Lectro-Tender™" (model #4051) was installed. The slaughter rate at the Sam Kane Beef Processors plant at the time of the installation of the "Lectro-Tender™" was 107 head/hr. Early in 1979, the LeFiell Company installed the first "Continuous-Trac Lectro-Tender™" (model #4055) in the Litvak Meat Company, Denver, CO; this accomplishment was significant because this electrical stimulator allowed for the application of electricity to carcasses in plants with slaughter rates exceeding 300 head/hr.

Because of the explosion of electrical stimulation technology and the deluge of requests for information and demonstrations on this subject, the

Department of Animal Science at Texas A&M University, through the joint efforts of researchers from the Texas Agricultural Experiment Station and meat specialists from the Texas Agricultural Extension Service, sponsored a symposium entitled "Electrical Stimulation for Improving Meat Quality" held in Corpus Christi, TX, in January 1979. This one-day symposium had a morning program dealing with the results of electrical stimulation research and an afternoon program consisting of demonstrations of electrical stimulation under industrial conditions at the Sam Kane Beef Processors plant along with evaluations in the cooler where test carcasses (one side had been electrically stimulated while the other side served as the nonstimulated control) had been ribbed after being chilled for 24 hr. This symposium was attended by over 100 meat industry personnel from many parts of the United States and several foreign countries; the U.S. participants represented companies that slaughter the vast majority of the beef in the United States. Soon after this symposium, many beef slaughtering plants in the United States began to order electrical stimulators and to adopt this process.

ELECTRICAL STIMULATION EQUIPMENT

There are two general classes of electrical stimulation equipment: high voltage and low voltage. Although results of high-voltage and low-voltage electrical stimulation appear to be closely related, the mechanisms of their actions and the methods used to apply the current to the carcass differ substantially. In order to better describe the electrical stimulation equipment manufactured and installed in U.S. meat packing plants, this discussion will be divided into two parts: high-voltage electrical stimulators and low-voltage electrical stimulators. The major portion of these discussions is adapted from the excellent review of electrical stimulators published by *Meat Industry* magazine and entitled "A Meat Industry Special Guide— Electrical Stimulation Equipment" (Meat Ind. 1982) and from promotional and informational materials supplied by the individual manufacturers.

High-Voltage Electrical Stimulators

The LeFiell Company. As mentioned earlier, the LeFiell Company was the first U.S. manufacturer of commercial electrical stimulators. LeFiell manufactures two basic types of high-voltage electrical stimulators: the bar-type ("Lectro-Tender™" model #4051—Fig. 7.1 and 7.2) and the continuous chain-type ("Continuous-Trac Lectro-Tender™" model #4055— Fig. 7.3 and 7.4). The LeFiell bar-type electrical stimulator uses an electrified bar that extends from the protective housing to stimulate a carcass and then retracts and undergoes sterilization before recycling to electrically stimulate another carcass. The number of carcasses that can be

7 INDUSTRIAL APPLICATIONS OF ELECTRICAL STIMULATION

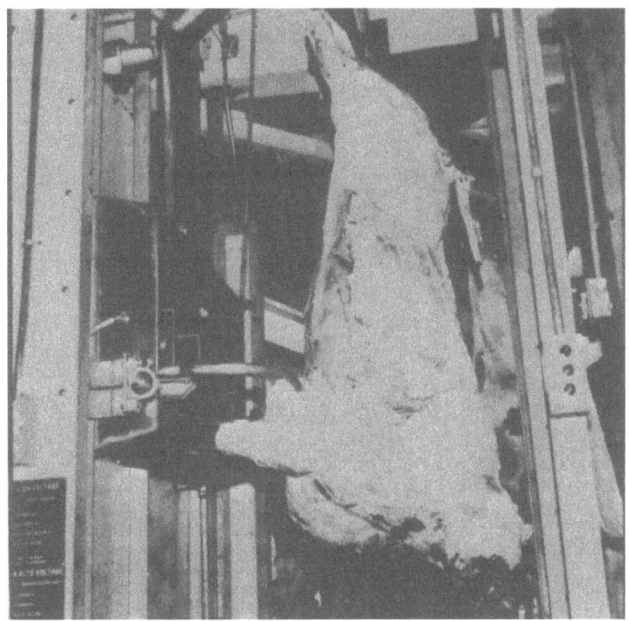

FIG. 7.1. LeFiell model #4051 "Lectro-Tender™."
Courtesy of the LeFiell Company and the National Provisioner.

electrically stimulated with the bar type of electrical stimulator depends on the number of bars present and the combined length of the bars—if the stimulator is equipped with one bar that is 2 m (7 ft) long, the capacity of the stimulator is 70 carcasses/hr; if the stimulator is equipped with two bars with a combined length of 4.6 m (15 ft), the capacity of the stimulator is 125 carcasses/hr; and if the stimulator is equipped with three bars totaling 5.5 m (18 ft) in length, the capacity is 180 carcasses/hr.

In order to accomplish electrical stimulation of carcasses in plants with a slaughter capacity greater than 180 head/hr, the "Continuous-Trac Lectro-Tender™" must be employed. The continuous-chain electrical stimulator is designed for plants with slaughter capacities of 155 to 390 head/hr. The feature of the continuous-chain electrical stimulator that allows its use in high-speed slaughter operations is the utilization of a continuously circulating chain that comes in contact with the carcass in the brisket region and travels with the carcass at the same speed as the slaughter rail conveyor (the speed of the continuous-chain conveyor can be adjusted to run at the same speed as the slaughter rail conveyor and will start and stop automatically with the slaughter rail conveyor). Because of the size of this unit (from 60 to 10.7 m or 20 to 35 feet long), the continuous-chain electrical stimulator can stimulate up to five moving carcasses

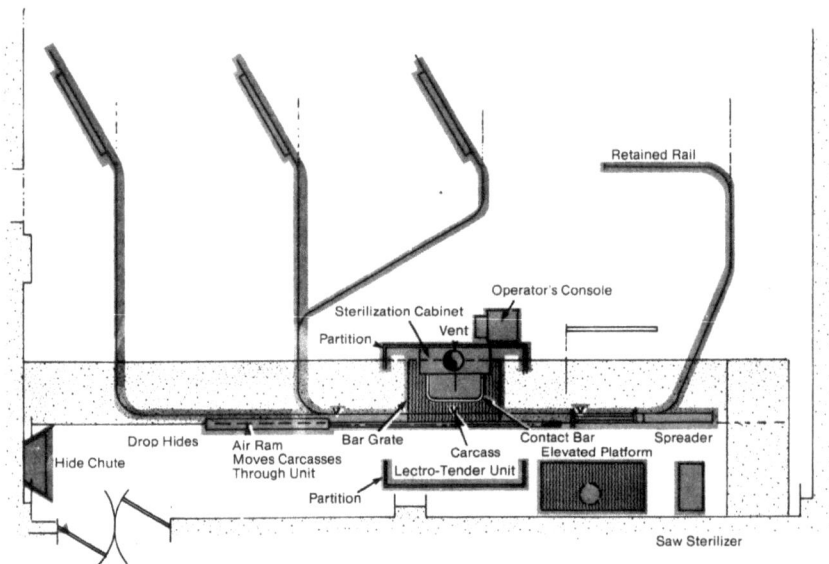

FIG. 7.2. Layout plan of "Lectro-Tender™" unit. Typical packing house installation of LeFiell single-bar "Lectro-Tender" electrical stimulation unit. Carcasses move via overhead conveyors after slaughter to hide puller (left), where hides are dropped. An air ram on the conveyor rail then pushes each carcass into the "Lectro-Tender" cabinet (center). U-shaped contact bar rotates down from the retracted position in sanitation cabinet to contact the carcass. After stimulation, bar retracts back to sanitation position, carcass moves on to evisceration, splitting, and further processing.
Layout from Morris (1979). Reprinted with permission of Food Engineering magazine.

at the same time. The continuous chain, which is made of high-strength plastic with stainless steel applicators, has high-voltage current delivered only on the front side of the circulating chain; no electrical current is delivered on the sides or on the rear of the circulating chain. A sterilizer tunnel is located on the rear of the stimulator for continuous sterilization of the circulating chain.

Koch-Britton. The Koch-Britton stimulators are manufactured by Britton Manufacturing Inc., of College Station, TX, and distributed by Koch Supplies of Kansas City, MO. The high-voltage models available from Koch-Britton include the model #350 and model #250 stimulators. These stimulators are designed for the small- to medium-sized slaughter plant and can handle slaughter rate capacities of 5 to 60 head/hr. These stimulators use a manual-probe system for administering the electrical current to the carcass. The manual-probe system (Fig. 7.5) is enclosed in a safety cabinet that has three doors: (1) a door for carcass entry, (2) a door

7 INDUSTRIAL APPLICATIONS OF ELECTRICAL STIMULATION 225

for carcass exit, and (3) a safety control door for the operator to enter and exit.

Electrical stimulation is applied by placing the carcasses in the cabinet; the operator (who enters and exits the cabinet through the safety control door where a series of photocells must be tripped in sequence) inserts the probe into the neck of the carcass. The operator passes his hand in front of another photocell and presses both start buttons, setting off a loud horn and flashing red lights while the electrical stimulation is in process for about 40 sec. Upon completion of stimulation, a horn sounds and the stimulator shuts off automatically and will not restart until these procedures are repeated.

FIG. 7.3. LeFiell model #4055 "Continuous-Trac Lectro-Tender™."
Courtesy of the LeFiell Company and Meat Industry *magazine.*

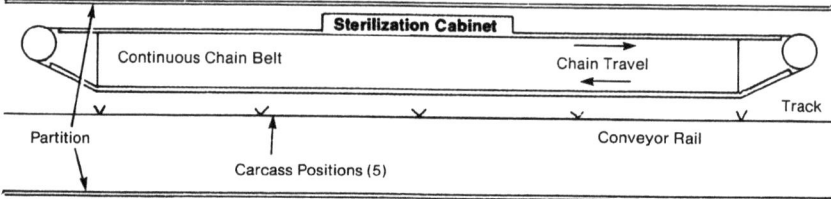

FIG. 7.4. Layout plan of "Continuous-Trac Lectro-Tender™" unit. Overhead view of the continuous electrical tenderizer illustrates continuous chain belt mounting stainless-steel plate electrodes which contact as many as 5 carcasses simultaneously for electrical stimulation. Carcasses move into cabinet via overhead conveyor rail and contact electrodes on chain belt for electrical stimulation. Chain is continuously sterilized with 180°F (82°C) water as it passes through sterilization cabinet on back side of travel.
Layout from Morris (1979). Reprinted with permission of Food Engineering magazine.

Requirements for the Koch-Britton high-voltage electrical stimulators include the need for 1.8 m (6 ft) of track section and sufficient space for installation of the safety cabinet. Electrical requirements include 220 volt, 60 cycle, single phase with a ground line from the rail to the outside. Safety features include a "lock-off" device that shuts the unit down if anyone attempts to enter the cabinet while stimulation is in progress, a buzzer system warning personnel the unit is about to activate, flashing red lights showing the unit is stimulating, barrier wall panels, and "Danger— High Voltage" signs.

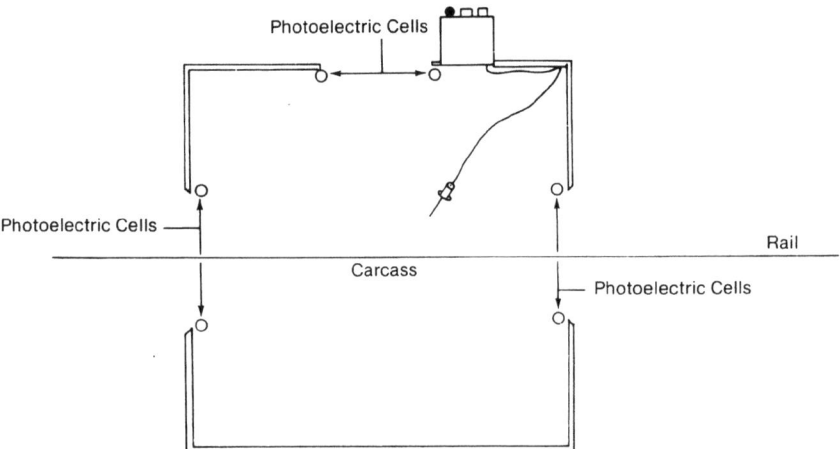

FIG. 7.5. Plan view of Koch-Britton model #350.
Courtesy of Koch Supplies.

7 INDUSTRIAL APPLICATIONS OF ELECTRICAL STIMULATION 227

Omeco-Boss. Omeco-Boss, Inc., of Omaha, NE (formerly Omeco-St. John), has three models of high-voltage electrical stimulators: model #625A, #625B, and #625C (detailed in Fig. 7.6 and 7.7). Models 625A and 625B are manual-probe, low-volume electrical stimulators, whereas model 625C is an automated, high-volume electrical stimulator. Designed for

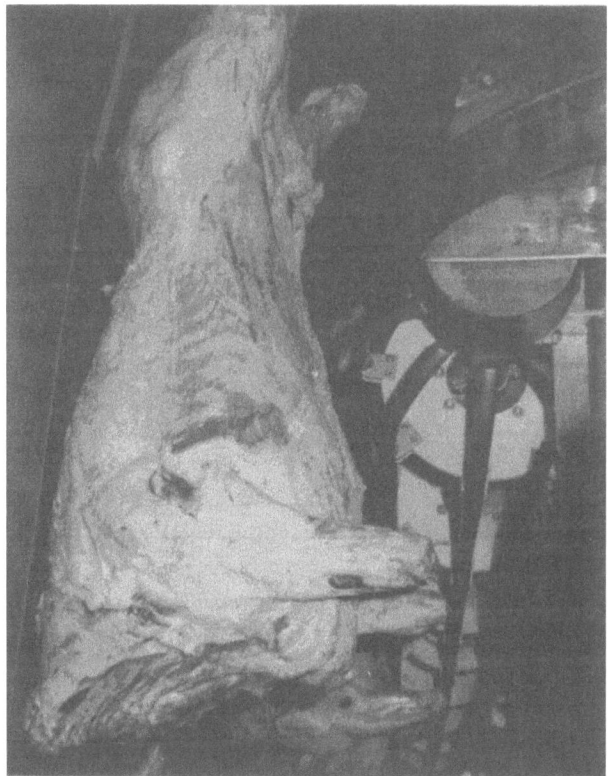

FIG. 7.6. Omeco-Boss (formerly Omeco-St. John) model #625C Rotating Rub Bar Stimulator.
Courtesy of Omeco-Boss and Meat Processing *magazine.*

small slaughter plants, model 625A has one entrance to the safety enclosure so that carcasses enter and exit through the same door (Fig. A of Fig. 7.7). Designed for small- to medium-sized slaughter plants, model 625B has double sliding doors at both ends of the enclosure so that carcasses enter through doors in one end and exit at the opposite end of the enclosure (Fig. B of Fig. 7.7). Safety interlock switches are mounted on the enclosure doors, and if all doors are not completely closed, a stimulation cycle cannot be initiated. If a door is opened while a stimulation cycle is in progress, the stimulus will be immediately interrupted.

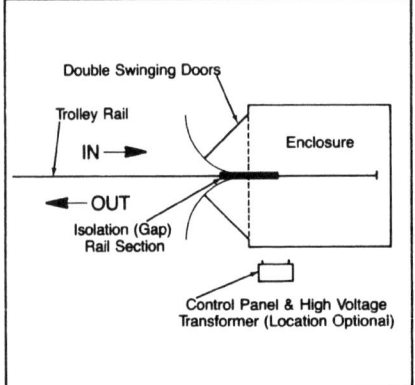

FIGURE A, MODEL #625A STIMULATOR with Manual Probe (Plan View)

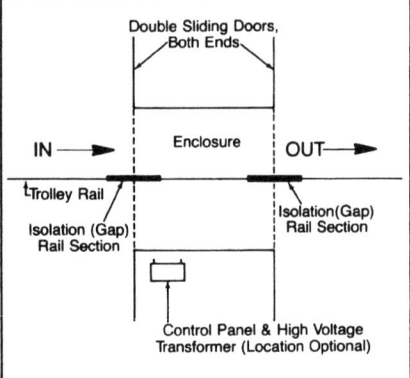

FIGURE B, MODEL #625B STIMULATOR with Manual Probe (Plan View)

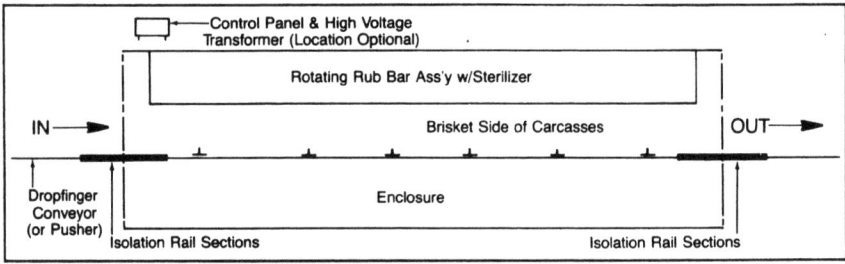

FIGURE C, MODEL #625C STIMULATOR with Rotating Rub Bar and Integral Bar Sterilizer (Plan View)

FIG. 7.7. Plan views of Omeco-Boss models #625A (Fig. A), #625B (Fig. B), and #625C (Fig. C).
Courtesy of Omeco-Boss.

For high-volume, automated electrical stimulation, the Omeco-Boss model 625C high-voltage unit consists of eight rotating bars. As each carcass moves through the stimulating enclosure, the brisket of each carcass contacts a sterilized bar that transmits the current during the time the moving carcass is rubbing against the bar (Fig. C of Fig. 7.7). After the carcass has received the electrical impulses, the bar rotates back into a partially enclosed sterilizing cabinet where it is sterilized with 82°C (180°F) water before it continues rotating back to stimulate another carcass. Capacity of the model 625C electrical stimulator ranges from 65 carcasses/hr [based on one unit of eight bars that are each 2.4 m (8 ft) long] up to 350 carcasses/hr [based on using several units of eight bars that are 2.4 m (8 ft) long].

Safety features of the model 625C include photo-eyes at the entrance and exit of the enclosure to automatically interrupt current if the photo-eye beam is broken, an emergency stop switch at each end of the enclosure, and an automatic shutoff of the high-voltage current if the overhead conveyor stops. An optional feature of this model is an automatic high-voltage circuit interrupter that automatically interrupts the high-voltage stimulus if some component in the circuit is shorted to ground. The interrupter module must be reset, the stimulator on-off switch must be pulled out completely, and the stimulation cycle start switch must be depressed before operation can be resumed.

Cervin. Cervin Automated Systems of Minneapolis, MN, have modified an electrical stunner to also perform electrical stimulation. The units contain output terminals for three stimulation probes as well as a stunning attachment. Isolated high voltage can be applied to each end of the carcass by means of one or two "cane hooks" that reach across the rounds of the carcass and a hand probe that can apply power to the neck of the carcass. The power panel of the stimulator/stunner is equipped with a power switch/breaker, seven-step rotary power control with indicators, a timer ranging from 0 to 5 min, and pilot and power-on lights.

Low-Voltage Electrical Stimulators

Koch-Britton. The Koch-Britton low-voltage electrical stimulators come in several models: model 150 LV is available for slaughter plants with capacities of 10 to 60 head/hr or in a high-speed version that stimulates in 15 sec and accommodates 40 to 150 carcasses/hr; the model 75 LV is available for slaughter plants with capacities up to 10 head/hr; and the model 100 LV, which is the company's export model and is designed for slaughter plants with capacities of 5 to 15 head/hr. All models mount on the wall near the bleeding area and electrical stimulation is administered to the animals by electrical lines extending from the units to a hook that can be attached to the nostrils on the head of the animal (Fig. 7.8). The 150 LV control unit measures about 40 cm × 50 cm × 75 cm (16 in. × 20 in. × 30 in.) so that it can fit into a small bleeding area in a slaughter plant. More than one unit may be installed to increase the number of carcasses handled, and each unit can be fitted with up to three nose hooks that conduct stimulation from the control unit.

The LeFiell Company. The LeFiell Company's low-voltage electrical stimulator is the "Lectro-Tender™" model #4051 LV, which is designed for plants with low slaughter rates. This model consists of a small stainless steel cabinet with operator controls and nose-clamp with a 9 m (30 ft) extension. The unit installs in the bleeding area of the plant. The cabinet, which plugs into a 120 volt outlet, has adjustments for voltage (up to 75 volts), cycle frequency, pulse duration, and stimulation time.

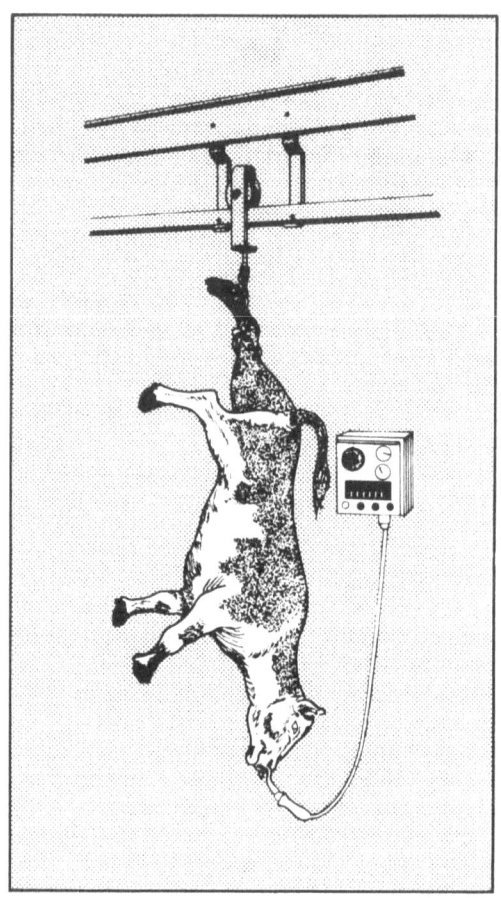

FIG. 7.8. Koch-Britton Stimulator 150 LV.
Courtesy of Koch Supplies.

Harneds. Harneds Inc., Wichita, KS, markets the Electro-Stim low-voltage electrical stimulator. This unit requires a 120 volt electrical hook-up and produces 45–48 volts of current at less than ½ amp. The Electro-Stim consists of a solid state control system in a stainless steel housing that measures 41 cm × 20 cm × 18 cm (16 in. × 8 in. × 7 in.), a spring-type nose clamp with a 7.5 m (25 ft) cord, and a 12 m (40 ft) ground cord. The control unit contains cycle and pulsation timers. The cycle timer automatically controls the duration of stimulation up to 150 sec. The pulsation timer has a dial marked from 0.5 to 3.5 sec. The recommended setting is 1 sec, which means that the current alternately is on 1 sec and off 1 sec for the duration of the cycle. During stimulation, a red light on the pulse timer and a red "stimulate" indicator lamp will flash on and off with the current.

7 INDUSTRIAL APPLICATIONS OF ELECTRICAL STIMULATION

Jarvis Corporation. Jarvis Products Corporation of Middletown, CT, is the supplier of the BV-80 low-voltage stimulator. This unit was developed in Argentina and has a capacity for 100 head/hr. The unit generates 21 volts RMS at 0.25 amp and utilizes a square wave form and a pincer in the nose for stimulation.

Van's International. Van's International, Silver Springs, MD, markets the "Stim-Ultima," a low-voltage electrical stimulator. This unit has both AC and DC capabilities and operates at 36 volts. Both nose clamps and rectal probes are sold with the unit.

INSTALLATIONS OF ELECTRICAL STIMULATORS WITHIN THE SLAUGHTER–DRESSING SEQUENCE

Like the many electrical stimulation units available, there is considerable variation with regard to the location of the unit installed within the slaughter–dressing sequence. The earliest location in the slaughter–dressing sequence where high-voltage electrical stimulation is being applied is the bleeding area. An advantage of this location is that additional blood forced from the carcass during stimulation can be readily processed. A surface-contact stimulator is required in this area to prevent contamination of the meat caused by penetration of the hide. Because contact is made with only the hide, no special sanitation of the equipment is necessary.

Certain disadvantages are associated with electrical stimulation in the bleeding area. Most plants shackle and suspend cattle by a single leg, and, because of this, there can be violent contractions of the free hindleg and subsequent damage to certain muscles. Secondly, because of the erratic jerking during electrical stimulation, some carcasses break the point of contact, causing some arcing of the electrical current and possible hide damage.

If low-voltage electrical stimulation is to be employed, it must be administered to carcasses in the bleeding area. Because low-voltage electrical stimulation causes the brain and central nervous system to elicit muscle contractions with a resultant decline in muscle pH, in order to be most efficacious, it must be applied to the animal within 10 min of stunning and bleeding. This time limit prevents the use of low-voltage electrical stimulation at any other location in the slaughter–dressing sequence. Low-voltage electrical stimulation does not require safety enclosures like those required for high-voltage electrical stimulation, and because the medium for administering electrical stimulation to the animal (a nose clamp in most instances) does not come in contact with the muscle tissue of the carcass, no special sterilization procedures are required.

The most common installations of high-voltage electrical stimulators involve locations in the slaughter–dressing sequence after hide removal but before evisceration, or after evisceration but before carcass splitting. Carcasses receiving electrical stimulation at these locations have their feet and heads removed, which prevents problems associated with the somewhat larger restricted areas required for administering electrical stimulation to cattle in the bleeding area. As alluded to in the "Safety, Installation, and Sanitation" section of this chapter, electrically stimulating carcasses after hide removal but before evisceration may cause some problems with fecal and urine contamination, although this can be prevented with proper preparatory practices. Whereas electrically stimulating carcasses after evisceration but before splitting appears to be the most ideal position (from a minimized problems standpoint), most slaughter facilities have limited space available in this area. Regardless of the position—after hide removal but before evisceration, or after evisceration but before carcass splitting—when electrical stimulation is applied, all probes or rub-bars that come into contact with the carcass must be sterilized between carcasses because the carcasses have not been officially inspected and passed by the USDA Meat and Poultry Inspection personnel at this point in the slaughter–dressing sequence.

The least common form of administering electrical stimulation is that applied to the split sides immediately prior to their entering the blast-chill cooler. Stimulating sides is not recommended because it is inefficient (there are twice as many to stimulate), it generally does not provide any advantage over stimulation of an unsplit carcass, and it is associated with violent contractions and lateral curling of the sides which can separate the vertebrae and pull the meat away from the feather bones. Slaughter plants that use this form of electrical stimulation usually do not have sufficient space available for installing an electrical stimulator at any other location in the slaughter–dressing sequence.

Although a variety of positions for locating electrical stimulators has been used by slaughter plants, the lack of uniformity in applying electrical stimulation to freshly slaughtered animals or to carcasses does not appear to lessen the positive benefits of using electrical stimulation for beef.

ELECTRICAL PARAMETERS

The electrical parameters for high-voltage electrical stimulation of beef used by the manufacturers of this equipment are almost identical. Manufacturers of electrical stimulators generally use electrical parameters based on those found most efficacious by McKeith et al. (1981). In their study, McKeith et al. (1981) evaluated 150 vs 550 volts, 0.5 to 5.0 amps, 17 (over 1 min) or 34 (over 2 min) impulses of 1.8 sec duration with a 1.8 sec interval between impulses administered to carcasses or sides at a variety of locations or stages in the slaughter–dressing sequence. They found that response in terms of tenderness increase was not closely associated with voltage but that lean color, freedom from heat-ring, and marbling scores appeared to be greatly improved by the use of the higher voltages. Neither number of impulses nor location within the slaughter–dressing sequence of administering electrical stimulation to the carcass had an impact on the results of their study. In general, most high-voltage electrical stimulation equipment has the following electrical parameters: 550–600 volts (AC), 5–15 amps, 60 cycles/sec, 15–20 impulses of electricity with a duration of 2.0 sec and an interval between impulses of 1.0 sec.

The electrical parameters for low-voltage electrical stimulators are less well defined, but the majority of low-voltage units operate in the range of 20 to 90 volts, usually 1 amp or less, and the current is applied to the carcass in pulses or in continuous form for about 15–20 sec.

SAFETY, INSTALLATION, AND SANITATION

The Food Safety and Inspection Service, in conjunction with the Occupational Safety and Health Administration, has developed safety and sanitation standards and requirements for the installation and use of electrical stimulation (ES) units or devices. According to Anderson (1980), electrical stimulators may be installed only upon approval by the Meat and Poultry Inspection's Facilities Group (who determine that there is adequate space to accomplish all of the slaughter floor functions, sufficient floor drains, and acceptable methods of construction) and Equipment Group (who assure that the stimulator is acceptable and who send the safety and sanitation operating procedures to the USDA Meat Inspector-in-Charge). Specific information regarding installation of electrical stimulators in federally inspected meat plants can be obtained from the Facilities, Equipment, and Sanitation Staff, Meat and Poultry Inspection, Food Safety and Inspection Service, United States Department of Agriculture, Washington, DC 20250.

Because of the danger involved with high-voltage electrical stimulation, persons who work near or operate such equipment must recognize the risk associated with high-voltage electrical stimulators, which produce lethal quantities of electricity. Some of the safety, installation, and sanitation concerns that must be followed by meat packers using electrical stimulation include the following (adapted from Anderson 1980): (1) in order to ensure worker safety, FSIS requires barriers at all openings to the stimulator, flashing or rotating lights, and audible signals to warn plant personnel that the unit is operating; (2) "Danger—High Voltage" signs must be displayed prominently, and emergency stop buttons to shut off the electrical current must be plainly labeled; (3) the power supply must be locked in the "off" position when not in use to prevent unauthorized personnel from turning on the stimulator; (4) a fail-safe system must be installed around the stimulator to prevent personnel from entering the area while the unit is operative; (5) all probes, bars, or contact surfaces by which electricity is applied to the carcass must be sterilized between carcasses; and (6) in electrically stimulating uneviscerated carcasses, care must be exercised to prevent fecal or urinary contamination of the carcass caused by the massive contractions—proper rimming and tying of the bung (and bladder in heifers) will normally alleviate any problems associated with feces or urine discharge.

TRADE NAMES AND PROMOTION OF ELECTRICALLY STIMULATED BEEF

Trade names for electrically stimulated meat products have evolved as commercial application has increased. Sam Kane Beef Processors, Corpus Christi, TX, call their product "ELECTRO-TENDER-aged®. They roller brand carcasses and use that trademark on their boxed beef, portion-controlled products, and trucks. Gooch Packing Company, Abilene, TX, calls its product Gooch Good'n Tender Beef® and makes point-of-purchase materials, such as gummed labels, posters and signs, and newspaper advertisement inserts available to retailers. Emge Packing Company, Anderson, IN, uses the trademark of Lectro-Tender® to identify their product and, like the Gooch Packing Company, uses many materials to aid retailers in selling electrically stimulated beef. Litvak Meat Company, Denver, CO, uses the name Electrolit® to identify their electrically stimulated beef.

Although there are other beef companies that promote electrically stimulated meat, many have the philosophy that because of the connotations to "shocking or electrocuting" animals (although dead) or carcasses, no mention to the sensitive consumer that carcasses have been electrically stimulated may be preferable.

INDUSTRY ADOPTION OF ELECTRICAL STIMULATION

Although the exact number of plants in the United States that use electrical stimulation would be hard to determine, it is safe to say that this process has been readily adopted by many plants with various slaughter capacities. Almost all of the major beef slaughterers in the United States have purchased and installed electrical stimulators in their plants, and several of these companies have made corporate decisions to adopt electrical stimulation in all of their facilities and to use this process as a sales and marketing tool. The adoption of low-voltage electrical stimulation began at a later time than that of high-voltage electrical stimulation, but low-voltage stimulation may already have surpassed high-voltage stimulation when measured by the number of units in operation. In this connection, though, the number of carcasses receiving high-voltage stimulation would be much greater than the number of carcasses receiving low-voltage stimulation due to the difference in slaughter capacities of plants that use high-voltage electrical stimulation versus the slaughter capacities of plants that use low-voltage electrical stimulation.

Even without knowing the actual number of beef carcasses in the United States receiving electrical stimulation, it is obvious that electrical stimulation has been an important development in the history of the United States beef packing industry. There may never be another technological advance in the meat industry that goes from research and development to the widespread industrial adoption stage as rapidly as has electrical stimulation.

SUMMARY

The incorporation of electrical stimulation into meat packing plants throughout the world has been largely due to the beneficial effect that electrical stimulation has on the quality and palatability of meat. With the interest shown by the U.S. meat industry regarding electrical stimulation, several companies have begun to manufacture commercial stimulators. The degree of complexity of commercial stimulators manufactured by these companies is varied; stimulators are made for slaughterers with capacities of 10 head/day up to 325 head/hr. The first commercially available electrical stimulator was the LeFiell Company's "Lectro-Tender™" model #4051 (installed in Sam Kane Beef Processors, Corpus Christi, TX, in November 1978). Since that time, hundreds of electrical stimulators have been installed in beef packing plants of all capacities in the United States. Manufacturers or distributors of electrical stimulation equipment include (listed in alphabetical order): Cervin Automated Systems, Harneds Inc., Koch-Britton, Jarvis Corporation, LeFiell Company, Omeco-Boss, and Van's International. Both high-voltage (550–600 volts) and low-voltage (20–90 volts) electrical stimulation units are available. Before electrical stimulators can be installed in federally inspected meat facilities, approval must be given by the Meat and Poultry Inspection's Facilities Group and Equipment Group. Several companies have developed campaigns to promote their brand of electrically stimulated meat. Few technological advances in the meat industry have gained widespread use as rapidly as has electrical stimulation.

ACKNOWLEDGMENTS

Appreciation is extended to the LeFiell Company, San Francisco, CA; Omeco-Boss, Inc., Omaha, NE; Britton Manufacturing Inc., College Station, TX; Koch Supplies Inc., Kansas City, MO; Cervin Automated Systems, Inc., Minneapolis, MN: Van's International, Silver Springs, MD; Jarvis Products Corporation, Middletown, CT; and Harneds Inc., Wichita, KS, for allowing the use of promotional and informational materials presented in this chapter. Also, the permission to use materials from the *National Provisioner*, Chicago, IL; *Meat Industry*, Mill Valley, CA; *Meat Processing*, Chicago, IL; and *Food Engineering*, Chicago, IL, is greatly appreciated.

REFERENCES

ANDERSON, R.W. 1980. Safety considerations and electrical stimulation. Proc. 26th Eur. Meet. Meat Res. Workers, Colorado Springs, CO, Vol. 2, K-5.

McKEITH, F.K., SMITH, G.C., SAVELL, J.W., DUTSON, T.R., CARPENTER, Z.L. and HAMMONS, D.R. 1981. Effects of certain electrical stimulation parameters on quality and palatability of beef. J. Food Sci. *46*, 13.

MEAT IND. 1982. A meat industry special guide—Electrical stimulation equipment. Meat Ind. (Feb.) 49.

MORRIS, C.E. 1979. Electrical tenderization. Food Eng. (Sept.).

SAVELL, J.W. 1979. Update: Industry acceptance of electrical stimulation. Proc. Recip. Meat Conf. *32*, 113.

SAVELL, J.W., SMITH, G.C., DUTSON, T.R. and CARPENTER, Z.L. 1980. Industry application of electrical stimulation in the United States. Proc. 26th Eur. Meet. Meat Res. Workers, Colorado Springs, CO, Vol. 2, K-2.

STIFFLER, D.M., SAVELL, J.W., SMITH, G.C., DUTSON, T.R. and CARPENTER, Z.L. 1982. Electrical stimulation: Purpose, application and results. Tex. Agric. Ext. Serv., College Station, TX, Bull. *B-1375*.

8

Cold Storage Energy Aspects of Electrically Stimulated Hot-boned Meat[1]

R. L. Henrickson[2] and A. Asghar[2]

Technologies of Hot Carcass Processing
Meat Yield and Losses
Quality Characteristics of Meat
Tenderness of Hot- and Cold-boned Meat
High Temperature Conditioning and Delay in Chilling
Electrical Stimulation
Innovation in Carcass Cutting
Functional Properties of Hot-boned Meat
Meat Curing
Microbiology of Hot- and Cold-boned Meat
Energy Conservation
Comparison of Energy Consumption During Chilling of Beef by
    Conventional and Hot Boning Methods
Effect of Plant Location on Transportation Energy
Industry Impediments
Summary
References

The continued increase (15 to 18%) in the number of animals slaughtered annually to meet the demand for beef products and the steady increase in the costs of meat processing strongly suggest a need to modify the existing processing methods to provide fabrication efficiency and energy conservation (Henrickson 1981). The prevalent commercial practices of chilling, freezing, and distributing carcasses consume tremendous amounts of energy (Rosoff and Ries 1976; Unger 1977). Consequently, the economic return is marginal. The energy requirements for processing per pound of live weight seem to vary with species, degree of processing, and plant type (Unger 1975). For example, in the case of beef, the energy expenditures vary from 750 to 1500 BTU/lb with an average of 910

---

[1]Journal Series Paper *4140* of the Oklahoma Agricultural Experimental Station. Financed in part by Station Project *2-4-2-1217*.
[2]Oklahoma Agricultural Experiment Station, Oklahoma State University, Stillwater, OK.

BTU/lb live weight. Processing of pork requires on the average 1650 BTU/lb live weight.

About 37 million cattle and over 3.2 million calves were slaughtered during 1983 in the United States. Processing of these carcasses requires a larger quantity of energy (over 99.3 trillion BTU) in different forms which accounts for about 12% of the energy used by the entire food industry in the United States (Unger 1975). The major shares of energy for the meat industry are contributed by natural gas (46%), electricity (31%), petroleum (14%), and coal (9%). This clearly indicates a need to conserve energy in meat processing. However, only a few attempts have been made to examine the economic impact that would motivate the meat industry to replace the prevalent processing practices with more efficient new technologies without risking any adverse effect on the quality characteristics of the meat. On the other hand, a vast amount of information is available on the potential advantages of new technological advances that could be adopted by the meat industry. Two of these are electrical stimulation and hot boning.

The significance of electrical stimulation and hot boning and the benefits associated with those technologies for the meat industry are presented in the initial sections of this chapter followed by energy conservation aspects.

TECHNOLOGIES OF HOT CARCASS PROCESSING

Traditionally, the beef carcasses, soon after dressing and splitting into sides, are transferred to a chilling room for a period sufficient to allow the completion of rigor and the conditioning process. The sides are then further divided into primal cuts for sale or are boned (cold) fabrication into meat products. Since the lowering of the carcass temperature from 100° to 40°F requires much space and inefficient use of energy, innovations in carcass processing technology were sought. One development in this direction emerged as "hot processing." This process involves the application of various technological operations to divide the prerigor carcass into primal cuts and/or to separate the muscle systems from extra bone, fat, and connective tissue prior to chilling, thereby substantially reducing the mass for cooling. This approach permits the excess fat to be rendered or used for sausage manufacture and the bones to be mechanically separated from the lean without undergoing the usual cooling–reheating cycle. Hot processing of carcasses is sometimes referred to as prerigor processing, prechill processing, hot boning, hot deboning, and accelerated meat processing. Extensive information has been reported on hot processing technology in relation to the quality characteristics of the meat. The important findings are summarized herein. More details on hot boning are also presented in Chapter 5.

MEAT YIELD AND LOSSES

Several workers have compared the yields and losses in hot- and cold-boned carcasses. Most of the research workers reported a significantly higher retail yield of salable meat by hot boning than by cold boning carcass sides (Brasington and Hammons 1971; Schmidt and Keman 1974; Taylor et al. 1981). Similarly, much smaller losses occurred in the weight of the hot rather than the cold-boned sides due to excessive evaporation in the latter case (Kastner et al. 1973; Falk et al. 1975; Kastner and Russel 1975). It has been shown that in an efficiently operated cooler the evaporative losses from carcasses may be in the magnitude of 2% (Lovett et al. 1976) or even 4%, of which 62% occurred during the initial 8 hr of cooling (Am. Soc. Heat., Refrig. Air Conditioning Eng. 1971B). Although partial dehydration of the carcass surface seems to be beneficial in decreasing the water activity and, hence, limiting the growth of some microbes (Leistner and Rodel 1975), it constitutes significant economic loss to the processor. In addition, the excessive dehydration of the meat surface adversely affects the appearance and color of the carcass.

The factors that account for evaporative losses from carcasses during chilling (Visser et al. 1976) are: (1) the difference in water vapor pressure between the surrounding air and the carcass surface, (2) evaporative losses induced by forced air circulation around the carcasses, and (3) evaporative losses due to heat transmission from the carcass surface. The contribution of all these factors can be minimized if, soon after dressing, the carcasses are severed into primal cuts and vacuum packed in moisture-proof films before subjecting them to a chilling temperature. The combined effect of controlled oxygen (1%) and carbon dioxide level (20%) developed in vacuum packs helps in inhibiting the growth of gram-negative aerobes (Seideman et al. 1976; Newton et al. 1977). The "boxed meat" technology originated from these concepts. Follett et al. (1974) observed 25% less loss in weight of hot-boned, vacuum-packed semimembranosus muscles than for the same cold-boned cut during 24 hr at 0°C.

The data in Table 8.1 summarize a detailed study on salable meat obtained by hot and cold processing of beef carcasses (Taylor et al. 1981). It can be seen that hot boning gave 2.6% more yield than cold boning. This difference was mainly due to evaporative losses, while improper trimming of fat, slightly heavier bones, and less dripping loss also contributed to the overall yield of hot-boned carcasses. This suggested that an effective method needed to be developed for the separation of lean from bone to realize the full potential of yield. At the same time, the industry also needed an improvement in fat trimming practices without additional labor expenditure.

Mandigo et al. (1977) compared the yield of hot and cold pork processed carcasses on a commercial scale and found no difference between the two methods of processing. They did not, however, use a film for packing the hot cuts to guard against the evaporative losses that can make the dif-

TABLE 8.1. Comparison of the Effects of Cold and Hot Deboning on Meat Yield from Opposite Sides of 16 Animals [Mean % of Hot Side Weight (with Standard Errors)]

	Cold deboning yield	Hot deboning yield	Significance of difference
Hot side	100.0	100.0	
Cold side	93.3 (0.08)		
Total usable meat	73.6 (0.38)	76.2 (0.28)	***
Bone	16.3 (0.23)	17.6 (0.22)	***
Fat trim	8.2 (0.37)	5.6 (0.27)	***
Evaporation loss during chilling	1.7 (0.08)		
Evaporation loss during boning, jointing, and packing	0.2 (0.06)	0.6 (0.06)	***

Source: Taylor et al. (1981).
***$P < 0.001$.

ference in yield between these two methods. Thus, proper packing of meat cuts in moisture-proof films such as polyvinyl chloride is important in order to realize the full yield potential of hot boning.

QUALITY CHARACTERISTICS OF MEAT

Many researchers have compared the quality characteristics of hot- and cold-boned meat (Weiner et al. 1966; Trumic et al. 1967; Kotter et al. 1968; Reddy and Henrickson 1969; Ungethüm 1971; Kowalski et al. 1972; Golowkin et al. 1973; Micyk et al. 1973; Proselkova and Bargajewa 1973; Pisula et al. 1973A,B, 1974, 1976). Generally, prerigor meat possesses significantly greater water holding capacity (WHC) than postrigor meat. The high WHC of prerigor meat is thought to be due to a high ATP content and, to some extent, to high pH (Hamm 1972; Honikel and Hamm 1978). This is probably why some workers also found hot-boned meat to be more juicy than cold-boned meat.

Taylor et al. (1981) reported that hot-boned meat was more uniform and had more color than cold-boned meat. The difference was ascribed to the greater exposure of surface area in the case of hot-boned meat, which also cooled relatively more evenly. This was particularly true for thick muscles, which otherwise cool at a slow rate. However, little difference has been reported for the juiciness, flavor, and overall acceptability of cooked meat from hot- and cold-boned carcasses (Schmidt and Keman 1974; Kastner and Russell 1975; Dransfield et al. 1976). The study by Strange and Benedict (1978) showed that, as compared with cold-boned meat, hot-boned meat had a higher free tyrosine value, indicating a higher proteolytic breakdown rate (Pearson 1968) and/or high microbial contamination (Strange et al. 1977), and a higher creatine content, which is negatively related to flavor development (Macy et al. 1979; Russell and Baldwin 1975). The hot-boned meat color, as measured in terms of $\triangle \% R$ ($\triangle \%R = \%R$ 630 nm $- \%R$ 580 nm; where %R at 630 and 580 nm is reflectance minimum of metmyoglobin and oxymyoglobin, respectively), was less acceptable than cold-boned meat color (Kastner and Russell 1975; Strange and Benedict 1978).

TENDERNESS OF HOT- AND COLD-BONED MEAT

The tenderness of meat as influenced by various ante- and postmortem factors has been reviewed in detail by Asghar and Pearson (1980) and Asghar and Henrickson (1982). There is general agreement among scientists that a marked decrease in tenderness does result if hot-boned meat is immediately subjected to a chilling temperature. The phenomenon commonly known as "cold shortening" is responsible for the toughening of prerigor meat when chilled at about 0°C before the onset of *rigor mortis* (Locker and Hagyard 1963). However, some muscles such as the biceps femoris and psoas major are not as adversely affected on prerigor chilling of the carcass as are the other muscles (Marsh *et al.* 1972; McLeod *et al.* 1973; Falk *et al.* 1975).

Biochemistry of Cold Shortening

The physiological contraction of muscle has been attributed to cyclic formation and breaking of cross-bridges between actin and myosin filaments according to the Huxley "sliding filament-moving bridge" model (Huxley 1974). In this process, ATP is enzymatically hydrolyzed by Ca^{2+}-activated actomyosin ATPase to provide energy for the contracting muscles. At the same time, the ATP content is restored for awhile from two main sources, creatine phosphate and glycolysis. This is an oversimplification of a very complex set of events, of which some are not yet completely understood. However, this aspect has been thoroughly discussed by Ashgar and Henrickson (1982) and is reviewed in greater detail in Chapter 6. Some excellent reviews on this subject by Cohen (1975), Squire (1975), Endo (1977), Frank (1980), and Adelstein and Eisenberg (1980) are available.

However, during cold shortening, the muscle fibers may decrease even up to 60% of their length depending on the amount of cold stress and the type of fibers. The red muscles, which contain high amounts of mitochondria, shorten more vigorously than white muscles. In the beginning, it was suggested that Ca^{2+} ions released from the sarcoplasmic reticulum into the myofibrillar regions at low temperatures were responsible for cold shortening of muscle (Pearson *et al.* 1973); but later studies revealed that the mitochondria release large amounts of Ca^{2+} ions at low temperatures and hence overload the sarcoplasmic reticulum. Consequently, an excess of free Ca^{2+} ions initiates cold shortening (Buege and Marsh 1975; Cornforth *et al.* 1980). Furthermore, a high pH value and high ATP levels are both prerequisites for cold shortening, whereas cold as such provides the direct stimulus for shortening (Asghar and Pearson 1980).

Since cold shortening was found to cause significant toughening of meat (Marsh and Leet 1966; Davey et al. 1976), the problem of cold shortening has been one of the serious drawbacks in hot boning technology despite many obvious advantages of hot processing for the meat industry. Several efforts were made to minimize the incidence of cold shortening in the carcass. Some of those approaches that have proved successful and feasible for the meat industry are: (1) high-temperature conditioning and delayed chilling, and (2) electrical stimulation.

HIGH TEMPERATURE CONDITIONING AND DELAY IN CHILLING

The temperature at which a carcass is stored greatly influences the extent and severity of rigor mortis and cold shortening, hence tenderness. Because of its high temperature coefficient, the rate of glycolysis increases with the temperature (Cassens and Newbold 1967). It has also been found that shortening of excised muscle increases as the cooling temperature decreases from 51° to 0°C (Locker 1960; Wilson et al. 1960; Goll 1968; Forrest et al. 1966). The shortening is least at temperatures between 15° and 20°C (Locker and Hagyard 1963). These studies have led to the concept of high temperature conditioning and delay in chilling of hot-boned carcasses. In view of the biochemistry of cold shortening, the following two technological approaches have been developed so as to increase the glycolytic rate and to encourage rigor mortis development at a temperature that will support the least cold shortening.

Conditioning of Hot Meat Cuts at High Temperatures

Some researchers have tried conditioning hot-boned meat for a period of time at temperatures ranging from 5° to 15°C. This time period allows rigor mortis to set in before chilling and helps to produce tender meat (Schmidt and Gilbert, 1970; Follet et al. 1974; Schmidt and Keman 1974; Weiner et al. 1966; Dransfield et al. 1976). However, this technique may suffer from some other serious drawbacks. The more exposed surfaces of cuts are likely to be contaminated and the high temperature conditioning may favor microbial growth, reducing the shelf-life of the meat.

Conditioning of the Carcasses at High Temperature Prior to Cutting and Cooling

Kastner et al. (1973, 1976), Falk et al. (1975), Kastner and Russell (1975) and Will and Henrickson (1976) have studied the conditioning of carcasses at 9°–15°C for 2 to 10 hr prior to boning and chilling. This approach also results in tender meat and minimizes cold shortening. The severity of contamination may be much less with this method as compared with the conditioning of hot meat cuts.

All these techniques have their place in minimizing cold shortening and improving tenderness, as does mechanical tenderization. However, the delay in chilling and holding the carcass at a high temperature is likely to favor rapid microbial growth on the surface. Contamination from the air and bacterial growth during the conditioning period will generally be greater in hot-boned meat handled in this manner.

ELECTRICAL STIMULATION

The application of postmortem electrical stimulation technology in meat science has helped to reduce the length of the delay in chilling time from 9–12 hr to 1–3 hr (Carse 1973; Chrystall and Devine 1978; Bendall 1976; McCollum and Henrickson 1980). This also reduces the risk of bacterial multiplication which would be likely during high temperature conditioning of the carcasses.

The historical events and philosophy that led to the application of electrical stimulation in the meat industry have been described at length by Asghar and Henrickson (1982). It is generally agreed that a high ATP content and a pH value at which Ca^{2+} ions are released from SR or mitochondria are the prerequisites for cold shortening. The underlying theme in the application of electrical stimulation to carcasses is to rapidly deplete the high energy compounds such as adenosine triphosphate (ATP), adenosine diphosphate (ADP), and creatine phosphate (CP) from the musculature while the carcass temperature is still high. Once the energy sources (ATP and CP) are exhausted, the muscle becomes inelastic and does not respond to cold stress. If the ATP has been depleted to 50% of the initial content, little cold shortening is expected (Bendall 1980). Incidentally, the pH of the muscle also drops rapidly due to conversion of glycogen into lactic acid during electrical stimulation, making the condition still more nonconducive for cold shortening.

A recent study by Rashid et al. (1983A,B) showed that excised semitendinosus (ST) muscle from electrically stimulated lamb carcasses shortened only 13.7% when rapidly chilled, whereas similar muscle from unstimulated carcasses shortened 20.4%. Similarly, Ca^{2+} ion-induced shortening for electrically stimulated ST muscle was 14% less than the control (20 vs 34%). Consequently, the meat from electrically stimulated carcasses was found to be 15% more tender than the control (Rashid et al. 1983B). Most investigators agree that postmortem electrical stimulation of carcasses generally produces a tenderizing effect on meat (Harsham and Deatherage 1951; Carse 1973; Chrystall and Hagyard 1976; Smith et al. 1979; Bouton et al. 1980; George et al. 1980; Ruderus 1980; Valine 1981). Several other quality factors have been reported to be associated with electrical stimulation of carcasses. For example, Pearson (1981) has quoted Texas workers as indicating that electrical stimulation of beef improved lean firmness score by 41%, panel tenderness by 26%, lean maturity by 23%, heat ring score by 23%, lean color score by 14%, marbling score by 11%, USDA quality grade by 8%, and flavor score by 6%. However, some of these claims have not been substantiated by several other studies as discussed by Asghar and Henrickson (1982).

According to Pearson (1981), Texas workers also reported that electrical stimulation greatly accelerated the cooling rate of carcasses. Even though the initial temperature of the electrically stimulated sides was shown to be higher than for nonstimulated sides, the former reached the cooler temperature about 2 hr earlier than the latter. If so, this will add to the benefits of electrical stimulation of carcasses as one of the sources in energy conservation by the meat industry. However, other workers have found little difference in cooling rates of electrically stimulated and nonstimulated carcass sides. Asghar and Henrickson (1982) have discussed these aspects of stimulation in detail.

According to the estimate of Hall et al. (1980), electrical stimulation costs $0.18/carcass; balanced against this expense is the saving in energy by shipping the carcasses a day sooner. This also results in a reduction in shrink loss. According to them, the biggest dollar-value benefit is in producing more choice meat by electrical stimulation of carcasses, while the energy use amounts to only about 2.5 kWh/head or a cost of about $0.15/head.

INNOVATION IN CARCASS CUTTING

Another serious problem associated with hot boning of carcasses is the preparation of conventional cuts with the desirable shape because of the pliable nature of the hot meat. Concern has also been expressed that British (Fig. 8.1A,B) and American (Fig. 8.2A) carcass cutting methodologies have relied more on the skeletal system to produce the desired cuts than on considering the muscle systems and their specific eating characteristics (Cuthbertson 1980).

The underlying principle of retail cutting in Britain is to bone out primal cuts followed by subdivision of the primals into muscle groups. These cuts hold together during cooling and can be carved across the muscle fibers to facilitate chewing (Strother 1975). The Meat and Livestock Commission (Meat Livestock Comm. 1974) emphasized the need to adopt a standard procedure that would provide better cutting efficiency and high quality finished products. Figure 8.1B illustrates the cutting lines of such a method superimposed on a skeletal diagram. Even with this method, some problems are associated with positioning the cuts that are parallel to the long axis of the side (Strother 1975).

There is quite a lot of similarity between British and American beef carcass cutting systems, especially with respect to the breakdown of the hindquarter and forequarter into primals (Fig. 8.2A). However, the American method is modified to facilitate the application of the band saw by following cutting lines that intersect the bones instead of following the contours of the joints. The obvious feature of American retail cutting methodology is the variety of steaks that are made for grilling. These methods increase the number of cut surfaces and hence increase the probability of drip loss and bacterial contamination. However, Gill and Newton (1978) remarked that growth of bacteria from both intact and cut muscle is likely to be similar since the membranes surrounding the muscles do not present a barrier to diffusion of substrates utilized by bacteria.

The French method of carcass cutting differs from the American and British methods in several ways (Fig. 8.2B). One of the major differences is in the preparation of retail items using boneless or partially boned cuts to ensure little variation in product quality due to different muscles. Besides this, another difference is the separation of the fore and hindquarter into the 5-rib forequarter and the 8-rib or "pistola cut" hindquarter (Strother 1975). The shoulder (*raquette*) is also removed from the forequarter. According to the French practice, more cuts are made along the seam between muscles rather than tangent to muscle fibers. For example, some muscles, such as the extensor carpi radialis and semimembranosus, are removed at their point of insertion, instead of cutting through them at the joint. However, despite certain attractive features of the French method of cutting, it is too costly in terms of man-hours of labor. For example, for retail cutting, the American method of cutting a beef side requires 1½ hr, the British method 3½ hr, and the French style 7 hr to prepare retail cuts.

Thus, keeping in view the merits of each cutting method, a modification of the existing carcass cutting procedures may help resolve some problems associated with hot cutting. Cuthbertson (1980) has expressed the view that the muscle seaming technique may prove particularly useful in hot processing because of the extensibility of the muscles and the softness of the fat in prerigor carcasses. Such innovations need to be explored.

FUNCTIONAL PROPERTIES OF HOT-BONED MEAT

Prerigor meat has been shown to possess a significantly higher emulsifying capacity for fat than postrigor meat (Saffle and Galbreath 1964; Acton and Saffle 1969; Froning and Neelakantan 1971), with more uniform and rounder fat globules. The greater extractability of salt-soluble proteins (myosin, actin, tropomyosin) in prerigor meat is thought to be responsible for this property (Johnson and Henrickson 1970). However, the binding strength of crude myosin, (extracted with 1.0 M NaCl and containing 0.15, 0.25, or 0.5% tripolyphosphate) from pre- and postrigor meat did not differ significantly (Turner *et al*. 1979). The high emulsifying capacity of prerigor hot-boned meat is advantageous in sausage production. Vogel (1963) claimed that the sausages made from hot-processed meat were of superior quality and possessed a longer shelf-life. Hamm and Potthast (1975) have also developed a procedure for sausage production from prerigor, freeze dried beef. The physiochemical and functional properties, such as WHC, color, flavor, and emulsion stability of the final product, have been reported as good as those from freshly prepared hot-

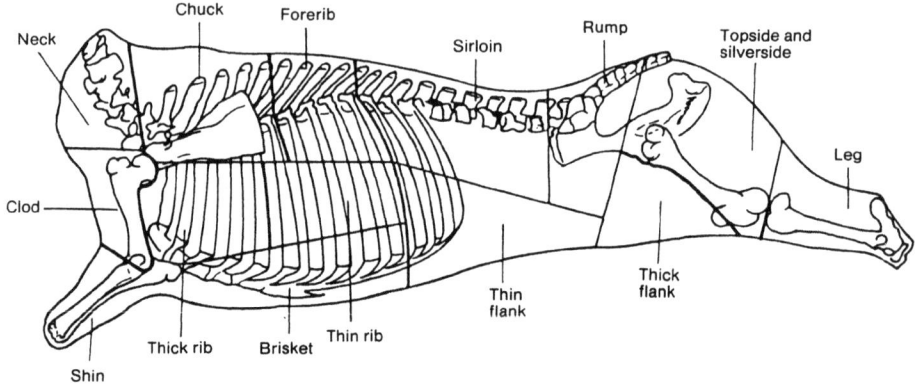

FIG. 8.1A. A British method of beef cutting.
From Strother (1975).

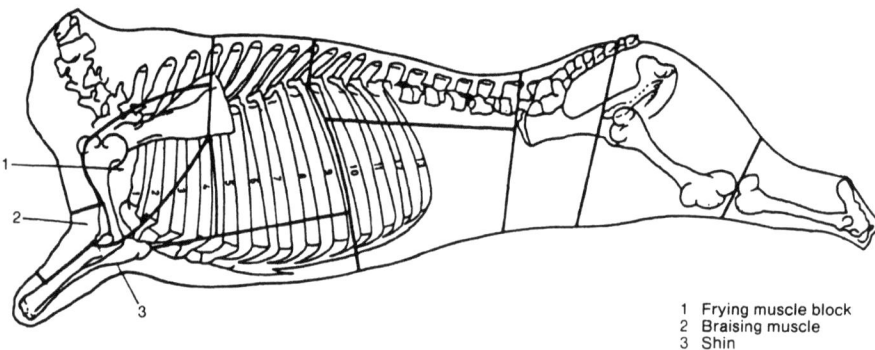

1 Frying muscle block
2 Braising muscle
3 Shin

FIG. 8.1B. Meat and Livestock Commission forequarter cutting method. (Note this side is quartered between the ninth and tenth ribs.)
From Strother (1975).

boned meat. Even those studies that found little difference in the quality characteristics of sausages made from hot- and cold-boned meat (Szaluszkowa *et al.* 1971, 1975; Tulejow and Jejeszchlebow 1973; Stilwell *et al.* 1978) should be regarded as a point in favor of using hot-boned meat in sausage production. There is, however, general agreement that the final yield of sausages made from hot-boned meat is 4 to 6% higher than that

8 COLD STORAGE ENERGY ASPECTS OF ES HOT-BONED MEAT

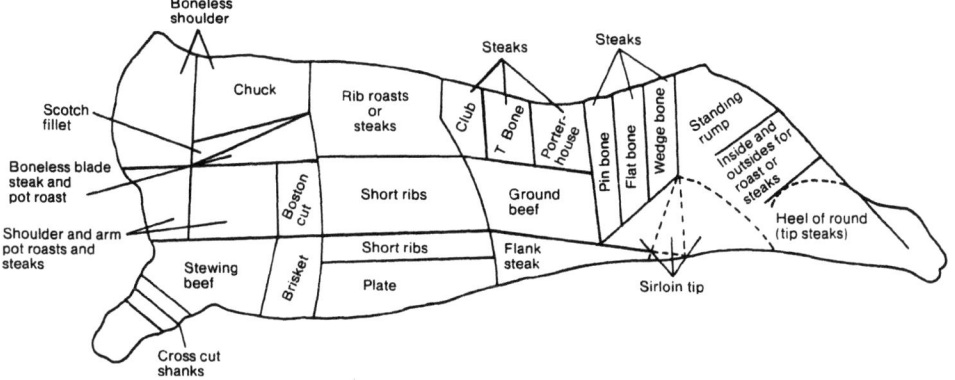

FIG. 8.2A. An American method of beef cutting.
From Strother (1975).

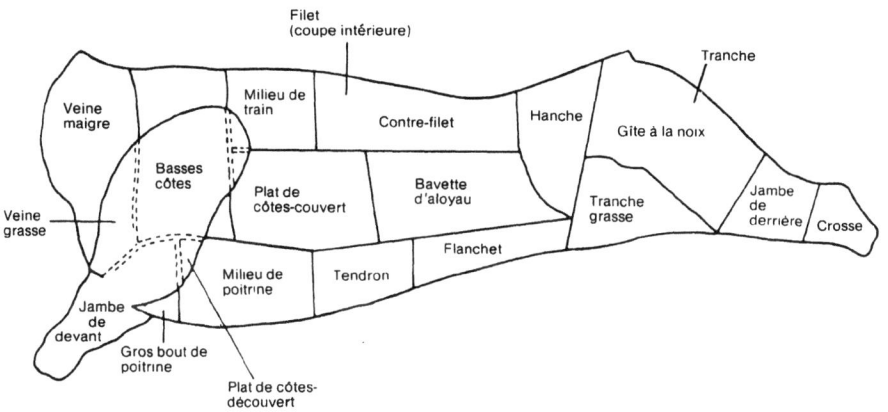

FIG. 8.2B. A French method of beef cutting.
From Strother (1975).

from cold-boned meat (Golowkin *et al.* 1973; Micyk *et al.* 1973; Tulejow and Jejeszchlebow 1973).

The difference in sausage yield between cold- and hot-boned meat seems to depend on the type of sausage. For instance, Pisula (1981) has reported that the yield of frankfurters (fine emulsion), zwyezajna (semifine emulsion), and szynkow (coarse emulsion) sausages was higher by 12, 7, and 3%, respectively, when prepared from hot-processed meat than from cold meat. Advantages in the yields for other fabricated products made by using hot-boned meat have also been reported. For example, Pepper and Schmidt (1975) obtained higher yields and better quality of beef rolls when using hot meat than cold meat. Similarly, Cross et al. (1976) and Jacobs and Sebranek (1980) considered higher cooking yields and less shrinkage of patties made from hot meat an advantage over those made from cold-boned meat.

MEAT CURING

Many studies have proved that the rate of diffusion of the curing ingredients into prerigor meat is much faster than in postrigor meat (Arganosa and Henrickson 1960; Jaroszewski 1974; Jablonski 1979; Pisula et al. 1973C; Wirth 1974). The ultimate yield of the product is either superior or similar to cold-cured meat (Mandigo and Henrickson 1966; Henrickson 1968; Cieloch and Turkowski 1968A; Pisula et al. 1974; Mandigo et al. 1977). Weiner et al. (1966) have reported that prerigor hams pumped with cold brine resulted in markedly lower drip and cooking losses than postrigor brined hams. If this is so, the ultimate yield is likely to be higher for prerigor cured meat.

The color of prerigor cured meat has also been claimed to be better and more stable than that of postrigor cured meat (Arganosa and Henrickson 1960; Henrickson 1968; Parr and Henrickson 1970). However, some workers did not find any difference in color stability, nitroso, and total pigments between hot- and cold-cured meat (Mandigo and Kunert 1973). On the other hand, Borkowska and Romanowski (1972) reported somewhat lower color scores for canned ham prepared from hot-boned pork with or without polyphosphate addition than for cold-processed ham treated similarly. In contrast to this, Skenderovic and Rankov (1975) and Hoes et al. (1980) noted a more marked, beneficial effect on addition of polyphosphates to prerigor than to postrigor meat.

MICROBIOLOGY OF HOT- AND COLD-BONED MEAT

Raw Meat

Pseudomonas and lactobacilli species generally cause spoilage of fresh meat at refrigeration temperatures under aerobic and anaerobic conditions, respectively (Ayres 1960; McMeekin 1975). The spoilage is confined mainly to the meat surface where low molecular weight, soluble components such as glucose and free amino acids provide the substrates for these microbes (Ockerman et al. 1964; Gill and Newton 1978). Log CFUs (colony forming units)/cm^2 or gram of meat are regarded as the bacterial spoilage index for raw meat.

Divergent reports have appeared on the storage life of hot-boned meat in comparison with conventionally handled cold meat. Some workers found that prerigor, vacuum-packed cuts, stored at 15°C for 48 hr, had more bacteria than cold-boned meat (Schmidt and Gilbert 1970). Follett et al. (1974) also observed a marked increase in bacterial counts in hot-boned as compared with cold-boned meat. In contrast, Falk et al. (1975) and Taylor et al. (1981) noted no real difference in total viable count between hot- and cold-boned meat. There was also no difference in the number of E. coli and Clostridium perfringens organisms in hot- and cold-boned meat (Taylor et al. 1981). Others have observed slower bacterial growth in hot than in cold-boned meat (Strange and Benedict 1978). Another study reported that aerobic plate counts (APC) for ground beef from hot-boned beef were either markedly lower or not significantly different from aerobic plate counts for ground beef from cold-boned carcasses (Emswiler and Kotula 1979). There were also no real differences of any practical importance in the most probable numbers (MPN) of coliforms and E. coli between hot- and cold-boned ground beef stored at 0°C.

Whether meat is hot- or cold-boned, the rate of cooling is the most important factor in determining the extent of microbial growth in meat. The work of Fung et al. (1980) strengthens this proposition. They have shown that hot-boned, vacuum-packed beef cuts had higher mesophilic and psychrotropic populations than conventionally chilled beef, which they attributed to the slower cooling rate in boxed meat. In a subsequent study, Fung et al. (1981) noted that hot-boned, vacuum-packed beef cuts cooled to 21°C in 3 and 5 hr or beef carcasses chilled at 2.2°C for 48 hr had lower total bacterial counts, and had more coliforms, more *Staphylococcus aureus,* more *C. perfringens,* and more streptococci organisms than the hot-boned meat that had been cooled in 9 to 12 hr to 21°C. A well planned study by Lara (1982), using different box sizes and three locations within each box, substantiated the findings of Fung et al. (1981). Hot-boned beef trimmings were cooled in boxes 4, 8, and 12 in. deep using an air temperature of 5°C with an air movement of 500 ft/min. Even though the microbial levels were high, the meat did not become putrid during 5 days storage at 5°C.

Electrically Stimulated Meat

Inconsistent reports have appeared in the literature on the microbiology of hot-boned meat from electrically stimulated and unstimulated carcasses. Whereas some workers found little difference in psychrotropic bacterial counts either initially or at the termination of display of electrically stimulated and unstimulated ground beef or steaks (Gill 1980; Hall et al. 1980; Jeremiah and Martin 1980; Taylor et al. 1981; Kotula 1981), other researchers have shown significantly lower bacterial growth on meat from electrically stimulated carcasses than from unstimulated ones (Raccach and Henrickson 1978; Mrigadat et al. 1980; Contreras and Harrison 1981).

Several explanations have been given for the significant differences in microbial growth on meat from electrically stimulated and unstimulated carcasses, based on the assumption of different redox potentials (Mrigadat et al. 1980) or the deleterious effects of electricity and of proteolytic enzymes (Riley et al. 1980). However, if not for those reasons, it could at least be expected that acceleration of postmortem glycolysis by electrical stimulation would permit rapid chilling of carcasses. This would help in extending the lag phase of microbial growth on the carcass surface, thereby resulting in more wholesome product with a longer shelf-life as compared with the product from unstimulated carcasses, which have been held at high temperatures for a period of time before cooling to avoid cold shortening (Asghar and Henrickson 1982).

Processed Meat

Controversial results have also been reported about the microbiology of processed meat products from hot- and cold-boned carcasses. For instance, Pulliam and Kelley (1965) found higher bacterial counts in hot-processed hams than in those that were cold processed. Similarly, the studies by Mandigo and Henrickson (1967) and Davidson et al. (1968) indicated higher mesophilic and psychrophilic bacterial counts in hot than in cold-processed pork sausages, although the peroxide value was lower in the former case.

In contrast, some workers have found the microbiological status of hot-processed meat to be either lower or similar to that of cold-processed meat (Barbe et al. 1966; Barbe and Henrickson 1967; Cieloch and Turkowski 1968A,B; Henrickson 1968; Proselkova and Bargajewa 1973; Pisula et al. 1974, 1976). Lin et al. (1979) have reported that prerigor sausages stored at 2° to 5°C had markedly lower total aerobic mesophile and lipolytic bacterial counts than similar sausages made from postrigor meat. Despite these differences, the acceptance scores were also surprisingly higher in the former case.

ENERGY CONSERVATION

In view of the rapid increase in energy costs of refrigeration and freezing, improved design of the cooling system based on the heat transfer characteristics of a food is a very important factor in keeping the processing costs low in the meat industry. In addition, the efficiency of cooling can also be improved by altering the geometry of the products without changing their quality characteristics. For the chilling of beef carcasses, both of these factors were duly explored in a series of studies at Oklahoma State University (Ferguson and Henrickson 1979). The important findings from these studies are summarized herein.

Cooler Design for Hot-boned Meat

The thermodynamic considerations of heat exchange suggest that a counterflow design is relatively more effective for better utilizing the energy source than other models. Two energy inputs are involved in this design. One energy source is for creating a temperature potential for energy transfer, and the second is for generating a pressure difference for maintaining the circulation of a given quantity of fluid at the required velocity (Ferguson and Henrickson 1979). On the basis of these considerations, a counterflow conveyorized cooler model was proposed by Ganni (1979). The counterflow conveyorized design is thought to have the highest effectiveness of any configuration with regard to optimum heat transfer, energy expenditure, and uniformity in the quantity and velocity of cooling airflow at minimum cost. Hence, this design is very close to the ideal system.

By using the counterflow conveyorized model, the relationship between heat transfer coefficient and energy transfer was determined for meat (Ferguson and Henrickson 1979). When the heat transfer coefficient was below 2 BTU/hr-ft^2-°F, the energy transfer was controlled by the heat transfer coefficient, whereas between 2 and 10 BTU/hr-ft^2-°F, the energy transfer was governed by both internal resistance and the heat transfer coefficient. It was further noted that the decrease in cooling period achieved by increasing the heat transfer coefficient beyond 10 BTU/hr-ft^2-°F was very small because the energy transfer was mainly controlled by the internal resistance of the meat.

Geometry of Carcass

One of the important factors that governs the rate of cooling is the thickness of the tissues. The average intact side of beef is 5 to 6 in. thick and requires about 25 hr (chilling room) to 48 hr (holding room) to reach the cooling air temperature (Am. Soc. Heat., Refrig. Air Conditioning Eng. 1971B; Bailey 1972A; Levy 1972). From engineering considerations, the thickness of the hindquarter not only limits the cooling rate, but the overall geometry of the side also restricts the optimum use of the chilling space. Furthermore, heavy duty refrigeration units must be employed to provide a uniform circulation around the sides. Ramsbottom and Strandine (1949) reported that the rate of cooling boneless beef cuts was faster than the rate of cooling the intact sides. About 10° to 15°F difference in temperature existed during the first 8 hr chilling period.

These facts led to the concept of hot boning technology as a means for energy conservation in the meat industry. In addition, the objectives of hot processing are to implement as much processing as possible on the hot meat before cooling and to avoid the interposition of a cooling phase for meat, which later on has to be heated for subsequent processing. On the

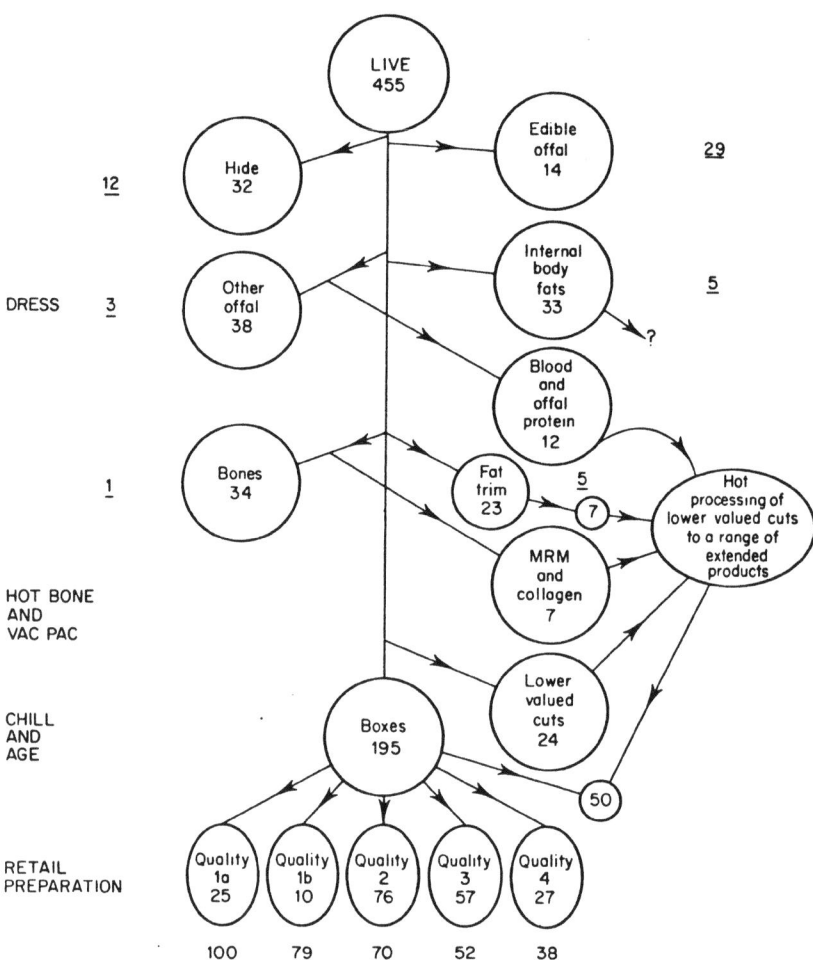

FIG. 8.3. Flow chart showing progressive breakdown of a 455 kg steer by a procedure that involves hot boning, followed by vacuum packaging of some primals, and hot processing other cuts incorporating recovered protein and fat. The figure within each circle is the weight in kilograms. The figure outside each circle is an index of value, with best steaks and roasts = 100. Total weight of meat product is 196 kg.
From Cuthbertson (1980).

basis of these considerations, a flow sheet diagram is presented in Fig. 8.3 for such a processing design.

From the viewpoint of heat transfer, two important factors that account for the fast cooling of hot-boned meat are the reduction in mass and the reduction in size of cuts. A study at Oklahoma State University, where various cuts of meat were grouped on the basis of thickness, indicated that 11 different cuts represented 59.8% of hot-processed yield (Table 8.2). The remaining 40.2% constituted the small pieces (lean) with an average thickness of 1.5 in. (group C). Among the 11 cuts, those having an average thickness of 2.8 (±0.3) in. constituted 29.3% by weight of the total yield (group A). The thickness of the other 7 cuts on the average was 1.9 (±0.3) in. and they composed 30.5% by weight of the total yield (group B).

TABLE 8.2. Modeling of Beef Cuts

Group	% by weight	Average thickness (in.)	Load factor 16 m/ft^2
A	29.3	2.8	15
B	30.5	1.9	10
C	40.2	1.5	8

Source: Ferguson and Henrickson (1979).
Loading factor (16 m/ft^2) = thickness × density. This is the weight of the product of a given group that can be loaded on a unit conveyor area.

Saving in Chilling Space

Hot boning the bovine carcass has numerous potential advantages for the meat industry. Most notable is the removal of excess fat and bone prior to chilling, thereby reducing the need for cooling energy. Storing a 600 lb carcass in a chill room requires valuable space. The cost for utilization of this area increases directly with the cooling period. The refrigerated space usually set aside for chilling a 600 lb carcass is 80 × 36 × 30 in. or 86,400 in.^{3} Space above and below the hanging carcass usually requires an additional 34,000 in.^{3}, making a total of 120,400 in.^{3} of space just for chilling. An equal amount of holding space (0°C) may also be provided, making a combined total space allocation of over 240,800 in.^{3}/carcass. Space allocation for fabrication must also be considered since this area is also cooled (12.8°C). The present practice is to chill the full carcass 24 hr before it is ribbed and prepared for shipment or fabrication. During this period, an attempt is made to reduce the internal temperature of the round from 50° to 38.9°C. An additional cooling period of 36 to 48 hr is required to reduce the internal round temperature to 0°C.

By contrast, the edible portion of a 600 lb carcass may be cooled using only 26,000 in.^{3} on a conveyor belt. Thus, the refrigerated space requirement for cooling the smaller pieces from the carcass may be more than 80% less than the present space allocation.

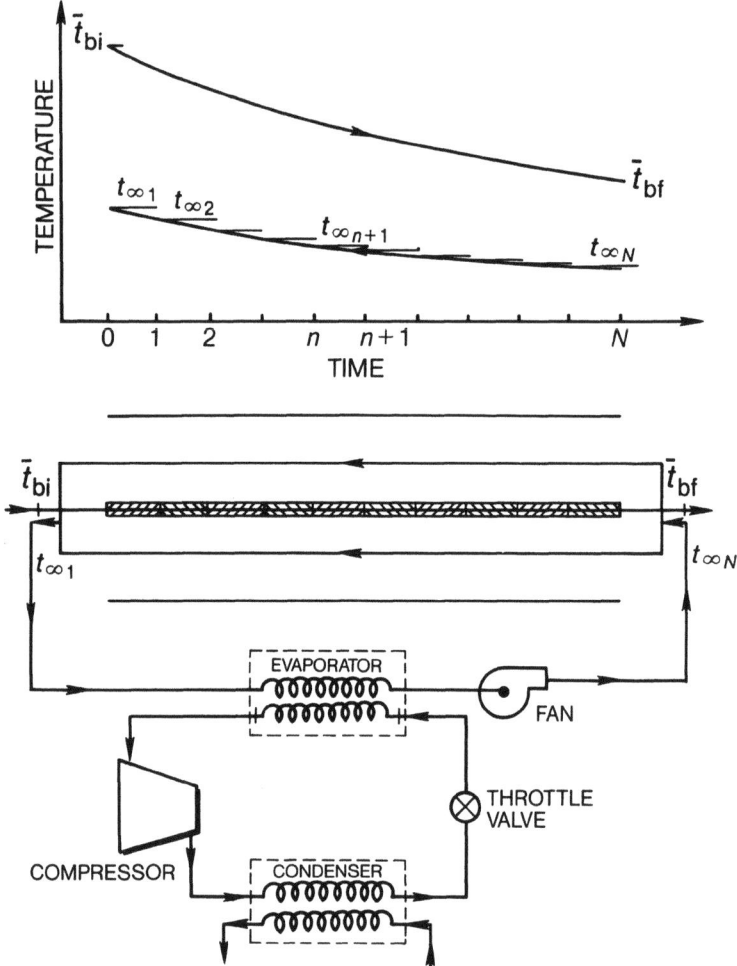

FIG. 8.4. Schematic of the counterflow system.

COMPARISON OF ENERGY CONSUMPTION DURING CHILLING OF BEEF BY CONVENTIONAL AND HOT BONING METHODS

Ferguson and Henrickson (1979) reported extensive studies on the energy consumption for the processing of intact beef sides by conventional cooling and by the hot boning method. For the purpose of comparing energy consumption, several assumptions were made. For instance, it was assumed that a 24 hr chilling period and a 24 hr holding period would be required to take the average carcass temperature of the intact side down

8 COLD STORAGE ENERGY ASPECTS OF ES HOT-BONED MEAT

TABLE 8.3. Specific Heat and Density of Lean, Bone, Fat, and Connective Tissue of Beef Carcass

Component	Specific heat (BTU/lb m-°F)	Density (lb m/ft³)	Thermal conductivity (BTU/hr-°F)	% weight of carcass
Lean	0.85	65.0	0.28	62.4
Bone	0.60	82.6	—	15.7
Fat	0.69	56.1	—	19.9
Tissue	0.70	—	—	2.0

Source: Ferguson and Henrickson (1979).
Average specific heat of carcass = (0.85 × 0.624) + (0.6 × 0.157) + (0.69 × 0.199) + (0.7 × 0.02) = 0.78 BTU/lb m-°F.
Note: The physical properties of beef are approximately constant above freezing, and the above properties are used over the temperature range 28°F < t < 102°F. The weight fraction of the various components of the carcass was established in Oklahoma State University's Meat Laboratory by studying 25 carcasses.

from 102° to 40°F (38.9° to 4.5°C), whereas equivalent chilling could be achieved in about 4 hr by hot boning the carcass and using a counterflow conveyorized cooling system (Fig. 8.4). The condenser water supply temperature was considered to be at 75°F (25°C) and all electric motors had an efficiency of 90%. The proportion of muscle, bone, and fat in an average beef carcass used in the calculations is shown in Table 8.3 along with the relevant physical properties (e.g., specific heat, density) of each type of tissue. The specific heat and the density of meat are constant in the temperature range of 28° to 102°F (−2° to 38.9°C). Although several heat transfer coefficient correlations have been suggested for calculating the heat transfer rate in beef (McAdams 1954; Earle 1966; Loginov 1969; Fleming 1971), Ferguson and Henrickson (1979) preferred Kay's 1966 equation because of its general nature in accounting for variation in the properties of beef.

$$ST_x Pr^{0.4} = 0.0295 \, Re_x^{-0.2}$$

Thus, for standard air with Prandtl number of 0.72 and specific heat of 0.24 BTU/lb m-°F, the heat transfer coefficient (h) can be derived as follows (Ganni 1979):

$$h_x = 8.0742E - 3*(\rho \, V)^{0.8}*(\mu/x)^{0.2}$$

$$h = (1/1) \int_{x=0}^{L} h \, dx$$

$$= 36.3*(\rho \, V)^{0.8}*(\mu/\text{liter})^{0.2}$$

For instance, if $t = 0°F$, and $L = 1$ ft, the heat transfer coefficient of beef would be

$$h = 0.52 \, (V)^{0.8} \, \text{BTU/hr-ft}^2\text{-°F}$$

*The design capacities of the chill cooler and holding cooler have been taken from Am. Soc. Heat., Refrig. Air Conditioning Eng. (1971A), Chapter 23, pp. 288–290.

TABLE 8.4. Physical Properties of Air in the Temperature Range of $-40°$ to $100°F$

Properties	Symbol	Value	Unit
Specific heat	c	0.24	BTU/lb m-°F
Prandtl number	Pr	0.72	
Density	ρ (t)	$(519.0 \times 0.0765)/(460.0 + t)$	lb m/ft^3
Absolute viscosity	μ (t)	$1.71865E - 8\, t + 1.11E - 5$	lb m/ft-sec
Kinematic viscosity	ν (t)	$4.6875E - 7\, t + 0.13E - 3$	(μ/ρ) ft^2/sec
Thermal conductivity	K (t)	$2.1875E - 5\, t + 0.0133$	BTU/hr-ft-°F

Source: Ferguson and Henrickson (1979).
Note: Some physical properties of air as a function of temperature were computed over the temperature range $-40°F < t < 100°F$ by the equations shown under the column "value."

The influence of temperature (between 4.5° and 38.9°C) on the physical properties of air (being used in the cooler) was computed as shown in Table 8.4.

The following example has been taken from Ferguson and Henrickson (1979), showing the comparison for energy expenditures of various forms between the conventional chilling process and the hot boning process. The number of carcasses in this example was assumed to be 520 head/day, each weighing 560 lb.

Load Calculation for Chilling of Beef by Conventional Method

The design cooling equipment* capacity of chill cooler = 1,073,000 BTU/hr or 89.4 tons

The design cooling equipment* capacity of holding cooler = 204,500 BTU/hr or 17.0 tons

Total cooling equipment capacity = 106.4 tons

Energy Calculations. The average operating conditions for condenser water inlet temperature condition of 75°F are obtained by simulating the equipment and are as follows

Refrigerant condensing temperature	= 93°F
Air to coil temperature	= 33°F
Air off the coil temperature	= 28°F
Evaporating temperature	= 21°F
Power to condenser water pump	= 0.75 hp
Input to chill cooler fan motor* = 45 hp/0.9	= 50 hp
Input to holding cooler fan motor = 10 hp/0.9	= 11.11 hp
Input to compressor motor	= 1.34 hp/ton

*See footnote on p. 255.

8 COLD STORAGE ENERGY ASPECTS OF ES HOT-BONED MEAT 257

Average total product load = 520 × 560 × 0.78 × (100 − 40)
= 1.3628 × 10^7 BTU/day or 47.3 tons

Circulating fan load = 161.11 × 2545
= 155,525 BTU/hr or 13.0 tons

a. Excluding the Building Load. Average total input to refrigeration system = [1.34 × (47.3 + 13.0)] + 61.86
= 143 hp or 106.4 kW

Energy required for processing 520 carcasses = 106.4 × 24
= 2553 kWh

b. Including the Building Load. Transmission, infiltration, personnel, and equipment heat load = 422,500 BTU/hr or 35.2 tons

Average total input to refrigeration system = [1.34 × (47.3 + 35.2)]
+ (61.86)
= 172.4 hp or 128.5 kW

Energy required for processing 520 carcasses = 128.5 × 24 = 3086 kWh

c. The Peak Power Demand. The peak power demand = 204.4 hp or 152.4 kW

Load Calculation for Chilling of Beef by Hot Processing Method

The same processing capacity as for cold processing is used in this example. The number of carcasses to be processed = 520 head/day.

Let the said capacity be handled in 8 hr shift.

Hot boning production = 520 × 560 × 0.624 = 181,709 lb m/shift
= 22,714 lb m/hr

Chilling capacity required for product to chill from 100° average to 40°F average = $\dfrac{22{,}714 \times 0.85 \times (100 - 40)}{12{,}000}$ or 96.5 tons

Building Load Estimation

Transmission, infiltration, personnel, and equipment heat in cold processing chill room = 291,000 BTU/hr

Circulating fan load = $\dfrac{45 \times 2545}{0.9}$ = 127,250 BTU/hr

TABLE 8.5. Hot-boned Beef Chilling Process Design Calculations

	Group A	Group B	Group C
Percentage by weight (Table 8.2)	29.3	30.5	40.2
Actual weight (lb m/hr)	6655	6928	9131
Cooling load (tons)	28.3	29.4	38.8
Chilling time (hr)	4	3	2.5
Loading factor (lb m/ft^2)	15	10	8
Conveyor area (ft^2)	1775	2078	2853
For 55°F condenser water			
Bhp/ton	1.25	1.15	1.15
Velocity of air on the conveyor (ft/sec)	16	14	13
Temperature of air onto conveyor (°F)	22	25	26
Ratio of heat capacity flow rates	0.25	0.25	0.25
cfm/ton	700	700	700
Temperature of air off the conveyor (°F)	36.5	39.5	40.5
Total cfm	19,810	20,580	27,160
Conveyor length-to-width ratio = 4	← 21 ft →	← 23 ft →	← 27 ft →
	↑ 6 in. ↓	↑ 6 in. ↓	↑ 8 in. ↓
(L is the length of the conveyor)	3 in.	2 in.	1.5 in.
	↑ 6 in. ↓	↑ 6 in. ↓	↑ 8 in. ↓
	L = 85 ft	L = 90 ft	L = 106 ft
Conveyor length-to-width ratio = 20	← 10 ft →	← 10 ft →	← 12 ft →
	↑ 12 in. ↓	↑ 15 in. ↓	↑ 17 in. ↓
	3 in.	2 in.	1.5 in.
	↑ 12 in. ↓	↑ 15 in. ↓	↑ 17 in. ↓
	L = 178 ft	L = 208 ft	L = 238 ft
Fan hp	1.5	1.5	1.5
For 79°F condenser water			
Bhp/ton	1.5	1.4	1.3
Velocity of air on conveyor (ft/sec)	18	16	15
Temperature of air onto conveyor (°F)	24	27	29
Ratio of heat capacity flow rates	0.25	0.25	0.25
cfm/ton	700	700	700
Temperature of air off the conveyor (°F)	38.5	41.5	43.5
Total cfm	19,810	20,580	27,160
Conveyor length-to-width ratio = 4	← 21 ft →	← 23 ft →	← 27 ft →
	↑ 5 in. ↓	↑ 5.6 in. ↓	↑ 6.7 in. ↓
	3 in.	2 in.	1.5 in.
	↑ 5 in. ↓	↑ 5.6 in. ↓	↑ 6.7 in. ↓
	L = 85 ft	L = 90 ft	L = 106 ft
Conveyor length-to-width ratio = 20	← 10 ft →	← 10 ft →	← 12 ft →
	↑ 11 in. ↓	↑ 13 in. ↓	↑ 15 in. ↓
	3 in.	2 in.	1.5 in.
	↑ 11 in. ↓	↑ 13 in. ↓	↑ 15 in. ↓
	L = 178 ft	L = 208 ft	L = 238 ft
Fan hp	2.0	2.0	2.0

Source: Ferguson and Henrickson (1979).
Total conveyor area required to chill as shown = 6706 ft^2.
To keep 1-day production on conveyor, conveyor area required = 18,222 ft^2.

8 COLD STORAGE ENERGY ASPECTS OF ES HOT-BONED MEAT

Transmission, infiltration, personnel, and equipment load excluding fan load in cold processing chill room = 163,750 BTU/hr

Taking one-half of the foregoing load as the load from transmission, infiltration, personnel, and the equipment excluding fan power (due to less required space) = 81,875 BTU/hr = 6.8 tons

The hot-boned beef chilling process design calculations should be done separately for each group of beef pieces. The grouping is given in Table 8.2. The design calculations are given in Table 8.5. The calculations are shown for both condenser water inlet conditions. In both cases, the chilling time of each group is kept constant to make comparisons of other parameters. All the design values are taken from the graphs by Ferguson and Henrickson (1979) in their earlier report. The other parameters are calculated using the design values and data from the example.

The design cooling equipment capacity required to handle the product load = 96.5 tons

The estimated design building load = 6.8 tons

Circulating fan load = $\dfrac{6 \times 2545}{0.9 \times 12{,}000}$ = 1.4 tons

Total cooling equipment capacity = 104.7 tons

1. Energy Calculations. For an 8 hr shift of this hot boning process, the chilling process takes 12 hr of operation. This is because the group A product loaded at the end of the eighth hour will require 4 more hours to complete the chilling process. The group B will take 11 hr and the group C will take 10.5 hr to complete the chilling process. For energy calculations, it is assumed that the total load should be averaged over 12 hr of chilling, even though a plant may not operate at full capacity for such a sustained period.

Average total product load = $\dfrac{181709 \times 0.85 \times (100 - 40)}{12 \times 12000}$ = 64.4 tons

Input to refrigeration system (BTU/ton, from Table 8.5)
= $(1.5 \times 0.293 + 1.4 \times 0.305 + 1.3 \times 0.402)$
= 1.4 Bhp/ton

a. Excluding the Building Load. Average total input to refrigeration system = $[1.4 \times (64.4 + 1.4)]$ = 92.12 hp
= 68.7 kW

Energy required for processing 520 carcasses = 68.7×12 = 824.4 kWh

Saving in energy = $\dfrac{2553 - 824}{2553} \times 100$ = 67.7%

b. Including the Building Load. Estimated design building load = 6.8 tons = 9.52 hp

Average input to the refrigeration system when the chilling process is on = (92.12 + 9.52) hp = 101.64 hp
= 75.8 kW

Average input to the refrigeration system when the chilling process is off = 9.252 hp = 7.1 kW

Energy required for processing 520 carcasses = (75.8 × 12) + (7.1 × 12)
= 995 kWh

Saving in energy = $\dfrac{3086 - 995}{3086} \times 100 = 67.7\%$

c. Peak Demand. The peak demand = 146.6 hp = 109.3 kW

Reduction in peak demand = $\dfrac{152.4 - 109.3}{152.4} \times 100 = 28.3\%$

EFFECT OF PLANT LOCATION ON TRANSPORTATION ENERGY

The Oklahoma study (Ferguson and Henrickson 1979) also evaluated the effect of plant locations on transportation energy consumption in hot boning as well as in conventional practice by using a transportation linear program model as shown in the following equation.

$$TC = \sum_{i=1}^{m} \sum_{j=1}^{n} (C_{ij} + P_i) \cdot X_{ij}$$

where TC = Total transportation cost, i = the amount of the commodity available for export from the ith surplus region, j = the amount of the commodity required by the jth deficit region, m = number of surplus regions, n = number of destination points, C_{ij} = the per unit cost of skipping from ith region to jth destination, P_i = the unit processing cost at surplus region i, X_{ij} = amount of the commodity shipped from ith origin to jth destination.

These data revealed that West Central, West South-Central, and Mountain regions of the United States are the major beef production areas, whereas Northeast, mid-Atlantic, Southeast, and Pacific are the beef-deficit regions. They produce less than demand because of high population density. Hence, about 50% of the annual beef production is transported from beef-producing to beef-deficient regions, and the interregional beef transportation accounts for over 70% of the total beef shipment costs. The meat industry in Kansas, Colorado, and Minnesota has the greatest incentive to switch to hot boning to curtail the transportation costs and to compete in the distant consumer markets.

This study also projected the transportation reduction in cost for 1980 to be approximately $139,595,554, with a reduction in energy consumption of 3.2604988×10^{12} BTU in the distribution of beef if the hot processing technology was used. This amounts to a *75% savings* in energy as compared with conventional practice because about 30% of the excess fat and bone no longer needs to be shipped to the retail point of sale. Instead, it will be processed in the plant as by-products (Ferguson and Henrickson 1979).

In the final analysis, hot processing of carcasses offers a great potential for energy conservation and lower operating costs for the meat industry. Figure 8.5 provides a summary of the projected energy savings should the beef industry in the United States fully adopt the suggested hot carcass processing technology. The findings discussed in the preceding sections

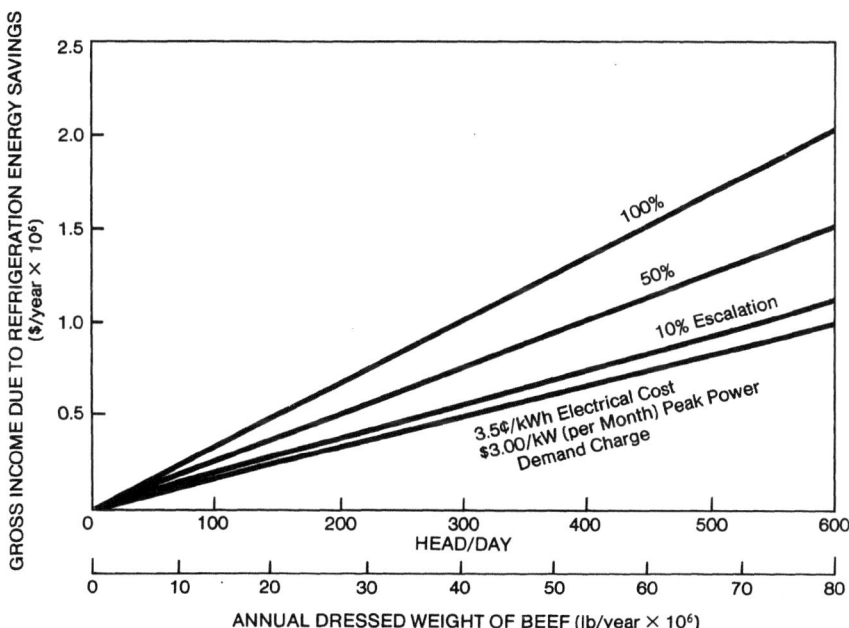

FIG. 8.5. Potential gross income due to savings in refrigeration electricity if hot boning process is used for cows and bulls.
From CENTEC (1980).

have been extrapolated for the 1978 U.S. beef slaughter data. Table 8.6 shows that 43,722,700 head of beef animals were slaughtered and inspected for human consumption. This represents a dressed weight of 2.4766×10^{10} lb of meat. Figure 8.6 provides a comparison of energy consumption if this amount of meat is cooled and transported according to conventional practice or hot processed.

Line 1 of the block diagram in Fig. 8.6 reflects the energy expenditures for conventional cold processing. Based on the energy requirement per pound for cooling the meat from 102° to 40°F, the gross energy consumption required as electricity for refrigeration amounted to about 2.69×10^{12} BTU. Line 2 denotes the energy consumption if waste heat is utilized in bagging of beef before chilling in a tunnel. Line 3 indicates the energy requirements for hot-boned carcass processing without making any use of waste heat, and line 4 represents the maximum potential for energy saving by using the hot carcass processing technology. It can be seen that a maximum reduction of 5.08×10^{12} BTUs of energy or a saving in energy of 1.62×10^7 BTU/head or 2.5 BTU/lb of processed beef can be achieved by hot processing of carcasses. Thus, an estimated 50% of the energy used in beef plant refrigeration can be considered if hot boning were included in the meat processing system. It should, however, be realized that this is a conservative estimate and does not include one of the most significant savings in energy that arises, namely, that from transportation.

In Britain, Harrington (1978) also made a detailed assessment of cost reduction by using the hot boning technique as compared with the conventional operation. His comparison was based on a plant design for 200 head/day or 30 head/hr. Table 8.7 provides a summary of various costs involved in the two methods of processing beef carcasses. It is obvious that the estimate of refrigeration space saving is much lower in Harrington's study than that perceived by Henrickson and McQuiston (1977) and Ferguson and Henrickson (1979). This disparity is due to Harrington's reliance on conventional cooling rooms even for hot-boned meat, whereas Henrickson and associates based their estimate on the conveyorized counterflow cooling design, which is believed to be much more thermodynami-

TABLE 8.6. Slaughter Statistics (Based on USDA Data for 1978)

Type	Head slaughtered	Average weight (lb)	Dressed weight (lb)
Veal and calf	4,170,200	123	5.1293×10^8
Steers and heifers	30,283,120[a]	645	1.9533×10^{10}
Cows	8,470,790[a]	489	4.1422×10^9
Bulls	798,590[a]	723	5.7738×10^8
Totals	43,722,700		2.4766×10^{10}

Source: Ferguson and Henrickson (1979).
[a] Animal category of nonfederally inspected was not given, assumed to have same distribution as federally inspected.

1. Cold Processing Method

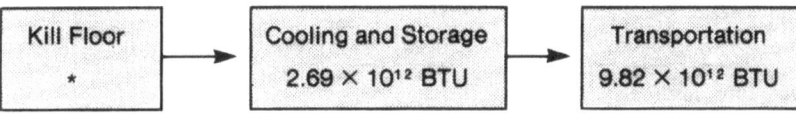

2. Hot Processing Method (using waste heat bagging shrink tunnel)

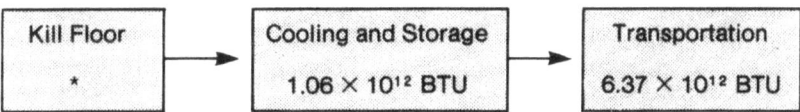

3. Hot Processing Method (using electrical bagging shrink tunnel)

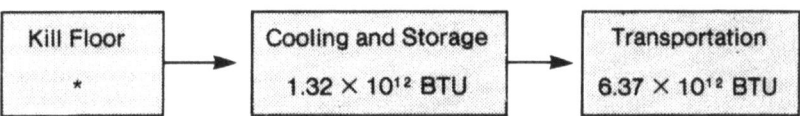

4. Maximum Potential Savings

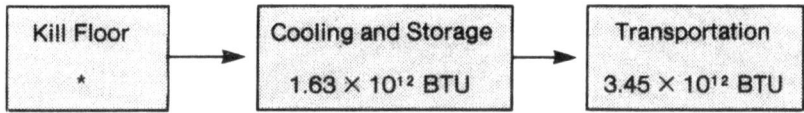

Grand total = 5.08 × 10^{12} BTU or 5.08 trillion BTU (source energy).

*It is assumed that kill floor energy requirements are the same under the hot and cold processing methods.

FIG. 8.6. Source energy block diagram.
From Ferguson and Henrickson (1979).

cally efficient. Besides, Harrington (1978) has shown a significant saving in labor, especially for boning and vacuum packaging, whereas Ferguson and Henrickson (1979) in one study did not find any difference in the labor requirement between the two processes. However, in a later report, Henrickson and Ferguson (1977) did note a 30% saving in labor requirements by hot boning. The savings in man-hours were shown to be about 15 min/head. This amounted to reducing the processing cost by $2.25/head. Despite these discrepancies, Harrington (1978) also projected a significant conservation in energy and a better return on capital if the hot boning technology were adopted by the meat industry.

TABLE 8.7. Costs and Returns for Two Model Beef Plants to Handle 50,000 Cattle/Year to Chilled Vacuum-packed Cuts Using Hot or Cold Boning (Costs Relate to the Spring of 1978)

	Hot boning (1000 £)	Cold boning (1000 £)
Operating costs		
labor: lairage, kill, dress bone,	150	150
vacuum pack management	145	180
	100	100
electricity	47	55
packing materials	475	475
other costs	200	200
	1,117	1,160
Gross margin		
cost	14,672	14,672
value of output	16,191	16,191
Margin	1,519 (9.4%)	1,519 (9.4%)
Margin before depreciation	402	359
Depreciation		
buildings (20 years)	40	45
plant (10 years)	20	20
equipment (5 years)	6	6
Total	66	71
Net margin	336 (2.1%)	288 (1.8%)
Annual return on invested capital	32.0%	25.0%

Source: Cuthbertson (1980).
Conversion rate: 1 British pound (£) = $1.94.

A study sponsored by the USDA (CENTEC 1980) for Packerland International has indicated that on-the-rail hot boning resulted in 35 to 40% saving in time as compared with table boning of cold carcasses. The yield of boneless cuts was higher by about 2%, with a resulting increase in product value per side.

Tables 8.8 and 8.9 summarize the minimum cost and energy requirements of the overall U.S. beef industry and different potential sources of annual savings in energy from adopting the hot boning and processing technology. It can be seen that the potential energy saving impact of hot

TABLE 8.8. Sources of Annual Savings in U.S. Beef Industry Resulting from Hot Boning Processing

Source	Energy savings		Dollar cost savings	
	(BTU)	(%)	($)	(%)
1976				
distribution	2.9184622×10^{12}	75.32	125,707,407	91.38
processing	0.9561898×10^{12}	24.68	11,860,780	8.62
distribution and processing	3.874652×10^{12}	100	137,568,187	100
1980				
distribution	3.2604988×10^{12}	75.61	139,595,554	91.44
processing	1.0518295×10^{12}	24.39	13,074,875	8.56
distribution and processing	4.3123283×10^{12}	100	152,670,429	100

Source: Ferguson and Henrickson (1979).

TABLE 8.9. Comparison of Cost and Energy Requirements for the U.S. Beef Industry with and Without the Hot Boning Technique

Annual cost item	Conventional	Hot boning	Potential savings
1976			
Minimum cost for distribution	$419,024,689	$293,317,282	$125,707,467
Minimum energy usage of distribution	9.7282072×10^{12} BTU	6.809745×10^{12} BTU	2.9184622×10^{12} BTU
1980			
Minimum cost of distribution	$465,318,514	$325,722,960	$139,595,554
Minimum energy usage of distribution	$10.8683292 \times 10^{12}$ BTU	7.6078304×10^{12} BTU	3.2604988×10^{12} BTU
1976			
Minimum costs of distribution and processing	$1,040,000,109	$902,431,922	$137,568,187
Minimum energy usage of distribution and processing	$12.4104233 \times 10^{12}$ BTU	8.5357713×10^{12} BTU	3.874652×10^{12} BTU
1980			
Minimum costs of distribution and processing	$1,152,913,532	$1,000,243,103	$152,670,429
Minimum energy usage of distribution and processing	$13.8188255 \times 10^{12}$ BTU	9.5064972×10^{12} BTU	4.3123283×10^{12} BTU

Source: Ferguson and Henrickson (1979).

boning technology originates from two major areas, namely, (1) the processing sector, and (2) the distribution sector. The data in Fig. 8.6 reveal that even though the total processing costs are far greater than the distribution costs, hot processing technology, if followed, would influence more the distribution costs than the processing costs. Since beef processing is labor-intensive and beef distribution is an energy-intensive operation, it is not surprising that major cost saving by hot boning will obviously be from the distribution sector. Data in Table 8.8 suggest that hot boning technology promises about 91% saving in distribution costs and 75% saving in distribution energy; it also projects about 9% reduction in processing cost, and 24% saving in processing energy as compared with conventional practices in the beef industry.

INDUSTRY IMPEDIMENTS

On-the-rail deboning of the beef carcass has not been readily accepted even though many potential advantages appear evident. A move toward hot processing of the beef carcass would cause great change in the slaughter and processing industry. At present, much of the livestock is slaughtered in one place and the product processed within another plant. This industry segmentation has caused the corporate leaders to be concerned about systems of operation, along with consumer satisfaction with the new products.

The greatest impediment faced by the industry is its financial status and the money needed to construct new facilities or to renovate old plants for handling the slaughter–processing system within the same plant. Since processing of hot meat will involve various alternative systems, one must consider these key systems in determining the size of the new modified facility. Processing may include bone-in or uniform boneless primals and subprimals; vacuum packaging; comminuted and restructured roasts, patties, and steaks along with the use of mechanically deboned meat; and curing and sausage manufacturing. New beef plants would need to be capable of handling over 400 beef/hr, whereas a hog processing plant would need to consider a capacity of 1000 head/hr. This slaughter capability would challenge present materials handling systems. It would require not only conveyorized cooling but also robotic handling of some products. These added new processing systems would encourage additional skilled labor to handle the unfrozen and frozen retail vacuum-packaged products.

A second impediment relates to the industrial tie to a grading system. Since grading is based on marbling, lean area, maturity, and fat cover, it is obvious that marbling level, the one factor difficult to measure in unchilled meat, is the limiting main concern. The price differential between the Choice and Good grades continues to encourage the importance of grading. These factors along with industry's concept of the sales value of the term "Choice" and the need to re-educate the consumer have caused delays in the acceptance of hot processing.

Warm sticky meat, due to high surface moisture, favors microbial growth. Therefore, greater awareness of facilities and handling hygiene will be required. Not only will there be more pieces to handle, but the cut surface area will be greater. A sanitation program must be synchronized with slaughtering, deboning, and processing lines. It is these materials handling systems that must be clearly outlined before hot boning can be fully implemented.

Packaging the deboned cuts or muscle systems to retain their shape and fiber orientation suitable for consumer acceptance has caused some concern to those considering hot boning of the beef carcass. The ease of placing the pieces within a vacuum barrier bag has presented a challenge to the package industry. Since the edible product would consist of several more pieces than the bone-in boxed beef, one must consider the additional handling costs.

On-the-rail hot deboning would provide a greater amount of fat to render and more bone to be placed through the mechanical deboner. Both of these products will cause a need for larger facilities with more capital outlay.

SUMMARY

From all these studies, it can be concluded that hot processing technology offers several energy- and cost-saving opportunities for the meat industry. These are summarized as follows.

1. Hot processing helps in reducing energy consumption since only edible parts of the carcasses are processed. As much as 65% energy can be saved by hot boning. There is a 37% reduction due to the elimination of extra bone and fat, which will reduce the cooling load by 32%. When these savings are combined with other process efficiencies, the saving could be well over 50% of that used in the conventional practice.
2. Hot processing provides better and more efficient use of the cooling space, since one-third of the chill cooler size is needed for hot-boned meat than is required for intact carcass sides. Moreover, no massive holding cooler would be required for hot boning.
3. Hot processing allows the use of more efficient cooling system designs, such as the conveyorized counterflow model. The conveyorized counterflow model system reduces the requirement for fan energy for circulating cooling air by 90%.
4. There is a four-fold saving in cooling time. The intact carcass sides require 48 hr to reduce the temperature from 102° to 40°F, whereas equivalent cooling of hot-boned meat can be achieved in less than 12 hr.

5. The peak power demand is reduced by 28% in the case of hot-boned meat.
6. Hot processing may reduce the building cooling load and meat distribution costs.
7. Of the total energy savings from hot boning, about 25% would result from processing and about 75% would be from distribution.

The major impediments to adoption of hot boning by the meat industry are listed as follows.

1. The cost of new buildings and equipment or the renovation of old facilities has been a major deterrent to adoption of hot boning and processing.
2. The problem of grading of hot-boned meat has also been an obstacle to adoption of hot boning.
3. Hot boning and processing require a high level of sanitation with a careful quality control program, beginning at slaughtering and extending through marketing.
4. Hot boning and processing result in more cutting and processing at the meat packing plant, thus increasing labor costs. However, lower labor requirements at the retail level will be less, thereby offsetting this cost.

Savings from hot boning and processing are sufficiently high to make them attractive to the meat industry. Increasing energy costs will in all probability force the meat industry to move toward hot boning and processing.

REFERENCES

ACTON, J.C. and SAFFLE, R.L. 1969. Preblended and prerigor meat in sausage emulsion. Food Technol. *23*, 367.
ADELSTEIN, R.S. and EISENBERG, E. 1980. Regulations and kinetics of the actin-myosin-ATP interactions. Annu. Rev. Biochem. *49*, 921.
AM. SOC. HEAT., REFRIG. AIR CONDITIONING ENG. 1971A. ASHRAE Guide and Data Book, Applications Vol. American Society of Heating, Refrigeration and Air Conditioning Engineers, New York.
AM. SOC. HEAT., REFRIG. AIR CONDITIONING ENG. 1971B. Meat products. *In* ASHRAE Guide and Data Book. p. 23:287. New American Society of Heating, Refrigeration and Air Conditioning Engineers, New York.
AM. SOC. HEAT., REFRIG. AIR CONDITIONING ENG. 1974. ASHRAE Fundamental Handbook. pp. 27.1–27.24. American Society of Heating, Refrigeration and Air Conditioning Engineers, New York.
ANON. 1974. A Need for Meat Cutting Efficiencies and High Quality Finished Products. Meat and Livestock Commission (MLC), London.
ARGANOSA, F.C. and HENRICKSON, R.L. 1960. Cure diffusion through pre- and post-chilled porsine muscle. Food Technol. *8*, 75.
ASGHAR, A. and HENRICKSON, R.L. 1982. Post-mortem electrical stimulation of carcasses: Effect on biochemistry, biophysics, microbiology and quality of meat. CRC Crit. Rev. Food Sci. Nutr. *18*, 1–58.

ASGHAR, A. and PEARSON, A.M. 1980. Influence of ante- and post-mortem treatments upon muscle composition and meat quality. Adv. Food Res. *26*, 53.
AYRES, J.C. 1960. The relationship of organisms of the genus pseudomonas to the spoilage of meat, poultry and eggs. J. Appl. Bacteriol. *23*, 471.
BAILEY, C. 1972A. An MRI view of commercial refrigeration design. In Symposium at MRI, Langford, U.K. International Institute of Refrigeration, Paris.
BAILEY, C. 1972B. Factors affecting rate of cooling and evaporation. In Symposium at MRI, Langford, U.K. International Institute of Refrigeration, Paris.
BARBE, D.D. and HENRICKSON, R.L. 1967. Bacteriology of rapid cured ham. J. Food Sci. *21*, 9.
BARBE, D.D., MANDIGO, R.W. and HENRICKSON, R.L. 1966. Bacterial flora associated with rapid-processed ham. J. Food Sci. *31*, 988.
BENDALL, J.R. 1976. Electrical stimulation of rabbit and lamb carcasses. J. Sci. Food Agric. *27*, 819.
BENDALL, J.R. 1980. The electrical stimulation of carcasses of meat animal. In Developments in Meat Science, Vol. 1. p. 27. Applied Science Publishers, London.
BENDALL, J.R., KETTERIDGE, C.C. and GEORGE, A.R. 1976. The electrical stimulation of beef carcasses. J. Sci. Food Agric. *27*, 1128.
BORKOWSKA, T. and ROMANOWSKI, J. 1972. Preparation and starting news of technologies of manufacturing seasoned products and sterilized pork preserved warm or partially cooled. Polish Meat Research Institute, Warsaw (*Cited by* Pisula 1981). (Polish)
BOUTON, P.E., FORD, A.L., HARRIS, P.V. and SHAW, F.D. 1980. Electrical stimulation of beef. Meat Sci. *4*, 145.
BRASINGTON, C.F. and HAMMONS, D.R. 1971. Boning carcass beef on the rail. U.S. Dep. Agric. ARS-52-63.
BUEGE, D.R. and MARSH, B.B. 1975. Mitochondrial calcium and post-mortem muscle shortening. Biochem. Biophys. Res. Commun. *66*, 478.
CARSE, W.A. 1973. Meat quality and the acceleration of post-mortem glycolysis by electrical stimulation. J. Food Technol. *8*, 163.
CASSENS, R.G. and NEWBOLD, R.P. 1967. Temperature dependence of pH changes in ox muscle post mortem. J. Food Sci. *32*, 13.
CENTEC. 1980. Energy Conservation by Electrical Stimulation and Hot-boning of Beef. A Report. CENTEC Corp., Reston, VA.
CHALCROFT, J.P. and CHRYSTALL, B.B. 1975. Current distribution in carcasses during electrical stimulation. Meat Ind. Res. Inst. New Zealand, Hamilton, N.Z. (MIRINZ) Annu. Rep. *32*.
CHRYSTALL, B.B. and DEVINE, C.E. 1978. Electrical stimulation, muscle tension and glycolysis in bovine sternomandibularis. Meat Sci. *2*, 49.
CHRYSTALL, B.B. and HAGYARD, C.J. 1976. Electrical stimulation of lamb—Laboratory study. Meat Ind. Res. Inst. New Zealand, Hamilton, N.Z. (MIRINZ) Annu. Rep. *32*.
CIA, G. and MARSH, B.B. 1976. Properties of beef cooled before rigor onset. J. Food Sci. *41*, 1259.
CIELOCH, J. and TURKOWSKI, T. 1968A. Is there a revolution in pork manufacturing? Meat Econ. (Gospodarka Miesna) *6*, 19 (*Cited by* Pisula 1981). (Polish)
CIELOCH, J. and TURKOWSKI, T. 1968B. The possibility of utilizating warm meat. Meat Econ. (Gospodarka Miesna) *8*, 9. (Polish)
COHEN, C. 1975. The protein switch of muscle contraction. Sci. Am. *233* (Nov.) 36.
CONTRERAS, S., HARRISON, D.L., KROPF, D.H. and KASTNER, C.L. 1981. Electrical stimulation and hot boning: Cooling losses, sensory properties and microbial count of ground beef in a model system. J. Food Sci. *46*, 475.
CORNFORTH, D.P., PEARSON, A.M. and MERKEL, R.A. 1980. Relationship of mitochondria and sarcoplasmic reticulum to cold shortening. Meat Sci. *4*, 103.
CROSS, H.R., BERRY, B.W. and MUSE, D. 1976. Sensory and cooking properties of ground beef prepared from hot and chilled beef carcasses. J. Food Sci. *44*, 1432.
CUTHBERTSON, A. 1980. Hot processing of meat: A review of the rationale and

economic implications. *In* Developments in Meat Science, Vol. 1. p. 61. A. Lawrie (Editor). Applied Science Publishers, London.

DALRYMPLE, R.H. and HAMM, R. 1974. Influence of sodium chloride on the breakdown of glycogen in minced beef muscle post-mortem. Fleischwirtschaft, 6. (German)

DAVEY, C.L., KETTEL, H. and GILBERT, K.V. 1967. Shortening as a factor in meat aging. J. Food Technol. *2*, 53.

DAVEY, C.L., GILBERT, K.V. and CARSE, W.A. 1976. Carcass electrical stimulation to prevent cold shortening toughness in beef. N.Z. J. Agric. Res. *19*, 13.

DAVIDSON, W.D., CLIPLET, R. and MEARDE, R.J. 1968. Post-mortem processing treatment on selected characteristics of ham and fresh pork sausage. Food Technol. *22*, 772.

DEVINE, C.E., GILBERT, K.V. and DAVEY, C.L. 1975. Carcass posture and electrical stimulation. Meat Ind. Res. Inst. New Zealand, Hamilton, N.Z. (MIRINZ) Annual Rep. *32*.

DRANSFIELD, E., BROWN, A.J. and RHODES, D.N. 1976. Eating quality of hot deboned beef. J. Food Technol. *11*, 401.

EARLE, R.L. 1966. The freezing of boneless beef in air blast freezers. 2nd Int. Congr. Food Sci. Technol. *4*, 79.

EMSWILER, B.S. and KOTULA, A.W. 1979. Bacteriological quality of ground beef prepared from hot and chilled beef carcasses. J. Food Protect. *42*, 561.

ENDO, M. 1977. Calcium release from sarcoplasmic reticulum. Physiol. Rev. *57*, 71.

FALK, S.N. and HENRICKSON, R.L. 1974. Feasibility of hot boning the bovine carcass. Okla. Agric. Exp. Stn. *MP-92*, 145.

FALK, S.N., HENRICKSON, R.L. and MORRISON, R.D. 1975. Effect of boning beef carcasses prior to chilling on meat tenderness. J. Food Sci. *40*, 1075

FERGUSON, E.J. and HENRICKSON, R.L. 1975. Energy conservation in the meat processing industry resulting from muscle boning of the unchilled bovine carcass. Okla. State Univ., Stillwater Energy Res. Dev. Admin. Res. Proposal ER 76-R-22.

FERGUSON, E.J. and HENRICKSON, R.L. 1979. Final report on energy conservation in the meat processing industry. U.S. Dep. Energy Contract *E4-76-5-05-5097*.

FLEMING, A.K. 1971. The numerical calculation of the freezing process. Proc. 13th Int. Congr. Refrig., Washington, D.C., 1970, 302.

FOLLETT, M.J., NORMAN, G.A. and RATCLIFF, P.W. 1974. The ante-rigor excision and air cooling of beef semimembranosus muscles at temperatures between $-5°C$ and $15°C$. J. Food Technol. *9*, 509.

FORREST, J.C., JUDGE, M.D., SINK, J.D., HOEKSTRA, W.G. and BRISKEY, E.J. 1966. Prediction of the time couse of rigor mortis through response to muscle tissue to electrical stimulation. J. Food Sci. *31*, 13.

FRANK, G.B. 1980. The current view of the source of trigger calcium in excitation-contraction coupling in vertebrate skeletal muscle. Biochem. Pharmacol. *29*, 2399.

FRONING, G.W. and NEELAKANTAN, S. 1971. Emulsifying characteristics of pre-rigor and post-rigor poultry muscle. Poult. Sci. *50*, 389.

FUNG, D.Y.C., KASTNER, C.L., HUNT, M.C., DICKEMAN, M.E. and KROPF, D.H. 1980. Mesophile and psychrotroph populations on hot-boned and conventionally processed beef. J. Food Protect. *43*, 547.

FUNG, D.Y.C., KASTNER, C.L., LEE, C.Y., HUNT, M.C., DICKEMAN, M.E. and KROPF, D.H. 1981. Initial chilling rate effects on bacterial growth on hot-boned beef. J. Food Protect. *44*, 539.

GALLOWAY, D.E. and GOLL, D.E. 1967. Effect of temperature on molecular properties of post-mortem porcine muscle. J. Anim. Sci. *26*, 1302.

GANNI, V. 1979. Design procedure for conveyorized chilling and freezing of hot boned beef. Ph.D. Thesis. Oklahoma State Univ., Stillwater, OK.

GAWRON, S., MISZTAK, M. and SABAT, H. 1972. A method for pickling meat, particularly bacon. Pol. Pat. 64628 kl. 53c, 2. Mar. 15. (*Cited by* Pisula 1981). (Polish)

GEORGE, A.R., BENDALL, J.R. and JONES, R.C.D. 1980. The tenderizing effect of electrical stimulation of beef carcasses. Meat Sci. *4*, 51.
GILL, C.O. 1980. The effect of electrical stimulation on meat spoilage floras. J. Food Sci. *43*, 190.
GILL, C.O. and NEWTON, K.G. 1978. The ecology of bacterial spoilage of flesh meat at chill temperatures. Meat Sci. *1*, 207.
GOLL, D.E. 1968. Resolution of rigor mortis. Proc. Recip. Meat Conf. *21*, 15.
GOLOWKIN, N., CYRENOW, B., WASILIEW, A. and SZITIKOW, J. 1973. Changes in beef with chilling and salting of the carcasses. Mjasnaja Industria SSSR *9*, 171. (Russian)
GOTHARD, R.H., MULLINS, A.M., BOULWARE, R.F. and HANSARD, S.L. 1966. Histological studies of post-mortem changes in sarcomere length as related to bovine muscle tenderness. J. Food Sci. *31*, 825.
HALL, L.C., SAVELL, J.W. and SMITH, G.C. 1980. Retail appearance of electrically stimulated beef. J. Food Sci. *45*, 171.
HALL, R., DICK, J. and CARSON, R. 1980. Electric stimulation tenderizes meat costs 18¢ per carcass, pays out in six months. Food Process. (Apr.), 120.
HAMM, R. 1966. The processing of beef frozen immediately after slaughter. Fleischwirtschaft *7*, 772. (German)
HAMM, R. 1972. Colloid chemistry of meat. Paul Parey Verlag, Berlin and Hamburg. (German)
HAMM, R. 1973. The significance of the water-binding abilities of meat in the production of hot sausage. Fleischwirtschaft *53*, 73. (German)
HAMM, R. 1977. Postmortem breakdown of ATP and glycogen in ground muscle: A review. Meat Sci. *1*, 15.
HAMM, R. 1978. The use of freeze-dried prerigor beef in sausages. Proc. Meat Ind. Res. Conf., Mar. 23, 1978, Univ. of Chicago, Am. Meat Inst. Food, 31.
HAMM, R. and Potthast, K. 1975. Freeze-dried hot-sausage meat. Fleischwirtschaft *55*, 87. (German)
HARRINGTON, G. 1978. *In* Symposium—New Technology in the Abattoir and Processing Plant. London Meat Trades Fair, Alexandra Palace, London, May.
HARSHAM, A. and DEATHERAGE, F. 1951. Tenderization of meat. U.S. Pat. 2,544,681.
HEIDMANN, R. 1964. The preservation of the roast during the cutting. Fleischwirtschaft *7*, 635. (German)
HENRICKSON, R.L. 1967. Pork can be processed before chilling. Okla. Agric. Exp. Stn. *MP-79*, 10.
HENRICKSON, R.L. 1968. High temperature processing effect on physical, chemical, microbial, and flavor properties of pork. Proc. Meat Ind. Res. Conf., 49.
HENRICKSON, R.L. 1975. Hot boning. Proc. Meat Ind. Res. Conf., 25. 1975.
HENRICKSON, R.L. 1979. Electricity and beef tenderness. Presentation at Am. Assoc. Meat Processors, 40th Am. Conv., Caesars Palace, Las Vegas.
HENRICKSON, R.L. 1981. Energy aspects of pre-rigor meat. Proc. 34th Recip. Meat Conf., 1981, *34*.
HENRICKSON, R.L. and FERGUSON, E.J. 1977. Energy conservation in the meat industry. Energy Res. Dev. Admin. Contract *EY 76-S-05-5097*. Prog. Rep. *ORO-5097-4*. Washington, DC.
HENRICKSON, R.L. and McQUISTON, F.C. 1977. A study of hot beef boning for energy conservation. Presented before Am. Soc. Heat., Refrig. Air Conditioning Eng., Feb. 16, 1977, Chicago.
HOES, T.L., RAMSEY, C.B., HINES, R.C. and TATUM, J.D. 1980. Yield and palatability of hot-processed phosphate-injected pork. J. Food Sci. *45*, 773.
HONIKEL, K.O. and HAMM, R. 1978. Influence of cooling and freezing of minced prerigor muscle on the breakdown of ATP and glycogen. Meat Sci. *9*, 181.
HUXLEY, A.F. 1974. Muscular contraction. J. Physiol. *243*, 1.
JABLONSKI, Z. 1979. Directions of progress of seasoned meat production. Meat Econ. (Gospodarka Miesna *1*, 2 (*Cited by* Pisula 1981). (Polish)

JACOBS, D.K. and SEBRANEK, J.G. 1980. Prerigor beef for frozen ground beef patties. J. Food Sci. 45, 648.
JAROSZEWSKI, Z. 1974. Preparation and starting new technologies of seasoned meat products made without cooling or from meat only partially cooled. Inst. Meat Prod. Inf. Bull. (Builetyn Informacyjny IPMs) 6, 4. (Cited by Pisula 1981). (Polish)
JEREMIAH, L.E. and MARTIN, A.H. 1980. The effects of electrical stimulation on retail acceptability and case-life of beef. Proc. 26th Eur. Meet. Meat Res. Workers, Colorado Springs, CO, 2, 30.
JOHNSON, R.G. and HENRICKSON, R.L. 1970. Effect of treatment of pre- and postrigor porcine muscles with low sodium chloride concentrations on subsequent extractability of protein. J. Food Sci. 3, 268.
KASTNER, C.L. and HENRICKSON, R.L. 1969. Providing uniform meat cores for mechanical shear force measurement. J. Food Sci. 34, 603.
KASTNER, C.L. and RUSSELL, T.S. 1975. Characteristics of conventionally and hot-boned bovine muscle excised at various conditioning periods. J. Food Sci. 40, 747.
KASTNER, C.L., HENRICKSON, R.L. and MORRISON, R.D. 1973. Characteristics of hot boned beef muscle. J. Anim. Sci. 36, 484.
KASTNER, C.L., SULLIVAN, D.P., AYAZ, M. and RUSSEL, T.S. 1976. Further evaluation of conventional and hot boned bovine longissimus dorsi muscle excised at various periods. J. Food Sci. 41, 97.
KAYS, W.M. 1966. Convective Heat and Mass Transfer. p. 239. McGraw-Hill Book Co., New York.
KOTTER, L., PRANDL, O. and TERPLAN, G. 1968. Testing of the effect of heating of meat on fluid retention. Fleischwirtschaft 4, 439. (German)
KOTULA, A.W. 1981. Microbiology of hot-boned and electrically stimulated meat. J. Food Protect. 44, 545.
KOTULA, A.W. and EMSWILER-ROSE, B.S. 1981. Bacteriological quality of hot-boned primal cuts from electrically stimulated beef carcasses. J. Food Sci. 46, 471.
KOWALSKI, Z., BYKOWSKI, W. and PANASIK, M. 1972. Preparation and starting new technologies of meat product manufacturing without cooling or only from partially cooled meat. Manuscript. Polish Meat Research Institute, Warsaw (Cited by Pisula 1981). (Polish)
LARA, R.O. 1982. Effect of box size and storage time on the total count, anaerobic count, and psychrotrophic count of electrically stimulated hot boned boxed beef. M.S. Thesis. Oklahoma State Univ., Stillwater, OK.
LEISTNER, L. and RODEL, W. 1975. The significance of water activity for microorganisms in meat. In Water Relations of Foods. p. 309. R.B. Duckworth (Editor). Academic Press, New York.
LEVY, F.L. 1972. Energy, time-temperature and weight loss during meat chilling. In Symposium No. 2—Meat Cooling Why and How. C.L. Cutting (Editor). p. 14.1–14.15. MRI Publishers, Langford, U.K.
LIN, H.S., TOPEL, D.G. and WALKER, H.W. 1979. Influence of pre-rigor and post-rigor muscle on the bacteriological and quality characteristics of pork sausage. J. Food Sci. 44, 1055.
LOCKER, R.H. 1960. Degree of muscular contraction as a factor in tenderness of beef. Food Res. 25, 304.
LOCKER, R.H. and HAGYARD, C.J. 1963. A cold shortening effect in beef muscles. J. Sci. Food Agric. 15, 787.
LOGINOV, L.I. 1969. Application of numerical methods for cooling process calculation. 12th Int. Congr. of Ref., Vol. 2. pp. 717–729, Madrid, 1967.
LOVETT, D.A., HERBERT, L.C. and RADFORD, R.D. 1976. Chilling of meat—Experimental investigation of weight loss. In Towards an Ideal Refrigerated Food Chain. p. 307. International Institute of Refrigeration, Annexe Sydney, Australia.
MACY, R.L., NAUMANN, H.D. and BAILEY, M.E. 1970. Water-soluble flavor and odor precursors of meat. J. Food. Sci. 35, 83.

MANDIGO, R.W. and HENRICKSON, R.L. 1966. Influence of hot-processing pork carcasses on cured ham. Food Technol. 20, 538.
MANDIGO, R.W. and HENRICKSON, R.L. 1967. The influence of pre-chill processing techniques on bacon. Food Technol. 21, 1262.
MANDIGO, R.W. and KUNERT, G.J. 1973. Accelerated pork processing: Cured color stability of hams. J. Food Sci. 6, 1078.
MANDIGO, R.W., THOMPSON, T.L. and WEISS, G.M. 1977. Commercial accelerated pork processing: Yields of cured ham, bacon and loins. J. Food Sci. 42, 898.
MARSH, B.B. and LEET, N.G. 1966. Studies in meat tenderness. III. The effect of cold shortening on tenderness. J. Food Sci. 31, 450.
MARSH, B.B., CASSENS, R.G., KAUFFMAN, G.R. and BRISKEY, E.J. 1972. Hot boning and pork tenderness. J. Food Sci. 37, 179.
McADAMS, W.H. 1965. Heat Transmissions, 3rd Edition. p. 249. McGraw-Hill Book Co., New York.
McCOLLUM, P.D. and HENRICKSON, R.L. 1980. The effect of electrical stimulation of the rate of post-mortem glycolysis in some bovine muscles. J. Food Qual. 1, 15.
McCRAE, S.E., SECCOMBE, C.G., MARSH, B.B. and CARSE, W.A. 1971. Studies in meat tenderness. IX. The tenderness of various lamb muscles in relation to their skeletal restraint and delay before freezing. J. Food Sci. 36, 566.
McLEOD, K., GILBERT, K.V., WYBORN, R.W., WENHAM, L.M., DAVEY, C.L. and LOCKER, R.H. 1973. Hot cutting of lamb and mutton. J. Food Technol. 8, 71.
McMEEKIN, T.A. 1975. Spoilage association of chicken breast muscle. Appl. Microbiol. 29, 44.
MEAT LIVESTOCK COMM. 1974. Beef Cutting. Meat and Livestock Commission, Milton Keynes, London.
MICYK, W., FIRGER, I. and STOPCZIK, R. 1973. Uskovennaja technologia proizwodatwa kopozienostiej iz parnoj swinmy. Mjasnaja Industria SSSR 3, 21 (Cited by Pisula 1981). (Polish)
MRIGADAT, B., SMITH, G.C., DUTSON, T.R., HALL, L.C., HANNA, M.O. and VANDERZANT, C. 1980. Bacteriology of electrically stimulated rabbit, pork, lamb and beef carcasses. J. Food Protect. 43, 686.
NEWBOLD, R.P. and SCOPES, R.K. 1967. Post-mortem glycolysis in ox skeletal muscle. Effect of temperature on the concentrations of glycolytic intermediaries and cofactors. Biochem. J. 105, 127.
NEWTON, K.G., HARRISON, J.C.L. and SMITH, K.M. 1977. The effect of storage in various gaseous atmospheres on the microflora of lamb chops held at $-1E$. J. Appl. Bacteriol. 43, 53.
OCKERMAN, H.W., CAHILL, V.R., WERSER, H.H., DAVIS, C.E. and SIEFKER, J.R. 1964. Comparison of sterile and inoculated beef tissue. J. Food Sci. 34, 93.
ORLINSKA, H. 1968. Emulsions and emulgators in seasoned meat manufacturing. Consumption Ind. (Przemysl Spozywczy) 9, 391. (Polish)
PARR, A.A. and HENRICKSON, R.L. 1970. Nitric oxide pigments in pre- and post-chill processed ham. Food Technol. 10, 118.
PARRISH, F.C., YOUNG, R.B., MINER, B.E. and ANDERSON, L.D. 1973. Effect of post mortem conditions on certain chemical, morphological and organoleptic properties of bovine muscle. J. Food Sci. 38, 690.
PEARSON, A.M. 1968. Application of chemical methods for the assessment of beef quality. 2. Methods related to protein breakdown. J. Sci. Food Agric. 19, 366.
PEARSON, A.M. 1981. What's new in research. Natl. Provis. 185 (25) 15.
PEARSON, A.M., CARSE, W.A., DAVEY, C.L., LOCKER, R.H., HAGYARD, C.J. and KIRTON, A.H. 1973. Influence of epinephrine and calcium upon glycolysis, tenderness and shortening of sheep muscle. J. Food Sci. 38, 1124.
PEPPER, F.H. and SCHMIDT, G.R. 1975. Effect of blending time, salt, phosphate and hot-boned beef on binding strength and cook yield of beef rolls. J. Food Sci. 40, 227.

PISULA, A. 1981. The technology of hot meat in processed meat and meat products. Presented at Int. Symp. Adv. Hot Meat Processing, Poznan, Poland, Aug. 17, 1981.

PISULA, A. and MROCZEK, J. 1976. Investigations on using warm pork in meat manufacturing. Warsaw, Agricultural Academy, 1976. (Polish)

PISULA, A., BELDYCKA, W. and WASILEWSKI, S. 1973A. Comparison of thermal contraction of meat during the pickling process applied to warm meat and to cool meat. Meat Econ. (Gospodarka Miesna) 3, 23. (Polish)

PISULA, A., GORECKA, E. and ABRAN, A. 1973B. A comparison of the water content in warm and cooled meat during the process of pickling and thermal treatment. Meat Econ. (Gospodarka Miesna) 9, 21. (Polish)

PISULA, A., POPIEL, J. and WASILEWSKI, S. 1973C. Investigation of modeling of nitrosodyes in pickled pork made of warm and cooled meat. Meat Econ. (Gospodarka Miesna) 5, 19. (Polish)

PISULA, A., MACH, L. and POPKO, T. 1974. Investigations on the usefulness of warm pork in manufacturing seasoned meat products and pasteurized preservation. 5th Sci. Session of KTi ChZ of the Polish Acad. Sci., Gdansk. (Polish)

PISULA, A., BOZYK, Z., GLOWACKI, A. and KLAMUT, A. 1976. Technologies of warm pork in meat manufacturing. Warsaw, Agricultural Academy, 1976. (Polish)

POTTHAST, K. and HAMM, R. 1976. Procedure for production of freeze-dried meat with preservation of water-binding ability of meat fresh from slaughter. W. Ger. Pat. 25,19,000. (German)

PROSELKOVA, T. and BARGAJEWA, G. 1973. The use of warm meat in the production of ham products. Mjasnaja Industria 5, 17. (Russian)

PULLIAM, J.D. and KELLEY, D.C. 1965. Bacteriological comparison of hot processed and normally processed hams. J. Milk Food Technol. 28, 9.

RACCACH, M. and HENRICKSON, R.L. 1978. Storage stability and bacteriological profile of refrigerated ground beef from electrically stimulated hot-boned carcasses. J. Food Protect. 41, 957.

RAMSBOTTOM, J.M. and STRANDINE, E.J. 1949. Initial physical and chemical changes in beef as related to tenderness. J. Anim. Sci. 8, 398.

RASHID, N.H., HENRICKSON, R.L, ASGHAR, A. and MORRISON, R.D. 1983A. Evaluation of certain electrical parameters for the stimulation of lamb carcasses. J. Food Sci. 48 (1) 10–14.

RASHID, N.H., HENRICKSON, R.L., ASGHAR, A. and CLAYPOOL, P.L. 1983B. The biochemical and quality characteristics of ovine muscles as affected by electrical stimulation. J. Food Sci. 48 (1) 136–140.

REDDY, G.S. and HENRICKSON, R.L. 1969. Quality of pre-chilled canned porcine muscles. Food Technol. 7, 81.

RILEY, R.R., SAVELL, J.W. and SMITH, G.C. 1980. Storage characteristics of wholesale and retail cuts from electrically stimulated lamb carcasses. J. Food Sci. 45, 1101.

ROSOFF, H.D. and RIES, K.M. 1976. Energy requirements and conservation in the meat packing industry. 1st Int. Congr. Eng., Boston, Aug. 12, 1976.

RUDERUS, H. 1980. Low voltage electrical stimulation of beef. Influence of pulse types on post-mortem pH fall and meat quality. In Proc. 26th Eur. Meet. Meat Res. Workers, Colorado Springs, CO, 2, 96.

RUSSELL, M.S. and BALDWIN, R.E. 1975. Creatine thresholds and implications for flavor of meat. J. Food Sci. 40, 429.

SAFFLE, R.L. and GALBREATH, J.W. 1964. Quantitative determination of salt-soluble protein in various types of meat. Food Technol. 18, 119.

SCHMIDT, G.R. and GILBERT, K.V. 1970. The effect of muscle excision before the onset of rigor mortis on the palatability of beef. J. Food Technol. 5, 331.

SCHMIDT. G.R. and KEMAN, S. 1974. Hot boning and vacuum packaging of eight major bovine muscles. J. Food Sci. 39, 141.

SEIDEMAN, S.C., VANDERZANT, C., SMITH, G.C., HANNA, M.O. and CARPENTER,

I.L. 1976. Effect of degree of vacuum and length of storage on the microflora of vacuum packaged beef wholesale cuts. J. Food Sci. *41*, 738.

SKENDEROVIC, B., and RANDOV, M. 1975. Investigations into the technological property of meat frozen pre- or post-rigor. Fleischwirtschaft *2*, 251. (German)

SMITH, G.C., DUTSON, T.R. and CARPENTER, Z.L. 1979. Electrical stimulation of hide-on and hide-off calf carcasses. J. Food Sci. *44*, 335.

SQUIRE, J.M. 1975. Muscle filament structure and muscle contraction. Annu. Rev. Biochem. Bioeng. *4*, 137.

STILWELL, D.E., MANDIGO, R.W., WEISS, G.M. and CAMPBELL, J.F. 1978. Accelerated pork processing. Frankfurter emulsion properties. J. Food Sci. *32*, 1646.

STRANGE, E.D. and BENEDICT, R.C. 1978. Effect of post-mortem boning times on beef storage quality during storage. J. Food Sci. *43*, 1652.

STRANGE, E.D., BENEDICT, R.C., SMITH, J.L. and SWIFT, C.E. 1977. Evaluation of rapid tests for monitoring alternations in meat quality during storage. J. Food Protect. *40*, 843.

STROTHER, J.W. 1975. The commercial preparation of fresh meat at wholesale and retail levels. *In* Meat. p. 183. D.J.A. Cole and R.A. Lawrie (Editors). AVI Publishing Co., Westport, CT.

SZALUSZKOWA, Z., SWIETOW, W., SLADKOWA, Z. and TARASOW, T. 1971. Affectiveness of the use of warm beef in sausage production. Mjasnaja Industria *4*, 30. (Russian)

SZALUSZKOWA, Z., TARASOW, T., SWIETLOW, W. and TKACZ, B. 1975. News in production of cooked sausage with warm meat. Mjasnaja Industria *4*, 23. (Russian)

TAYLOR, A.A., SHAW, B.G. and MacDOUGALL, D.B. 1981. Hot deboned beef with and without electrical stimulation. Meat Sci. *5*, 109.

TRAUTMAN, J.C. 1964. Fat emulsifying properties of prerigor pork protein. Food Technol. *18*, 1065.

TRUMIC, Z., PETROVIC, N. and RISTIN, V. 1967. Uticaj razlicitin temperatura vode i rastvora natrijumchlorida i polifosfata na hodraciju usitujenog gevedeg mesa. Technologia Mesa *5*, 130.

TULEJOW, E. and JEJESZCHLEBOW, A. 1973. The production of sausages from defrosted meat with the addition of salted warm meat. Mjasnaja Industria *3*, 29. (Russian)

TURNER, R.H., JONES, P.H. and MacFARLAND, J.J. 1979. Binding of meat pieces: An investigation of the use of myosin-containing extracts from pre- and post-rigor bovine muscle as meat binding agents. J. Food Sci. *44*, 1443.

UNGER, S.G. 1975. Energy utilization in the leading energy-consuming food processing industries. Food Technol. *14*, 35.

UNGER, S.G. 1977. Energy conservation goals for the food industry in the 1980's. Presented before Am. Soc. Heat., Refrig. Air Conditioning Eng., Chicago, 1977.

UNGETHÜM, W. 1971. Water-binding abilities of meat. Part I. Fleisch *4*, 108. Part II. Fleisch *5*, 128. (German)

U.S. DEP. AGRIC. 1984. Livestock Slaughter Statistics. Statistical Reporting System, Crop Reporting Board, Washington, DC. (1983 Statistics)

VALINE, C. 1981. Tenderizing meat by electrical stimulation. Recherches *12* (122) 612. (French)

VISSER, K., AIRAH, A.M. and IIR, A.M. 1976. Current refrigeration practices in Australian abattoirs. *In* Towards an Ideal Refrigerated Food Chain. pp. 249–304. Refrigeration Science and Technology, Sydney, Australia.

VOGEL, P.W. 1963. Process for sausage manufacture. Belg. Pat. 636,004. (French)

VON HOOF, J. and HAMM, R. 1972. Influence of sodium chloride on the breakdown of adenosine triphosphate in minced beef muscle post-mortem. Z. Lebensm. Unters. Forsch. *150* (5) 282. (German)

WEIDEMAN, J.F., KAESS, G. and CARRUTHERS, L.D. 1967. The histology of pre-

rigor and post-rigor of muscle before and after cooking and its relation to tenderness. J. Food Sci. *32*, 7.
WEINER, P.D., KROPF, D.H., MACKINTOSH, D.L. and KOCH, B.A. 1966. Effect of muscle quality of processing pork carcasses within one hour post-mortem. Food Technol. *20*, 541.
WILL, P.A. and HENRICKSON, R.L. 1976. The influence of delay chilling and hot boning on tenderness of bovine muscle. J. Food Sci. *41*, 1102.
WILSON, G.D., BROWN, P.D., POHL, G., WEIR, C.E. and CESBRO, W.R. 1960. A method for the rapid tenderization of beef carcasses. Food Technol. *14*, 186.
WIRTH, F. 1974. Sausage production today: Water-binding, fat-binding, structure. Fleischwirtschaft *1*, 9. (German)

9

Electrical-Stimulation Research: Present Concepts and Future Directions[1]

B. Bruce Marsh[2]

General Considerations
Specific Aspects
Conclusions
References

Many discoveries in science have lain unused and dormant for years until their significance was recognized and they could be used in practical situations. Recent and continuing advances in electronics technology and the spectacular progress anticipated in genetic engineering exemplify this "learn now, apply later" category, for, without prior knowledge on which they could be constructed, these subjects would not exist at all.

In other fields of scientific endeavor, the sequence is reversed. An empirical, cause-and-effect modification or procedure is put into use before there is any real comprehension of what is happening, or why it is effective, or how it might be further refined. Elucidation of the mechanism may come quite soon, yielding new knowledge on which to base subsequent improvements, or it may be long delayed simply because little incentive exists to study an already satisfactory process. This "use now, explain later" approach has yielded, and continues to yield, results of inestimable value; consider, for instance, the sad state of the human race if surgical procedures and medical treatments could be applied only when every biological detail of their actions was known and fully understood.

Meat science, for the most part, does not fall into the first of these categories. No great store of preexisting fundamental knowledge (or, at least, of knowledge specifically labelled Basic Meat Science) has long been available to guide the meat investigator, and no great urge has arisen to amass such a store as an end in itself. Furthermore, the very success of

---

[1]Muscle Biology Manuscript *176*. Contribution from the College of Agricultural and Life Sciences, University of Wisconsin, Madison, WI.
[2]Muscle Biology Laboratory, Department of Meat and Animal Science, University of Wisconsin, Madison, WI.

early efforts to explain (if not to greatly improve) various existing technologies has acted as a further disincentive to study the more basic aspects. It is not surprising, therefore, that meat science still relies heavily on the try-it-and-see approach, for it has grown and thrived to the present on an almost exclusively empirical diet.

There is nothing at all wrong, of course, with a *modus operandi* that seeks application before there is full understanding and use before there is total knowledge. We would still be waiting for the first aspirin tablet if its discovery had depended on a detailed perception of biological pyrogenesis, and for the first frankfurter if its formulation had been contingent upon a complete theory of protein–salt interaction. Yet there are several very good reasons why, without deserting trial-and-error methodology, we should direct a little more of our attention in the future toward basic aspects than we have in the past. First, there comes a time when the returns from empiricism diminish rapidly and eventually become vanishingly small. The problems already solved have been the relatively easy ones, amenable to cause-and-effect investigation; those that remain are still with us simply because they have withstood this form of attack and will persist until their true nature is better understood. Second, masses of highly relevant information have been produced in recent years, particularly in the disciplines of physiology and biochemistry; they are already available if we are but willing to relate them to our own domain. Third, crises of which we presently have no suspicion will almost certainly arise in the future, and our ability to deal with them quickly and effectively will depend very much on the extent of the basic knowledge we acquire before their eruption.

We can bring these rather vague generalizations into sharper focus by considering the origins of some of our most significant knowledge of meat. The current stockpile of information on myofibrillar proteins (whose properties and interactions are major determinants of meat quality) owes very little to meat studies; it stems almost entirely from biochemical and physiological investigations of muscle as a contractile tissue. Rigor mortis and its relationship to pH, fluid retention, and the PSE condition (summarized by Bendall 1973) are understood largely because the initial studies were undertaken, not to solve an urgent problem, but rather to explain a common observation in terms of the underlying mechanism. The phenomenon of cold shortening (Locker and Hagyard 1963) was not discovered in a crisis-generated study of toughness but in a long-term study of meat tenderness that had been initiated some years before (though it is worthy of note that the knowledge so obtained was put to large-scale practical use very soon afterward to avoid excessive toughness in lamb). Because of the stores of information produced by these and similar studies, we are now in a far stronger position to understand, to modify, and to enhance the properties and qualities of meat.

These broad concepts have been dealt with right at the start, and at some length, because I believe that advances in the use of electrical stimulation (ES) in the meat industry will depend at least as much on the acquisition of "background" information as on the empirical development of techniques and applications. At this still early stage of ES implementation, therefore, it is appropriate to examine both our comprehension and our lack of knowledge of the process and its mechanism so that future efforts may be directed toward a more profitable interplay between understanding and practice.

GENERAL CONSIDERATIONS

Almost by definition, an assessment of tomorrow's research needs is an evaluation of today's ignorance and can be made only after critical appraisal of yesterday's learning. Development of future strategies depends upon the exposure of present deficiencies, so a devil's-advocate role must be adopted to probe the recent history of ES research and development.

Past Contributions to Present Knowledge

Almost 40 years have elapsed since ES was first seriously examined in relation to the practical problem of meat toughness. In his fascinating history of the technique's development, Deatherage (1980) has described studies, undertaken during World War II, that led to the filing of a patent application in 1947 and the later granting of patent protection (Harsham and Deatherage 1951). Despite its demonstrated great effects, however, the process received so little attention that its reappearance, more than 20 years later (Carse 1973), was regarded almost as a brand new technique. Indeed, even the New Zealand group responsible for this second disclosure was unaware of the original work until shortly before its paper was submitted for publication (W. A. Carse: personal communication).

The long dormancy of ES, between its birth and its vigorous growth phase, is readily accounted for: most meat scientists have at least a nodding acquaintance with the literature of their own specialized field, but very few routinely examine patent specifications. To explain the sudden and intense interest that was aroused by Carse's paper, I believe we have to invoke the earlier detection of cold shortening and its consequences. Appearing about midway between the first and second published discoveries of ES, the work of Locker and Hagyard (1963) exerted a unique stick-and-carrot influence, forcing the researcher to recognize the importance of the early-postmortem period while offering in return an exciting and potentially very valuable area for exploration and exploitation. With its emphasis on the prerigor condition, cold shortening turned ATP and muscular contraction into acceptable—even desirable—topics for laboratory discussion, thus paving the way (and ensuring an enthusiastic reception) for the reentry of ES to the meat scene.

Now we must turn to the postdormancy period of ES study and use. If progress were measured by activity, and knowledge by publications, we would have to conclude that the present is a golden age of advancement in meat-quality research and development; but the extent of progress and the acquisition of knowledge are not always proportional to enthusiasm and printer's ink. The critical question must be asked: How big, really, is the store of information generated by all this vigor and exertion?

It is fortunate that the Harsham-Deatherage patent was prepared with such thoroughness for it provides a quite precise benchmark with which present knowledge may be compared. Consider these excerpts (Harsham and Deatherage 1951):

> The invention is directed particularly to an improved process for tenderizing meat in much less time than heretofore has been required, and at much less cost.
>
> ... The invention contemplates a procedure wherein the animals, after being killed and bled and preferably, but not necessarily, within approximately 15 to 30 minutes thereafter, are contacted with electrodes to produce galvanic or faradic responses therein with attendant contraction of the muscles.
>
> ... Voltage must be sufficiently high to overcome the resistance of the tissues and permit an activating current to flow therein. For instance, low voltages, e.g., 40–50 volts, or even much lower, produce a stimulating effect, but higher potentials of from 100 to 3000 volts are preferred because of the better current distribution over the cross-section of the carcass and the corresponding greater certainty which attends their use.
>
> ... Since it is desirable to perform stimulation as soon as possible after death, this treatment is conducted preferably before skinning.... The production, within the muscles, of a medium favorable to catabolic enzyme activity while the temperature of the carcass is at or slightly above the normal temperature hastens the action of the acids and enzyme systems on both the muscle fibers and the connective tissues.
>
> ... It is found that the pH drops to 6.0–6.2 within one hour and may fall to as low as 5.6–5.9 in approximately 12 hours. Thus, soon after slaughtering, the muscles and connective tissue are caused to reside in the environment of an acid medium which aids or favors the action of certain enzymes upon both the connective tissue and the muscles.
>
> ... Treatment—affects tenderization to a more or less uniform level of approximately 7 to 7.5 [10 = very tender]—The process exhibits greater effect on tough tissue.

Other sections of the patent specification describe the effects of varying frequency, the rhythmic jerking that occurs "to an undesirable degree" at frequencies below 10 Hz, the preferred method of applying the electrodes, and the several ways in which appropriate periodic electrical discharges could be produced. Further quotation is unnecessary to make the point that the discoverers of the technique did a remarkably thorough job. A complete reading of the original document today, a hundred or so published studies later, is indeed an edifying experience.

Many significant technological advances have been built upon the foundation provided by the original specification. Repeated confirmations of the beneficial effects of ES on meat quality have come from many laboratories. Numerous other real or supposed advantages of the process have been unearthed. Alternative methods of administering the treatment have been devised: live rails, rub bars, spring-loaded pan electrodes, nostril clips, pithing-rod electrodes, chain contacts, and a probe apparatus permitting *per rostrum ad rectum* stimulation via natural orifices. Low-voltage systems have been developed to ensure maximum operator safety. The effectiveness of ES, and of ES coupled with various other recognized tenderizing treatments, has been demonstrated. It is because of these applied studies and their generally favorable results that large segments of the meat industries of several countries have incorporated ES into routine practice. As measured by commercial installation and application, therefore, ES developmental progress has been very substantial.

These large and quite rapid advances in technology, however, have come about by application and extension of the original specification and for the most part have not been accompanied by significant gains in our understanding of the process itself. Granted, if the present state of ES technology is so advanced and so complete that no need exists for further modification, then the principal incentive for additional in-depth study is eliminated (even though a strong interest would remain to elucidate the physiological mechanism). Still, if the practical employment of ES is not yet perfect—if results from its use are variable, for instance, or if improvements in one attribute are accompanied unaccountably by deterioration in another—then a greater knowledge of underlying mechanisms is an absolute prerequisite to further progress. Without it, a long period of futile cycling of observations cannot be avoided.

Past contributions to our knowledge of ES have thus been almost entirely technological in nature, virtually unaccompanied by significant advances in understanding of the process or of its consequences. While application of the Harsham-Deatherage discovery has been intensive, elucidation has been negligibly small; we still lack the fuller comprehension needed to account for present observations and to point the way to further innovations. An intentional channeling of enthusiasm and effort toward cause, rather than effect, is urgently needed.

Meat Quality and Carcass Stimulation

Although widely separated in time, the studies of Harsham and Deatherage (1951) and Carse (1973) were prompted by the same need: to tenderize meat. Many other beneficial effects of ES have since been reported, including improved color, decreased heat-ring incidence, greater ease of early grading, reduced refrigeration needs, increased practicality of hot boning, improved retail-display appearance, extended retail-caselife, lower shrinkage, increased flavor, and improved marbling appearance.

In view of this truly remarkable array of near-panacean abilities, it is not really surprising that tenderization per se is not the principal inducement for stimulator use in U.S. meat plants (Terrell *et al.* 1982). In most countries, less emphasis is placed on the cosmetic features of marbling and color (and, perhaps, correspondingly more on direct consumer reaction) than in the United States, and tenderness is the primary—and often the only—reason to apply the ES process. (This is particularly so in countries that are major meat exporters, for significant adverse consumer reaction to imported meat can lead quite rapidly to changed patterns in international trade, loss of export markets, and perturbations in national economies.) In this country, however, the major enticement to use ES is the process's effect on appearance, since marbling largely determines, and color may influence, the quality grade on which carcass value depends. Indeed, if tenderness enhancement, without change in visual characteristics, were the only inducement, it is likely that industrial use of ES in the United States would still be very limited. It is thus the potential for improvement *in these quality indicators,* rather than in tenderness itself, that has encouraged the quite rapid adoption of the ES technique. (This cosmetic motivation would no longer exist, of course, if marbling ceased to be the basis of quality grading—a possibility that certainly cannot be ignored in view of the commendable study (Dolezal et al. 1982) of subcutaneous fat thickness as a predictor of palatability.)

It is no mere exercise in semantics to distinguish quality as estimated visually from quality as perceived organoleptically, for the great variability in eating satisfaction among identically evaluated carcasses is only too well known. Even though such a differentiation introduces the vexed and controversial subject of quality grading, it is necessary to do so at this point in order to focus attention on the major research need in the ES field: an explanation of the mechanism(s) by which stimulation exerts its quality-enhancing action(s). Already, without this understanding, several potential fallacies have crept into ES doctrine: that the rate of pH decline is necessarily a measure of tenderizing efficiency, for instance, and that the brighter color of ES-treated meat is an indicator (albeit a weak one) of higher eating quality. It is a matter of some urgency, therefore, to acquire a basic understanding of ES before these and other misconceptions become ineradicable articles of faith.

Marbling and (to a lesser extent) color have long been used to provide an estimate of the eating quality of beef. The extent to which the former really predict the latter is still far from clear, despite the great number of studies that have been undertaken over many years, but we may be fairly confident it is quite small. It is on this tenuous, unexplained, and possibly inexplicable relationship that the U.S. quality-grading system for beef is largely based. (In fairness to its originators, developers, and administrators, it must be added that no clearly superior system has yet appeared.) Leaving aside for the present any further discussion of the merits and deficiencies of grading as presently practiced, let us assume that these visual characteristics do indeed offer some guidance to eating-quality differences among carcasses, even those of identical maturity, and that intramuscular fat-deposition patterns and myoglobin content really do exert a significant influence on eating satisfaction.

Now, although this quality-grading system was devised before the introduction of ES, we might (at first sight) suppose that it would perform no more poorly with ES carcasses than with those receiving no stimulation. After all (it could be argued), ES usually increases both tenderness *and* visual characteristics, the former often to a significant extent and the latter sometimes enough to elevate the grade as assessed one day postmortem. So why should the quality indicators be any less predictive of eating quality in ES beef than they are in the non-ES product?

The reason, though neither direct nor obvious, is nonetheless definite and important. If two non-ES steer or heifer carcasses of the same age do not receive the same quality grade, it is because they differ in intramuscular-fat content/distribution and possibly in color. There is thus some basis for implying a difference in eating quality, if our stated assumption—that visual characteristics are a guide to eating quality—is valid. But if two beef sides differ in these visual indicators simply because only one of them has been stimulated, this basis disappears entirely. Neither fat content nor fat distribution nor myoglobin has been changed at all; ES has merely created the illusion that these variables now differ between the sides.

How, then, are we to account for the parallel improvements in eating quality and grading indicators that so often follow ES treatment if at the same time we are to explain the numerous observations in which no such relationship can be detected? The answer appears to lie in the multiplicity of effects that are produced by the ES treatment, for there can be no doubt that several consequences result directly from stimulation. If all of them always occurred, and to the same extent, it would be entirely legitimate to relate one to the other: brighter color, for instance, to greater tenderness. Unfortunately, however, these various effects may occur to widely differing degrees, depending both on the characteristics of individual carcasses and on the mode of administering the stimulating current; parallel responses in appearance and palatability are then no longer observed. Savell *et al.* (1982), for instance, have demonstrated an almost complete separation of cosmetic improvement (which was considerable) from palatability enhancement (which was negligible) when ES was applied to carcasses of young bulls. This divorce of visual appraisal from eating quality must raise serious doubts of the validity and basis of the quality grading of ES carcasses in general. In addition, as we shall see later, visual characteristics can be uncoupled quite readily from eating quality, even in steers, by a fairly simple change in the method of applying the ES treatment. Although it would be both pointless and far from practical to use this modification on a large scale, the result nevertheless again calls into question the soundness of applying non-ES quality standards to ES carcasses.

Yet this added uncertainty that ES has introduced into an already dubious grading system is not the most important reason for seeking an early understanding of carcass stimulation. There are other motives, more significant and less transient than grading facilitation, for elucidating ES actions. Without such knowledge, we cannot know if the empirically derived physical parameters now in use are really producing optimal results; we shall remain unaware of potentially great improvements in process efficiency or product quality that might arise from presently unknown concepts. Rather more concrete and of greater consequence is the distinct possibility that (in the absence of a fuller understanding) ES will continue to be applied in a way that is appropriate for one set of circumstances but much less suitable for another. A rather simple example of this situation may have arisen already; it concerns the relative quality-enhancing merits of high-voltage and low-voltage usage in ES installations, and it will be discussed later.

The extent to which ES improves quality thus depends largely on the meaning we choose to bestow on the word "quality." Even in nonstimulated carcasses, visual (or grading) quality provides only a weak measure of eating quality, and in the electrically stimulated product the potential exists for an even poorer relationship between the two. It is fortunate that stimulation as generally applied often results in improvements that are both visually *and* organoleptically detectable, but there is evidence that distinct and separate mechanisms are responsible for these quite different effects. If these mechanisms remain obscure, and if they cannot be individually identified with improvements in specific properties, we must be content to administer ES as a sort of shotgun therapy, hoping that at least one pellet will hit and activate each of the quality-enhancing processes. On the other hand, if the mechanisms can be elucidated, and if each can be associated with a particular benefit, then it should be possible to devise appropriate routines that will strengthen those actions that are beneficial and eliminate those that do not help. There are, after all, many variables already available for manipulation, and perhaps others of which we are presently unaware. Our eventual understanding of how each of them affects each of the desirable attributes will put us in a much better position to formulate the most beneficial treatment for each set of given conditions.

The Dynamic Nature of Early-postmortem Muscle

Finally, before turning from general considerations to more specific aspects, we must remind ourselves of the extreme complexity of muscle. Even when completely at rest, muscle *in vivo* is very far from being a static tissue. Proteolytic and protein-synthesizing reactions are still going on, regardless of contractile activity; glycolysis and respiration continue their course, contributing energy for the maintenance of concentration gradients, turnover mechanisms, and body temperature; blood flow is maintained, bringing oxygen and substrates to the tissue and removing end-products or metabolites for use elswhere. When a nervous stimulus reaches this already highly dynamic system, almost every component is altered, at least momentarily: membrane permeabilities change, sodium and potassium migrate, calcium ions leave the organelle(s) previously entrapping them, regulatory proteins change their configuration, contractile proteins are provoked into movement, glycolysis and respiration are stimulated to produce more immediately available energy, and blood flow increases manyfold to cope with both the greater inflow of required substrates and the disposal of larger amounts of waste products.

Not all of these systems and mechanisms are activated, of course, when external ES treatment is imposed on an early-postmortem carcass, since blood flow (and hence inward and outward transport of metabolites) has ceased some minutes before. Other changes, *not* taking place *in vivo*, replace those that can no longer occur, ensuring that stimulation after death is at least as complex as in life: membrane deterioration, lactic-acid accumulation, declining temperature, reduced ATP-resynthesizing ability, and a totally abnormal activation of part of the central nervous system by the acts of stunning, decapitation, and carcass splitting. Still further complications are introduced by the need to use voltages that are hundreds of times higher than the body was designed to utilize, often causing membrane destruction, supraphysiological contraction, and extensive fracturing of contractile fibers and connective tissues.

ES exerts its desirable effects *only* in the early-postmortem period; to delay application until the musculature approaches the rigor state is to lose every benefit the process can provide. It follows that significant progress will depend not only on a thorough knowledge of muscle structure and composition but also on a continual awareness of the dynamic and delicately balanced state of the prerigor (and near-to-living) tissue.

SPECIFIC ASPECTS

My emphasis up to now has been on the need for a more basic approach to ES research if we are to understand, control, and (eventually) optimize the process in its practical applications. It is necessary at this point to become more specific, adding flesh to this conceptual skeleton in order to justify and encourage investigations of a more fundamental nature. It is not only future studies that merit consideration; many past projects, I believe, would yield further valuable information if they were reexamined in terms of basic mechanisms. Stimulation investigations have been reported at such a fast pace in recent years that the task of assimilating them and integrating their results into an overall picture has become almost impossible. As a consequence, many of them will necessarily remain only isolated observations, unrelated and unrelatable to each other simply because, with our present limited knowledge, we lack the ability to comprehend their findings.

The first of the three topics to be discussed in this section, early-postmortem muscle temperature, may appear at first sight to be only distantly related to the overall theme. It is true that muscle temperature varies very little among carcasses up to the time of stimulation, for the process is always applied within a few minutes (certainly less than an hour) of slaughter, an interval too brief to allow appreciable temperature change. Later temperature conditions, however, can (and frequently do) vary over a wide range; and, through their ability to influence glycolysis and proteolysis, they can (and probably do) exert significant effects on quality. It is likely that some of the unknowns and inconsistencies in the ES field are due to our failure to recognize the significance of muscle temperature during the first few critical postmortem hours, and it is at least possible that they might be resolved if greater attention were given to the monitoring of this important variable.

Early-postmortem Muscle Temperature

The strong influence of temperature on prerigor muscle behavior and ultimate meat quality has been examined extensively (though by no means exhaustively) in recent years, particularly in relation to rigor pattern, shortening, and tenderness. It is widely recognized that none of these effects is due *solely* to temperature; rather, it is the interaction between temperature and other variables, particularly glycolysis, that is responsible for the wide range of observed responses. Yet there remains a tendency to overlook the magnitude of the differences in time–temperature relationships among different studies. This is particularly so when apparently quite similar investigations are undertaken in facilities having widely varying cooling efficiencies and on carcasses of highly disparate cooling susceptibilities.

The data in Table 9.1 are presented to illustrate the appreciable effects of several variables on early-postmortem muscle temperature. They are derived from recent studies in Wisconsin (Lochner et al. 1980) and Brazil (unpublished data: B. B. Marsh, G. Cia, O. Corte, and G. Takahashi). Air velocities were 90–120 m/min in all three groups.

The effects of the variables—fat cover and content, weight, and chiller temperature—on muscle cooling rate are not unexpected. Fat is well recognized as an effective thermal insulant (more than twice as efficient, in fact, as the same thickness of lean tissue); simple carcass mass is obviously a major determinant of heat loss; and ambient temperature clearly influences the rate of heat transfer from carcass to environment. Table 9.1,

TABLE 9.1. Effects of Air Temperature and Carcass Size and Fatness on Muscle Cooling Rate

Breed	Live wt. (kg)	Fat cover (cm)	Marbling	Chiller air (°C)	Longissimus temperature at 5 hr postmortem
Angus	570	2.6	Slightly abundant	−2	26
Angus	430	0.5	Small	−2	13
Nelore (Zebu)	430	0.1	Devoid	+5	13

thus, reveals no novel information at all, and indeed is not intended to do so. Its sole purpose is to demonstrate a straightforward (though usually overlooked) fact: that early-postmortem temperature is a function of a *number* of variables, of which ambient temperature is but one. It is not the magnitude of these several effects that gives cause for surprise; rather, it is the extent to which they are ignored. It is rare indeed for the measurement of muscle temperature (as distinct from air temperature) to be included in meat-quality studies. This is not to say, of course, that it should be monitored routinely in every investigation, for in many of them it would be clearly irrelevant; but if there is a likelihood—or even a possibility—that muscle temperature might influence the result (or might affect the outcome if the study were to be repeated elsewhere), then periodic measurement of muscle temperature is essential.

Part of my reason for placing such heavy stress on early-postmortem temperature is obvious, for (depending on its rate of decline) it can initiate cold shortening, change proteolytic rate, influence color, and exert appreciable effects on weight loss and microbial growth. In addition, its ability to regulate several of these processes is magnified by the power it exerts concurrently over glycolysis. A higher muscle temperature promotes a faster rate of ATP splitting; the consequent increase in ADP accelerates pH fall and rigor onset. pH decline in excised bovine muscle, for instance, is almost four times as fast at 37°C as it is at 17°C. Besides its direct effects on ongoing processes in muscle, temperature thus exercises an important indirect influence. For activities such as proteolysis that may be highly sensitive to both temperature *and* pH, a powerful cascade form of control is exerted.

It is not only in relation to our basic understanding of muscle that early-postmortem temperature is important; it is also critical to our comprehension of meat-quality attainment in general and of the role of ES in particular. Clearly, to optimize the beneficial influence of stimulation, we must first know which of its several effects is (or are) really responsible for the desired end result. The usually quoted list of possibilities—not by any means an exhaustive catalogue—consists of cold-shortening prevention (through acceleration of glycolysis and rigor onset), proteolytic-enzyme release and activation (through rapid acidification while muscle temperature remains high), and physical rupture of the fibrous elements of the tissue (through extreme contraction in some zones of the fiber and compensating stretch and fracture in others). Under quite widely differing experimental conditions, all three of these effects—elimination of cold shortening, release of lysosomal enzymes, and tissue fracture—have indeed been observed, though (to my knowledge) never simultaneously. This failure to detect all of the consequences of ES in a single sample is no cause for dismay; in a system as complex as prerigor muscle, quite minor differences among carcasses might be responsible for quite significant differences among tissue responses to stimulation. The lack of synchronism, however, does not help to clarify the role of ES in meat-quality enhancement, particularly when it is accompanied by a distinct lack of agreement among researchers undertaking apparently similar investigations. Thus, several studies (e.g., George et al. 1980) have revealed no significant structural damage in the musculature of ES carcasses, whereas others (e.g., Sorinmade et al. 1982) have found extensive rupturing. Similarly, the release of lysosomal enzymes by stimulation has been demonstrated by Dutson et al. (1980), who ascribed the effect to rapid ES-induced pH fall while carcass temperature was still high, but Marsh et al. (1981) found no evidence that rapid acidification per se caused any tenderizing at all; indeed, it was accompanied by a significant toughening. Such major discrepancies—in fact, frank contradictions—in results from different laboratories gravely complicate our present understanding of the actions of ES, and their early elimination is a matter of some urgency.

It is here, I believe, that early-postmortem temperature must enter the picture, for it appears capable of explaining many of the conflicting results and of reconciling most of the diverging views. Let us suppose that two investigators undertake very similar studies in which only one side of each carcass is stimulated; following this treatment, both sides are chilled. Each observer uses cattle and cold rooms that are immediately at hand and are thus familiar to him and typical of his geographical region. But X's cattle are well finished, and his chillers are relatively mild, whereas Y uses very lean cattle, and chillers that are very severe. For X, the possibility of cold shortening in the control (unstimulated) sides is almost negligibly small since the slow cooling conditions ensure both a fairly rapid glycolytic rate and a very slow advance of the "cold front" into the musculature. X is thus completely justified in concluding that ES has nothing to do with cold-shortening prevention, and he deduces that fiber rupture and possibly acid-activated cathepsins must be responsible for the small but significant difference he observes in eating quality. The very rapid muscle cooling in Y's study, by contrast, not only ensures a fast temperature fall into the cold-shortening range but also retards glycolysis and postpones rigor onset in the control sides—conditions that will almost certainly provoke extensive cold shortening in the non-ES musculature. Y is as justified as X in coming to a quite different conclusion: that the large quality improvement evident in the stimulated sides is due almost entirely to the treatment's elimination of cold shortening. We cannot fault the inference of either of the observers; in each case the deduction is appropriate and the conclusion is correct. What *can* be faulted is the supposition that X or Y will almost certainly make: that his results and recommendations are directly applicable elsewhere without further investigation of the local conditions.

The interpretation one places on ES effects thus depends very much on where one's studies are undertaken, for location largely determines the magnitude of the temperature influence. If the location is the United States, with its generally fatter animals and often quite "lenient" chillers, then cold shortening (at least to a great enough extent to affect tenderness significantly) is unlikely, even without stimulation; but if it is in one of the many countries where cattle are leaner and chillers more severe, then appreciable shortening is highly probable in the unstimulated sides, and prevention of this shortening is almost certainly the predominating mode of ES tenderizing action. These two scenarios, of course, portray only the extremes; quite lean carcasses do indeed sometimes encounter quite vigorous chillers in this country, and the converse no doubt holds true in other lands. In general, however, carcass cooling rates tend to congregate in a bimodal pattern, some toward the fast end of the possible range and others toward the slow.[3]

Early-postmortem muscle temperature is thus a major determinant of the tenderizing effectiveness of ES; its importance appears to be due, not to its influence on the stimulated carcass or side (from which any cold-shortening tendency will have been eliminated already by the electrical treatment), but rather to the continued temperature sensitivity of the *un*stimulated material. It is, after all, the tenderness *difference* between ES and non-ES meat that is the customary measure of treatment efficiency, so the degree of cold shortening in the unstimulated musculature largely determines our evaluation of the tenderizing powers of the ES process. Since this cold-shortening extent is so dependent on a number of factors, each of which can exert a very significant influence, it is really not surprising that results and deductions vary so much among widely separated laboratories.

One other aspect of early-postmortem muscle temperature merits discussion: its role in explaining the widely varying tenderizing effectiveness of low-voltage stimulation. Comparisons of studies in this area are very difficult to make since (through either choice or necessity) almost every investigator selects a combination of experimental conditions—carcass characteristics, chilling rate, electrical parameters, etc.—that is different from all others. Yet a vague pattern is beginning to emerge from the chaos, allowing the tentative expression of a concept that reconciles a number of apparently conflicting results: The tenderness improvement that is brought about by low-voltage ES is proportional to the extent of shortening-induced toughening in the *un*stimulated (control) carcass or side. Since cold shortening in this non-ES situation is dependent on chill-

---

[3] Reasons for this curious distribution are not hard to find. First, many countries must now conform to stringent European Economic Community (EEC) regulations on cooling rate; the United States exports little or no beef to Europe and so does not need to comply with the EEC's requirements for rapid chilling. Second, lean carcasses suffer a significantly smaller evaporative weight loss if chilling is very fast; the greater fat cover of U.S. beef quite effectively reduces evaporation so weight-loss reduction by increasing the cooling rate is not a strong incentive. Third, the grading systems of some countries (e.g., Sweden: Ruderus and Fabiansson 1980) favor the production of lean (and therefore more easily cooled) carcasses; the U.S. quality-grading system encourages the deposition of fatty insulation, at least to the point where it has a very marked influence on cooling rate. Finally, feed supplies are a major controller of carcass fatness, in most countries placing a severe limit on the extent to which adipose tissue can be deposited economically.

ing rate, it follows that early-postmortem muscle temperature is the factor determining the effectiveness of low-voltage stimulation. Thus we would expect that, in situations where lean carcasses and severe chilling conditions are the rule, low-voltage treatment would produce a large tenderness improvement through cold-shortening elimination, and this indeed appears to be true in Sweden (Ruderus and Fabiansson 1980) and Australia (Taylor and Marshall 1980; Bouton et al. 1980). Similar results have been obtained in this country when carcass-cooling rates were uncharacteristically rapid (McKeith et al. 1980). The study of Eikelenboom et al. (1981) in the Netherlands was particularly informative, for it included a demonstration that appreciable cold shortening occurred in the control sides; 85 volts (peak) caused a significant tenderizing, the extent of which was not much less than, and not significantly different from, that produced by 300 volts. On the other hand, the decline in muscle temperature under typical U.S. conditions is usually too slow to cause much cold shortening (as mentioned earlier), because of greater carcass fatness and/or a milder chilling routine, so there is little or no shortening-induced toughening to be overcome. In these circumstances, low-voltage ES produces only an insignificant tenderness improvement relative to the unstimulated control (McKeith et al. 1981).

It is highly pertinent to this discussion that low voltages (32–85 V), despite their relatively small "punch," can cause quite rapid declines in muscle pH (Ruderus and Bergquist 1980; Bouton et al. 1980; Taylor and Marshall 1980); indeed, Eikelenboom et al. (1981) showed that 85 V was as effective as 300 V in accelerating glycolysis (though the methods of administering the two treatments differed somewhat). Thus, in combatting cold shortening, there is no reason why a low-voltage treatment should exert much less influence than one employing a high voltage. The great toughening effect of the length change will much exceed and largely conceal that due to other causes, and its elimination (even if no other tenderizing mechanism is activated) will be very readily detected as a major quality improvement. If, however, because of a slower temperature decline in the musculature, cold shortening cannot occur to a significant extent, then considerably more than a mere acceleration of glycolysis will be needed to further tenderize the product; the applied stimulus must now be powerful enough to activate other mechanisms (one of which, fiber fracture, will be discussed later). Therefore, fatter and slower-cooling carcasses would be expected to respond much less positively to a low-voltage system. The seemingly contradictory effects reported by different observers—many of them apparently overlooking the great temperature influence—are thus seen to be no more than a consequence of differing cold-shortening tendencies brought about by widely dissimilar cooling rates.

Finally, in this temperature-centered section, reference must be made to the very widespread use of pH decline as a measure of ES effectiveness. No problem arises when this indicator is employed on rapidly cooling carcasses, for the rate of pH fall is an appropriate marker of declining cold-shortening propensity. Nor is there any difficulty with slowly cooling carcasses if cosmetic improvement is the only goal, for visual (quality-grading) characteristics are certainly enhanced by rapid pH decline while the carcass is still hot. However, if increased palatability—primarily tenderness—is a major objective in these non-cold-shortening carcasses, then pH fall is of negligible value, and (in admittedly quite contrived circumstances) may even indicate the reverse of the true situation. Although McKeith *et al.* (1981) failed to measure pH changes, we may safely assume (in the light of numerous other studies already cited) that their 150 volt treatment greatly stimulated glycolysis; yet it caused only very small tenderness improvements in grain-finished steers. As for the reversal effect, Marsh *et al.* (1981) showed that the use of 2 Hz current produced both a rapid pH decline and a distinct *toughening* of the meat. Rate of pH fall is thus not absolutely coupled to ES effectiveness, and, in studies of the influence of stimulation on the palatability of quite slowly cooled carcasses—a not unusual subject of research in this country—pH change could be a highly misleading measure of quality enhancement.

This section may appear to many to have done no more than restate the obvious since muscle temperature and electrical stimulation have been widely acknowledged for some years as major determinants of meat quality, but perhaps it is the obvious that needs restating. The strong tendency to the present time has been to concede the importance of both factors, temperature and stimulation, but to treat them as separate entities, with little or no recognition of the great modifying influence exerted by the first on the actions of the second. The root of the problem is obvious: Of necessity, we assess the efficacy of ES by reference to an unstimulated control that is itself highly variable, for it is on this control that temperature exerts its sometimes very large and sometimes negligible effects. We should not be disconcerted, then, to find that ES-induced quality differences fluctuate so much from one laboratory to another; it is not the effects of stimulation on the treated side or carcass that are responsible for this variability but the effects of *temperature* on the *untreated* sides. As demonstrated by the high-voltage/low-voltage example, apparently conflicting observations and conclusions can sometimes be reconciled by taking this very important temperature factor into consideration. There is thus some reason to suppose that a thorough monitoring and/or control of muscle temperature, coupled with a continual awareness of the strong temperature influence on quality enhancement, will lead to significant progress in both comprehension and application.

ES-induced Changes in Early-postmortem Metabolism

Electrical stimulation can produce two broad types of effects in early-postmortem muscle: accelerative and disruptive. These effects do not invariably occur together, and by altering the characteristics of the applied current—certainly the frequency (Marsh et al. 1981), and probably the voltage (as described earlier)—it is quite easy to maintain the first and suppress the second. With this evidence that glycolytic acceleration and structural disruption can be readily uncoupled from each other, we are justified in considering the two effects separately—though of course with a continuing awareness that each may impinge significantly on the other. The low-frequency study also demonstrates again that we are not dealing with a simple cause-and-effect situation; much more is required than a naïve concept that ES *directly* determines the quality characteristics of meat. Instead, stimulation must be regarded as the instigator of a host of metabolic and structural responses, and it is these, rather than the initial electrical provocation, that are immediately responsible for the several quality changes ultimately observed. These intermediate effects, occupying a central position in the chain of events, are thus to be viewed as both result and cause—the result of prior ES application, and the cause of later quality modification.

Although muscle in its early-postmortem, prerigor state is still very much alive in terms of its contractility and metabolic activity, it is nevertheless irreversibly dying. Very marked differences separate the living and the immediately postmortem tissue, whether we consider the resting or the stimulated condition, and all of them are due to a single incident: the cessation of blood flow at the time of death. This one trauma results in a great cascade of interrelated events that start with the onset of anaerobiosis and end with the attainment of rigor mortis.

Despite the complexity of this chain of activities, one major simplication can be made right at the start: There is no reason to suppose that the final composition of a previously stimulated muscle, once it is in rigor, differs in any way from that of a corresponding nonstimulated one in the same "set" condition. (There are, of course major *structural* differences, to be dealt with later.) The great acceleration of glycolysis by stimulation may well be accompanied by a momentary accumulation of glycolytic intermediates, just as has been observed (on a more extended time scale) when glycolysis is accelerated by homogenizing the early-postmortem tissue. But when the stimulus ceases and the glycolytic rate returns to (or toward) its normal resting pace, pressure on the rate-limiting enzyme(s) declines, and the brief pile-up of metabolites at the bottleneck is quite rapidly depleted. By the time rigor is complete, therefore, the tissue has returned to a normal (non-ES) composition. Gratifying though this may be (in the sense that no abnormal constituent has accumulated as a result of the treatment), the return to compositional conformity is frustrating for the investigator, since it totally prevents the later measurement of any ES effect on the early-postmortem pH/temperature profile. In elucidating this important relationship and its influence on shortening and (probably) proteolysis, it is thus not enough to look only at the before-and-after states of the tissue; it is necessary also to assess the situation *during* the transition from muscle to meat.

It is the rate, therefore, and not the nature or the extent of postmortem metabolism, that is of major interest in relation both to comprehension and to consequences. If we could restrict our attention to ES-provoked effects on this rate, without regard to what might be happening in the unstimulated controls, a reasonably complete and consistent pattern would emerge, for stimulation very appreciably speeds glycolysis while the current is flowing, and probably (though not necessarily: see Bendall 1980) results in continued rapid acidification after it has ceased. However, no such helpful restriction can be imposed; it is the *difference* (in rate, or extent, or any other measure) between stimulated and unstimulated carcasses that is important, as was demonstrated earlier for tenderness, since we have no other way of determining ES effectiveness. It is thus as necessary to determine glycolytic rate in the unstimulated tissue as it is in the stimulated, for without this control value, no assessment of ES-induced acceleration can be made. Yet here a major problem arises; even without ES inducement, the pace of glycolysis varies widely, both among corresponding muscles in different carcasses (Tarrant and Mothersill 1977; Bendall 1978) and within a single muscle of one carcass (Marsh et al. 1981). This is a surprising discovery; it was assumed for many years that muscle temperature was the one determinant of glycolytic rate in the early-postmortem material, as indeed it still appears to be in the special case of early-excised muscles. The practice of hot boning, however, is still the exception rather than the rule, and until it becomes more widespread, we must continue to concentrate on the much more common procedure of cooling beef as sides. The question thus arises, specifically for non-ES beef: Apart from temperature, what determines the rate of early-postmortem glycolysis (and perhaps of other changes dependent on this rate) in muscles that remain skeletally attached?

There are several reasons why this matter is important and relevant to the present discussion, even though it is not directly ES-related. First, although the variability in glycolytic rate is often small and sometimes negligible (both among and within muscles), on some occasions it is remarkably large; in carcasses held at 37°C to avoid temperature-difference complications, we have found 3 hr longissimus pH values spanning 1.3 units among muscles and 0.5 unit within a single muscle. Second, if pH-sensitive proteolysis contributes to meat palatability, then the rate at which the tissue traverses the pH range of 7.0–5.5 could be a significant determinant of eating quality, particularly when the magnifying effect of early-postmortem temperature on enzyme activity is taken into account. Similar reasoning applies, of course, to cold shortening since rapid glycolysis (no matter if it occurs spontaneously or is induced by ES treatment) effectively prevents the detrimental length change. Appreciable differences in eating quality might thus be anticipated among carcasses that differ widely in their "natural" (unstimulated) rates of muscle glycolysis. Within single muscles, where the range of glycolytic rates among sites is smaller (though in some cases by no means insignificant), we would expect this influence on quality to be less; when associated with temperature gradients in the cooling musculature, however, it might still be great enough to contribute to the tenderness variability sometimes encountered within a single steak or roast. Third, it is important to note that all of the muscles in our study attained ultimate pH values of below 5.7. If prerigor pH sampling had not been undertaken, therefore, this very curious behavior would have escaped detection entirely, and any effects it might have exerted on quality would necessarily have been ascribed to interanimal variability.

Now, although this phenomenon of unexpectedly fast glycolysis occurs in non-ES carcasses, there seems to be no reason why its effects would be any different from some of those produced by stimulation. (The word "some" is included because tissue disruption is an unlikely consequence of the spontaneous acceleration of pH fall; fiber rupture appears to occur only with the development of supercontracted zones, the formation of which does not depend on rapid glycolysis.) In particular, the suppression of cold shortening and its toughening consequences, together with the cosmetic improvements that depend on rapid pH fall—brighter appearance, earlier grading readiness, heat-ring elimination, etc.—should be achieved as quickly and as effectively in these "naturally" fast-glycolyzing muscles as in those that are stimulated.

Several aspects of this non-ES acceleration of pH decline appear worthy of further consideration. First, our total inability to account for it with our present fund of knowledge is another indication of the complexity of muscle and of how far we are from fully comprehending the tissue. There is obvious scope here for the basically inclined investigator, for even if stripped of its meat-quality implication, the phenomenon still presents a provocative challenge to understanding. Second, the effect bears at least a superficial resemblance to the rapid early-postmortem glycolysis sometimes observed in porcine muscle. No distinctly PSE condition has been observed in beef that has gone through the rigor process very quickly, and no indications of stress have been seen in the animals that later glycolyzed very rapidly; the analogy is thus by no means a strong one. Nevertheless, a parallel does exist, and even though the root causes of the two conditions may be quite different, their *immediate* causes (once the putative aberrant signals have reached the musculature) could well be very similar. Third, the fast-glycolysis effect is obviously triggered at about the time of slaughter, so there is some slight reason to suppose that the act of stunning may be responsible. No matter how standardized the stunning method may appear to be, many different sequelae could result from quite minor variations in the strength and position of the blow, just as in human brain-injury victims a wide range of consequences may follow. Perhaps the rapid muscle glycolysis occasionally observed in beef has its parallel in the extreme hyperthermia sometimes observed in human head-trauma survivors; even when it is behaving normally, muscle is a significant contributor to body-heat maintenance, so a severalfold increase in its metabolic rate would almost certainly result in a major temperature elevation. Finally, and more directly relevant to the topic of stimulation: if we could account for this occasionally observed rapid glycolysis, it might prove to be a relatively simple matter to devise some means (a modified stunning method, for instance) that would initiate it routinely, thereby largely eliminating the need for carcass stimulation.

Yet, despite its obvious interest and probable importance to meat-quality studies, this rapid-glycolysis phenomenon occurs far too sporadically to merit specific investigation at this time. Rather, it is to be regarded as an effect that might be gradually elucidated through secondary observations made in the course of other studies; in its simplest form, such a side investigation would require nothing more than a pH and a temperature measurement at each of two standard early-postmortem times. Besides surveying the incidence of fast glycolysis, this routine would facilitate the interpretation of studies involving stimulated/control comparisons where the unrecognized occurrence of rapid pH decline in the control tissue could conceal any beneficial effect produced by stimulation.

My omission from this discussion of several other ES-induced effects on muscle metabolism does not indicate any lack of appreciation of their significance. The consequences of altering various electrical parameters, for instance, or of applying the stimulation at different times postmortem, may be very pronounced, strongly influencing glycolytic rate, cosmetic appearance, and eating quality. Much has been achieved already in these areas, and what remains to be done can best be accomplished through a straightforward empirical approach that is not dependent on further basic understanding. For the immediate future, I believe it will be much more profitable to direct our limited research resources along new avenues (or along old ones that remain blocked, despite all our past efforts), leaving the necessary processes of optimizing and fine-tuning to the development sector of the research and development partnership.

ES-induced Changes in Muscle Structure

Stimulation-induced effects on the eating quality of meat (as distinct from its visual quality) are obviously structural in nature, whether we are considering the prevention of cold-provoked shortening, the promotion of tissue disruption, or the proteolysis of individual building blocks. Architectural considerations must thus play a large role in future studies of ES and its actions, so it is appropriate to gather together and interrelate several concepts that have stood in comparative isolation to the present.

The contraction-provoking effects of various stimuli on prerigor muscle are generally quantified by using the initial excised-muscle lengths as a base on which to calculate the extent of shortening. Imprecise though it is, this measure serves quite well in allowing us to compare results from different sources and from widely differing types of experiments. When these comparisons are made, a rather striking common feature appears:

Several structure-related events occur at about 35–40% shortening. In the well-known relationship between shortening and toughening in beef, for instance, peak toughness is found at 35–40% shortening for both cold- and thaw-shortened muscles; toughness falls away steeply on either side of this value (Marsh and Leet 1966). The exudation of "drip" from thaw-shortened beef (Marsh and Leet 1966) and lamb (Marsh and Thompson 1957) increases quite suddenly as the extent of shortening passes the 40% mark. Cold-shortened bovine muscle retains a uniform (though, of course, diminishing) sarcomere periodicity as shortening increases to about 40%, but supercontracted zones and extensive internodal fractures appear when this degree of contraction is exceeded (Marsh et al. 1974). Frog muscle fibers, subjected to repeated electrical stimulation, enter the "delta state" at about 35% shortening; they fail to return spontaneously to their initial lengths and develop subnormal isometric tension when stimulated again after forcible reextension (Ramsey and Street 1940).

There is thus considerable evidence that a major "discontinuity" intervenes when shortening attains about 35–40% of the initial muscle length, resulting in very extensive changes in both muscle structure and meat properties. It could perhaps be argued that these quite dramatic alterations—in muscle architecture on the one hand and in meat quality on the other—are not necessarily related, mere coincidence being responsible for their occurrence at the same shortening degree; but it is much more logical and constructive to believe that all of them, the physiological and the qualitative, are the consequences of one event. It is easy to view the paradoxical (high shortening) tenderizing, the suddenly increasing "drip" flow, and the tension-development failure as direct results of a single cause: disruption.

By referring these several effects back to one origin, this explanation simplifies the picture considerably, but it also raises fresh questions: why should the disruption take place at all, and if it *has* to occur, why does it do so at 35 or 40% shortening? One possible answer (though probably not the only one) is based upon calculations and conjectures made by Marsh and Carse (1974) in a theoretical consideration of the cold-shortening phenomenon. From known thick- and thin-filament lengths, they deduced that, at about 35% contraction, the thick filaments (or, rather, those that have successfully penetrated the Z-line) just reach the near ends of the distal thin filaments of the longitudinally *adjacent* sarcomeres (Fig. 9.1). If the muscle were to pass into rigor at this length (as sometimes happens), the filaments would be continuously and uniformly crosslinked throughout the fiber length, resulting in a structure that, after cooking, would be perceived as very tough. However, shortening does not necessarily cease at the 35% point; once it has gone this far, it is much more likely to increase further, for the juxtaposition of the ends of the two filament species from neighboring sarcomeres has readied the contractile mechanism for another burst of activity. The original paper should be consulted for further discussion.

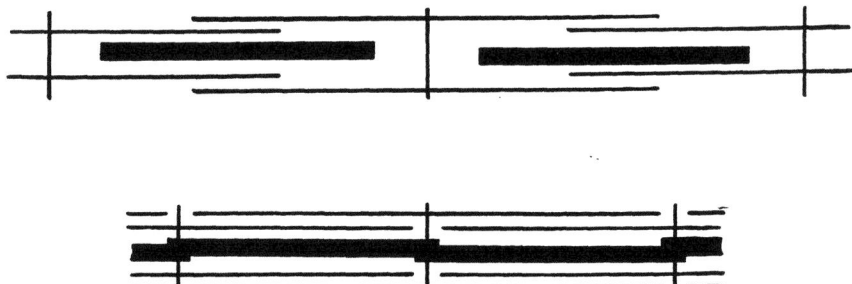

FIG. 9.1. Highly schematic diagrams of two sarcomeres in the unshortened configuration (upper) and the 35%-shortened state (lower). The thick filaments have been offset (lower) to illustrate their overlap at high shortening.

Now, these thick-to-thin filament intersarcomere connections are made over a relatively long time interval since we are no longer dealing with a highly brain-synchronized and nerve-orchestrated system. The chilling provocation travels through the tissue by heat conduction and not by nervous transmission, so it activates some areas sooner than others. In those regions where 35% shortening (and thus trans-Z-line interfilament contact) is attained more quickly, the second phase of contraction commences early, whereas in other regions at this time, the ends of the two filament species of neighboring sarcomeres have not yet reached each other. The stage is thus set for the act of disruption; areas in which the filaments have interacted quickly are able to start their vigorous second-stage contraction and do so at the expense of those areas where the interaction was more tardy. (Fibers in these latter areas, of course, are not only still unprepared for renewed contraction; they also lack the interfilament connections between sarcomeres that would help them to resist extension.) The net result of all this activity is to produce zones of super-contraction ("contraction nodes") interspersed among zones of stretch and (if the nearby contractions have been great enough) of fracture. From the viewpoint of the meat-quality researcher, it is these reextended or broken areas that are responsible for the increasing tenderness as overall shortening proceeds through the 40–60+% range.

Inaccurate and oversimplified though it may be, this hypothesis accounts reasonably well for the whole of the shortening/toughening structural relationship in cold-shortened muscle, and it appears to be equally applicable to the thaw-shortening phenomenon. My primary purpose in presenting it here in some detail, however, is not to resuscitate a theory that is concerned only with cold-induced changes. Rather, it is to suggest that essentially the same concept can account quite rationally for the second tenderizing action of ES—the disruptive process whose effects are often observed in the stimulated tissue. Several major differences exist, of course, between the cold-triggered and the ES-provoked systems. The time scales are separated by several orders of magnitude; the electrical treatment (in contrast to the imposition of cold) is applied briefly and is terminated while relaxation can still occur; and the shortening phases in the two treatments take place at very different temperatures. Furthermore, a critical feature of the model (as applied to cold-induced events) is the development of temperature gradients that result from the invariably uneven cooling of the muscle mass; nothing parallel to this slow-conduction consequence can be invoked to account for the ES-produced fracturing since passage of the current is virtually instantaneous.

Yet the same underlying structural changes can indeed be postulated for both processes, and evidence supporting the proposed mechanism has been available for well over 50 years. Cooper and Eccles (1930) showed that muscle twitches gradually fuse into a tetanus as the frequency of the current is increased, the tension meanwhile rising to a value several times higher than that produced by a single stimulus. The transitional frequency range (from discrete twitches to a smooth tetanus) varies among species; for bovine muscle, 10 Hz appears to be within it (Bendall et al. 1976). Thus, if the time interval between successive stimuli is large, say more than about 0.25 sec, the muscle has ample time to relax completely before it is next provoked. Its glycolysis is greatly accelerated because ATP is being split rapidly to provide the energy needed for contraction; but the shortening is completely reversible because it does not have sufficient time to traverse the whole 0–35% span needed to initiate second-stage contractions. If stimuli follow each other rapidly, however—as they certainly do when 50 or 60 Hz frequency is applied—there is no time for relaxation to occur; each succeeding twitch is superimposed on those already past. Provided the stimulus is strong enough, therefore, shortening can proceed to and beyond 35%, the thick filaments of one sarcomere interacting with the remote thin filaments of the next. Areas in which this happens are then able to supercontract at the expense of others that are slightly weaker or that fail by a millisecond or two to attain the critical configuration needed for second-stage shortening. It is these latter areas that yield under the influence of their neighbors' strength, being either stretched back toward their unshortened (and tender) length or forcibly broken. With smaller stimuli, as in low-voltage ES, too few of the fibers are able to supercontract to initiate tearing within the tissue so that acceleration of glycolysis is not accompanied by disruption.

Support for this hypothesis comes from the elegant light-microscope studies of other physiologists (Nageotte 1937; Ramsey and Street 1940), who used prolonged tetanizing stimulation to produce "alternating bands of close and widely separated striations." (These classical papers deserve greater recognition by meat investigators, who took almost 40 years to rediscover the contracture-producing ability of tetanizing current.) The proposed mechanism is also in line with the relative tenderizing effects of high and low voltages, as discussed earlier: the former are able to cause disruption, and therefore tenderizing, doing so even if carcass cooling is too slow for cold shortening to occur, but the latter merely prevent the shortening and its toughening consequences. Perhaps the most directly meat-related evidence for the twitch-tetanus concept comes from recent studies showing that 2 Hz stimulation of beef sides produces no electron microscope (EM)-detectable breaks at all in the musculature, whereas 60 Hz current, applied in precisely the same manner, causes extensive tissue fracture (G. Takahashi, S. M. Wang, J. V. Lochner, and B. B. Marsh: to be published). This observation explains why, in the study of Marsh *et al.* (1981), the low-frequency (and thus nontetanizing) current had no tenderizing effect at all in circumstances that allowed no possibility of cold shortening.

This hypothesis has been described in some detail (and perhaps at excessive length) for several reasons. First, it demonstrates that a single reasonably plausible theory can be formulated to account for the consequences of a number of different quality-affecting early-postmortem treatments: cold exposure, freeze-thawing, and electrical stimulation. (Probably heat shortening can be included also; supercontraction and tearing occur when beef is cooked rapidly in a prerigor state, but not when it receives the same treatment after rigor onset: Cia and Marsh 1976.) Second, it accounts for the seemingly contradictory effects sometimes observed when widely differing voltages are used, or when high and low frequencies are compared. In particular, it indicates that, if *either* of these readily altered variables is low, cold shortening and its associated toughening are eliminated but disruption-created tenderizing cannot take place; if both of them are high, however, a distinctly advantageous fracturing is added to the shortening-elimination process. Third, it stresses the relevance and importance to present meat-quality studies of long-past physiological investigations that were not even remotely food-related when they were undertaken. A wealth of information lies hidden in the older literature of the basic life sciences; to ignore it is to cut ourselves off from an invaluable resource.

My principal reason for dwelling so long on this hypothesis is to demonstrate the need for (and the advantages of having) a model that provides some sort of visualization of structural changes that may be taking place. I am very much aware that the hypothesis as presented may be far from correct and indeed might prove quite soon to be totally irrelevant. Even if it were completely demolished in the near future, however, this prototype would have served a useful purpose, for it brings together a number of effects and concepts into a form that can be readily attacked or defended. All efforts to test it (even if the intent is wholly destructive!) will be welcome, for every addition to or subtraction from it will inevitably take us closer to comprehension and control.

By no means can all of the tenderness changes undergone by meat be ascribed to the promotion or prevention of shortening or to microscopically detectable disruption. Sometimes the "no-visible-cause" changes are quite pronounced, and because they clearly must involve structural damage of some sort, we are forced to postulate that submicroscopic weakening or damage can take place and is able to do so to widely differing extents in apparently identical material. Of particular interest to us here is the possible influence of ES on the course of these proteolytic processes.

Because they cannot be detected visually, the effects of proteolysis are much more difficult to examine than those causing length changes or fractures. In addition, the field is extremely complex, with a bewildering array of enzyme candidates (each with its own unique set of pH, temperature, and ionic optima) from which to choose, and a wide (and still-increasing) field of structural proteins on which the enzyme(s) may work; desmin (Young et al. 1981) and connectin (the "gap" filaments: Locker 1982) have been added recently to the list of aging-sensitive components whose degradation might be expected to affect tenderness significantly. The sad truth is that, even with full access to the vast literature of proteolysis from several specialized fields of research, we are still a long way from even a quite elementary understanding of the process as it may operate in meat. It is not merely that we remain ignorant of the enzyme(s) specifically causing the tenderizing effect; we cannot even be sure which of the structural-protein substrates that undergo proteolytic modification are really responsible for the observed quality changes. Both for the whole area of meat-quality enhancement and for the narrow segment of it that is directly concerned with carcass stimulation, proteolysis and its effects merit high research priority.

Even before an enlarged knowledge of the subject becomes available, however, one particular question relating closely to ES and proteolysis needs urgent attention: Is proteolytic tenderizing aided, hindered, or unaffected by rapid, early-postmortem pH decline? It is a problem of obvious basic interest because it concerns pH/activity relationships of proteolytic enzymes, and an answer would provide very real clues to the identities of the tenderizing proteases; but it is also of immediate practical concern. We can largely control the rate of pH fall by appropriate manipulation of ES-current characteristics and stimulation times, and it is thus within our power to either encourage or discourage the rapid attainment of low-pH conditions; but we do not know which of these possible routes is the more beneficial. Clearly, if cold shortening occurs to a marked extent, rapid rigor onset must be the goal, regardless of the additional good or bad effect that may be exerted by the concomitant rapid decline of pH. However, in the many situations where cold shortening is negligible (because of carcass fatness and/or lenient chilling), it becomes very important to determine if a high or a low rate of glycolysis will more readily promote proteolysis.

The question is one of recent origin. During the several years when ES was believed to exert its benevolent tenderizing effect solely by preventing cold shortening, it was logical to postulate that rapid pH decline was desirable, for pH fall is an excellent guide to rigor progress and hence to deteriorating cold-shortening ability. With the later demonstration (Savell et al. 1977) that ES improves tenderness even if there is no shortening-induced toughening to be overcome, two additional mechanisms were proposed to account for the beneficial effects of stimulation: proteolytic activity (Savell et al. 1977) and tissue disruption (Savell et al. 1978). ES, of course, still accelerates glycolysis even if (because of milder chilling) cold shortening cannot take place; so a fast pH fall would be expected to accompany *any* stimulation-induced tenderizing effects, whatever their mechanisms. However, much more than simple accompaniment was envisioned in this association; it seems to have been assumed that the rapid pH decline was *responsible* for the tenderness improvement, rather than merely consorting with it. From this dubious deduction, it was but a small step to the suggestion that it had to be the proteolytic activity, and not the disruptive effect, that was aided by the rapid pH fall since disruption was fairly obviously not dependent on fast glycolysis. In this whole series of seemingly logical concepts, it appears to have been overlooked entirely that the rapid pH fall might be merely a companion, and not a cause, of the tenderizing process.

With the recognition of this potentially fallacious reasoning, a very interesting possibility arises: that disruption might be the principal (or even the sole) means by which ES exerts its rapid tenderizing effect in a non-cold-shortening situation, the accompanying fast acidification being without influence on tenderness or perhaps even harmful to it. There is, after all, good evidence that disruption can produce a dramatic tenderization when shortening is great (Marsh et al. 1974), but there are no comparably strong grounds to indicate that rapid pH fall (or the fast attainment of a low pH) can do so.

Evaluation of this concept requires not only the elimination of any cold-shortening ability, but also the total abolition of *either* disruption *or* fast pH decline, in order that the effects of only one of these two latter mechanisms can be observed. As described earlier, this suppression can be achieved by the use of 2 Hz stimulation, which encourages rapid glycolysis while completely preventing disruptive changes in the tissue (Marsh et al. 1981). With this technique, it was shown that loin steaks from beef sides receiving low-frequency ES were at least as tough as (and usually tougher than) those from the paired sides receiving no electrical treatment at all. The simplest, and perhaps the only possible, explanation of this result is that the rapid attainment of a low pH, when isolated from cold shortening and disruptive complications, does not promote tenderization at all and probably impedes it.

This observation suggests, but of course does not prove, that neutral proteases are involved and that they are more effective tenderizers than those proteolytic enzymes that are activated by low pH. It thus revives interest in the frequently denied possibility of early-postmortem (prerigor) aging, and in addition offers a reasonable explanation of several quite puzzling earlier observations (listed and summarized by Marsh et al. 1981) that suggested an association between tenderness and either slow glycolysis or high ultimate pH. [A contrary view has been expressed recently by Dutson et al. (1982), who found that ES had no tenderizing effect when the ultimate pH was about 6.6 (i.e., when no rapid pH decline could occur). Since even the *un*stimulated material was remarkably tender, however, no additional ES-induced tenderizing could have been expected (Harsham and Deatherage 1951; Savell 1979). Indeed, the fact that the untreated meat was so tender actually *supports* the concept developed so far, for it suggests that a long dwell time in the high-pH region is distinctly beneficial to tenderness.] It is certainly to be hoped that further studies in this area will be undertaken very soon, both to give a positive clue to the proteolytic enzyme(s) that might be involved in tenderizing and to provide guidance for the optimal design and use of stimulation equipment.

The subject of possible prerigor aging is a facet of meat-quality research that has been neglected for far too long. We in meat science are in a very curious situation concerning this topic. We acknowledge the role played by proteases in living muscle, whose proteins are continually and quite rapidly turning over by concurrent processes of catabolism and resynthesis. We are equally aware that proteolytic activity is almost certainly responsible for much of the slow tenderizing that takes place when meat is aged for a week or two beyond rigor onset. Yet we cling to the belief that protein breakdown does not occur in the first few postmortem hours, thereby perpetuating a rather absurd scenario: that proteolysis is an active process to the instant of death, ceases entirely from that moment until rigor completion, and then is restored to vitality with the attainment of the ultimate pH. I know of no evidence for such a reversible shutdown mechanism in meat, or for a sudden inhibition of one protease at slaughter and the abrupt activation of another at rigor completion, and I can only assume that the concept originated a long while ago, before meat researchers were made aware of the considerable proteolytic activity of living muscle. It is unfortunate that the on-off-and-on-again protease myth has been given a new lease on life by being referred to, in several recent papers, as if it were a self-evident truth.

My criticism does not depend on merely negative evidence (that is, on the absence of any substantial grounds for denying prerigor proteolysis); rather, it is based on several quite positive indications that the process may indeed be able to take place in the early-postmortem period. Consider these facts: (1) Muscle contains proteolytic enzymes that are active at near-neutral pH. (2) These enzymes are not oxygen-dependent, so the sudden onset of anaerobic conditions at (or just after) slaughter should be without effect. (3) A high early-postmortem temperature, whether due to carcass fatness or to imposed physical conditions, promotes tenderness to a greater extent than can be accounted for by the mere elimination of cold shortening. (4) In the absence of disruption and cold-shortening complications, acceleration of pH fall and rigor onset (by low-frequency ES) causes no tenderizing at all and in fact often results in toughening. (5) Meat of high ultimate pH is usually more tender than meat of normal pH if other conditions are similar. None of these items, or combinations of them, prove that prerigor proteolysis *does* occur, but they certainly present an irresistible temptation to suggest that it *might* do so, and to a significant extent. I believe that an investment of research effort in this area, in relation to both normal and ES-modified early-postmortem events, might yield substantial returns.

CONCLUSIONS

I have attempted in this chapter to indicate both the limits of our present knowledge of carcass stimulation and the areas of current uncertainty (and frank ignorance) toward which research might be profitably directed. The article is broader in one sense than its title implies, for ES cannot be treated in isolation; it has been necessary to introduce other topics—temperature effects, grading concerns, and wider aspects of proteolysis, for instance—that, while not an integral part of stimulation, cannot be profitably or logically separated from it. In another sense, however, I have dealt with the subject very narrowly, confining the discussion fairly rigorously to research as distinct from technology or development. While this restriction may not be well received, I believe the time has come when we must turn our attention increasingly to *understanding* (and not merely using) electrical stimulation and its actions, and—even more so—to seeking an integrated comprehension of meat quality (of which ES is but one aspect). The attractiveness of employing the novel technique of carcass stimulation and the economic advantages to be gained from its early industrial use have proved much more enticing than the task of slowly accumulating background knowledge, yet it is only by means of this knowledge that the full potential of the stimulation technique will be realized.

My emphasis on the need for a greater understanding is an attempt to correct the serious imbalance that has developed in the recent past and an incitement to augment the pool of information on which further technical progress will rely. To the present, the empirical approach has served us well, albeit with declining effectiveness; in the future, practical development will depend much more on the acquisition of new knowledge than on the further manipulation and recycling of the old.

REFERENCES

BENDALL, J.R. 1973. Postmortem changes in muscle. *In* The Structure and Function of Muscle, Vol. 2, 2nd Edition. p. 243. G.H. Bourne (Editor). Academic Press, New York.

BENDALL, J.R. 1978. Variability in rates of pH fall and of lactate production in the muscles on cooling beef carcasses. Meat Sci. 2, 91.
BENDALL, J.R. 1980. The electrical stimulation of carcasses of meat animals. In Developments in Meat Science, Vol. 1. p. 37. R.A. Lawrie (Editor). Applied Science Publishers, London.
BENDALL, J.R., KETTERIDGE, C.C. and GEORGE, A.R. 1976. The electrical stimulation of beef carcasses. J. Sci. Food Agric. 27, 1123.
BOUTON, P.E., SHAW, F.D. and HARRIS, P.V. 1980. Electrical stimulation of beef carcasses in Australia. Proc. 26th Eur. Meat Res. Workers Conf. 2, 23.
CARSE, W.A. 1973. Meat quality and the acceleration of postmortem glycolysis by electrical stimulation. J. Food Technol. 8, 163.
CIA, G. and MARSH, B.B. 1976. Properties of beef cooked before rigor onset. J. Food Sci. 41, 1259.
COOPER, S. and ECCLES, J.C. 1930. The isometric responses of mammalian muscles. J. Physiol. 69, 377.
DEATHERAGE, F.E. 1980. Early investigations on the acceleration of postmortem tenderization of meat by electrical stimulation. Proc. 26th Eur. Meat Res. Workers Conf. 2, 2.
DOLEZAL, H.G., SMITH, G.C., SAVELL, J.W. and CARPENTER, Z.L. 1982. Comparison of subcutaneous fat thickness, marbling and quality grade for predicting palatability of beef. J. Food Sci. 47, 397.
DUTSON, T.R., SMITH, G.C. and CARPENTER, Z.L. 1980. Lysosomal enzyme distribution in electrically stimulated ovine muscle. J. Food Sci. 45, 1097.
DUTSON, T.R., SAVELL, J.W. and SMITH, G.C. 1982. Electrical stimulation of ante-mortem stressed beef. Meat Sci. 6, 159.
EIKELENBOOM, G., SMULDERS, F.J.M. and RUDERUS, H. 1981. The effect of high and low voltage electrical stimulation on beef quality. Proc. 27th Eur. Meat Res. Workers Conf. 1, 148.
GEORGE, A.R., BENDALL, J.R. and JONES, R.C.D. 1980. The tenderizing effect of electrical stimulation of beef carcasses. Meat Sci. 4, 51.
HARSHAM, A. and DEATHERAGE, F.E. 1951. Tenderization of meat. U.S. Pat. 2,544, 681. Mar. 13.
LOCHNER, J.V., KAUFFMAN, R.G. and MARSH, B.B. 1980. Early-postmortem cooling rate and beef tenderness. Meat Sci. 4, 227.
LOCKER, R.H. 1982. A new basis for meat tenderness, in terms of gap filaments. Proc. 28th Eur. Meat Res. Workers Conf. 1, 117.
LOCKER, R.H. and HAGYARD, C.J. 1963. A cold shortening effect in beef muscles. J. Sci. Food Agric. 14, 787.
MARSH, B.B. and CARSE, W.A. 1974. Meat tenderness and the sliding-filament hypothesis. J. Food Technol. 9, 129.
MARSH, B.B. and LEET, N.G. 1966. Studies in meat tenderness. III. The effects of cold shortening on tenderness. J. Food Sci. 31, 450.
MARSH, B.B. and THOMPSON, J.F. 1957. Thaw rigor and the delta state of muscle. Biochim. Biophys. Acta 24, 427.
MARSH, B.B., LEET, N.G. and DICKSON, M.R. 1974. The ultrastructure and tenderness of highly cold-shortened muscle. J. Food Technol. 9, 141.
MARSH, B.B., LOCHNER, J.V., TAKAHASHI, G. and KRAGNESS, D.D. 1981. Effects of early-postmortem pH and temperature on beef tenderness. Meat Sci. 5, 479.
McKEITH, F.K., SMITH, G.C., SAVELL, J.W., DUTSON, T.R., CARPENTER, Z.L. and HAMMONS, D.R. 1980. Electrical stimulation of mature cow carcasses. J. Anim. Sci. 50, 694.
McKEITH, F.K., SMITH, G.C., SAVELL, J.W., DUTSON, T.R., CARPENTER, Z.L. and HAMMONS, D.R. 1981. Effects of certain electrical stimulation parameters on quality and palatability of beef. J. Food Sci. 46, 13.

NAGEOTTE, J. 1937. The extreme contraction of skeletal muscles in vertebrates. Z. Zellforsch. Mikrosk. Anat. *26*, 603. (French)

RAMSEY, R.W. and STREET, S.F. 1940. The isometric length-tension diagram of isolated skeletal muscle fibers of the frog. J. Cell Comp. Physiol. *15*, 11.

RUDERUS, H. and BERGQUIST, A. 1980. Industrial application of low voltage electrical stimulation in Sweden. Ann. Technol. Agric. *29*, 659.

RUDERUS, H. and FABIANSSON, S. 1980. Research on low voltage electrical stimulation of beef carcasses in Sweden. Ann. Technol. Agric. *29*, 581.

SAVELL, J.W. 1979. Electrical stimulation of meat provides many advantages. Natl. Provis. *180* (23) 49.

SAVELL, J.W., SMITH, G.C., DUTSON, T.R., CARPENTER, Z.L. and SUTER, D.A. 1977. Effect of electrical stimulation on palatability of beef, lamb and goat meat. J. Food Sci. *42*, 702.

SAVELL, J.W., DUTSON, T.R., SMITH, G.C. and CARPENTER, Z.L. 1978. Structural changes in electrically stimulated beef muscle. J. Food Sci. *43*, 1606.

SAVELL, J.W., McKEITH, F.K., MURPHEY, C.E., SMITH, G.C. and CARPENTER, Z.L. 1982. Singular and combined effects of electrical stimulation, post-mortem aging and blade tenderization on the palatability attributes of beef from young bulls. Meat Sci. *6*, 97.

SORINMADE, S.O., CROSS, H.R., ONO, K. and WERGIN, W.P. 1982. Mechanisms of ultrastructural changes in electrically stimulated beef longissimus muscle. Meat Sci. *6*, 71.

TARRANT, P.V. and MOTHERSILL, C. 1977. Glycolysis and associated changes in beef carcasses. J. Sci. Food Agric. *28*, 739.

TAYLOR, D.G. and MARSHALL, A.R. 1980. Low voltage electrical stimulation of beef carcasses. J. Food Sci. *45*, 144.

TERRELL, R.N., CORREA, R., LEU, R. and SMITH, G.C. 1982. Processing properties of beef semimembranosus muscle as affected by electrical stimulation and postmortem treatment. J. Food Sci. *47*, 1382.

YOUNG, O.A., GRAAFHUIS, A.E. and DAVEY, C.L. 1981. Postmortem changes in cytoskeletal proteins of muscle. Meat Sci. *5*, 41.

Index

A-band, 28, 30–34, 37, 187–188, 192–195, 208
Abattoir, 1, 39, 97, 99, 114, 127–128, 136, 178, 237, 261, 268, *See also* Packing plants
  beef slaughtering, 149–153, 219–222, 229, 234–235, 262
  freezing works, 3
  lamb slaughtering, 128
Accelerated processing, 160. *See also* Boning
Acetyl-CoA, 196
Acetyl-coenzyme A. *See* Acetyl-CoA
Achilles tendon, 91, 188
Acid, 280
Acidification, 287–288, 293, 301
Actin, 28, 31–32, 57, 189, 196, 201–202, 206, 245
  binding site, 202
  filaments, 85, 189, 193–195, 198, *See also* Thin filaments
  interaction sites, 195–196
  molecule, 195
  -myosin bonds. *See* Myosin-actin, bonds
Actomyosin, 32–33, 38, 66
  formation, 66
  gel, 32
  linkages, 32–33
Adenine nucleotides, 140
Adenosine diphosphate. *See* ADP
Adenosine monophosphate. *See* AMP
Adenosine triphosphate. *See* ATP
Adipose tissue, 289. *See also* Fat
A-disc, 187
ADP, 196–197, 213, 243, 287
Aerobic, 17, 196, 248
Aerobic plate counts, 174, 249–250
A-filament, 30–32, 34, 37
Age. *See* Aging, physiological

Aging, 1, 6, 9–10, 19–20, 24, 26–30, 34–35, 37, 45, 53, 63–64, 75–76, 78–81, 101, 104–106, 113, 129, 133, 136–137, 139–140, 162–163, 165–167, 172, 198, 211–213, 252, 300–302, *See also* Conditioning
  extended, 20
  high temperature, 34–35, 37–38
  physiological, 45, 50, 78, 130
  prerigor, 301–302
  rapid, 20
  -sensitive components, 300
Air, *See also* Atmosphere; Velocity
  liquid, 13
  properties of, 256
  temperature, 286–287
Aitchbone, 24. *See also* Suspension
Alkaline rigor. *See* Rigor, mortis
α-actinin, 36, 57
α-helix, 32, 37
α-red fibers. *See* Muscle Fiber, intermediate
α-white fibers. *See* Muscle fiber, white
Alternating polarity. *See* Polarity
American cutting system, 244–247
Amino acid, 186, 249
AMP, 196
Amperage, 230, 233
  intermediate, 124
  low, 124
Anaerobic, 18, 28, 76, 197, 248
Anaerobic condition, 16, 202, 302
Anaerobiosis, 18, 292
Animal, age, *See* Aging, physiological; Maturity
  condition, 115
  exhaustion, 77, *See also* Stress
  husbandry, 2

307

308　INDEX

Animal (cont.)
　live, 21–22, 39, 173, 186, 196–197, 219, 252, 285, 302
　meat-producing, 186
　number slaughtered, 237
Anoxia, 18
Anterigor excision, 160. See also Boning; Muscle, excised
Argentina, 231
Artificial conditions, 18
Asia, 161
Aspirin, 278
Atmosphere, air, 18
　nitrogen, 16, 18
ATP, 14–19, 76–77, 101, 196–197, 199, 201–202, 205, 213, 241–243, 279, 285, 287, 298
　breakdown, 9, 196, 213
　dephosphorylation of, 77, 101
　intramitochondrial, 205
　physiological concentration, of, 28
　reduction of, 50
　synthesis, 197, 285
ATPase, 18, 196, 201, 205
　activity, 63
　mitochondrial, 205
Australia, 2–3, 91, 96–97, 121, 123–124, 139, 164, 290
Autolysis, 36, 46, 51, 53, 58, 66. See also Cold-reduced autolysis
Azide, 16

Back, broken. See Spine, broken
Bacteria, 173, See also Microbial; Microbiology
　gut, 13, See also Microorganisms
　surface, 173. See also Microbiology; Microorganisms
Bacterial, See also Microbial, spoilage; Microbiology
　counts, lipolytic, 250
　growth, 75, 106, 174, 177, 243, 245, 250
　spoilage index, 249
Balaena, 8, 13
Banding patterns, 187–189, 195, 298
Bands, contracture, 199, 208–213
Bar. See Rub bar
Basic research. See Research, basic
Beef, See also Carcass, beef; Muscle, beef
　forage-fed, See Feeding regimens
　freeze dried, 245
　grain-finished. See Feeding regimens

Beef rolls, 248
Belgium, 123–124
Best and Donovan Hog Stunner, 124, 126, 128–129, 221
β-galactosidase. See Lysosomal enzymes
β-glucuronidase. See Lysosomal enzymes
β-red fibers. See Muscle fiber, red
Bidirectional pulse train. See Pulse, train, bidirectional
Bimodal pattern, 289
Binding properties, 178
Binding strength, of myosin, 245
Biochemical changes, 76
Biochemical response to stimulation, 90
Biochemistry, 13, 73, 90, 278
　of cold and thaw shortening, 14–19
Biopsies, human, 12
Bladder, 234
Blade tenderization, 136–137, 139. See also Mechanical, tenderization
Bleeding, 135, 231, 280
Bleeding area, 229, 231–232
Block-beef, 137
Blocking agents, of, See also Inhibitors
　electron transport, 18
Blood, flow, 284–285, 292
　loss, 151
　processing, 231
　recovery/removal, 151, 231, 252
　staining (of carcass), 151
　vessels, 13, 186–187
Bone, 179, 252–253, 255, 261, 267
Boneless cuts, chilling of, 251
Boning, 20, 101, 105–106, 152, 164, 263
　early, 105–106, 165
　hot, 24–25, 39, 97, 106–107, 138–139, 160–180, 242–268, 281, 293
　history of, 161–164
　of pork, 178
　hot vs cold, 164–180, 239–241, 246–250, 254–257, 260–268
　lines, 177–178, 267
　mechanical, 267
　on-the-rail, 179, 264, 266–267
　summary of research, 169–171
　table, 264
Bound water. See Water
Bovine serum albumin, 19
Boxed beef, 234, 249, 252, 267
Boxed beef distribution system, 159
Boxed meat, 106, 174, 178
Brain, 231
Brain-synchronized system, 297
Breaking. See Cutting; Primal cuts; Sub-primal cuts
Brine, cold, 248
Brine pipes, 3

Brisket, 223, 228
Britain, 2–3, 123–124, 127, 139, 164–165, 262
British cutting system, 244–246
Buffer, 19
Bung, 234
By-products, 159, 261

Ca^{2+}, 16–19, 36–37, 201–206, 243, 284
  accumulation, 202–205
  -binding, 15–19, 202–203, 206, 215
  capture, 202
  chelators, 201, 203
  cold-induced release of, 16
  pCa^{2+}, 16
  physiological concentration, 28
  pump, 18–19
  release, 18, 202–203
  -uptake, 19
CAF. See Calcium-activated factor (CAF)
Calcium, 28, 201, 284
Calcium binding, 191
Calcium pump, 206
Calcium release, 191
Calcium-activated actomyosin ATPase, 84
Calcium-activated factor (CAF), 28, 36, 56, 58
Calmodulin, 28
Calpain, 28
  caseinolytic activity of, 36
  proteolytic activity of, 36
Calpastatin, 28
Canada, 123–124
Cane hooks, 229. See also Probe; Electrode
Capacitive discharge unit, 92
Capillaries, 186–187
Capital, 263, 267
  fixed, 179
  invested, 263
Carbon dioxide, 176, 196
Carcass, 84, 89, 162, 165–167, 178–179, 234, 264, 280, 282–283, 285–286, 291, 299, 303
  beef, 21, 22, 81–82, 92–97, 103–106, 110–113, 115, 126–127, 130–136, 139–141, 142–150, 165–168, 185, 199–201, 206–207, 214, 220–235, 237–240, 243–247, 251–258, 262–268, 288, 293, 301
  Choice, 131–132, 142–143, 166, 220, 244, 266–267
  cutting, 244–247
  dressing, 2, 25–26, 76, 107, 110, 112, 252, 261–262
  drip, 151–153
  fatness, See Finish
  frozen, 6–10, 113, See also Frozen, lamb, mutton
  geometry, 251–253
  Good, 131–132, 142–143, 166, 266
  lamb, 2–3, 6–8, 14, 22, 24–26, 75–76, 81, 89, 91, 98–101, 103–104, 107–110, 112–115, 127–131, 165, 199, 206, 243, See also Frozen, lamb
  pork, 142–143
  Prime, 131–132
  processing, 73, 76, 79, 96, 99, 101, 103, 111–115, 159–160, 173, 222–232, 237–240, 243–247, 251–268
  quality traits, 148–153, See also Cosmetic improvement; Grading; Lean scores; Quality
  shipping, 159, 179, 244, 253
  shrink, 127, 152–153, 168, 179, 244, 281
  split, 134–135, 139, 149–150, 167, 185, 191, 251, 254–255
  Standard, 131–132
  suspension, See Posture; Suspension
  swine, 22, 138, 147, 162, 238–239
  thawing of, 10
  value, 282
  weight, 104, 130, 261–262, 286
Cartilage, 187
Cartoned meat. See Boxed meat
Caseinolysis, 36
Caselife. See Retail, caselife
Catabolic enzyme. See Enzyme
Cathepsin, 28–29, 37, See also Lysosomal enzymes
  acid-activated, 288
  -C, 208
Cathodic stimuli, 92–93
Cause-and-effect investigation, 278
Cells, 197
Centralized breaking-point, 159–160
Centrifugates, 201
Centrifuge cuts, 17
Chain, continuous, 223, 224–226
  hanging, 111–112
Chain contacts, 281
Chewing, 208, 244
Chiller. See Refrigeration, cold room
Chilling, 10, 18, 19–21, 45–57, 75–77, 81–82, 84, 99–101, 103–106, 114–115, 121–122, 142–143,

310 INDEX

Chilling (cont.)
  145, 151–153, 161, 163–167, 170–174, 176–179, 202, 206, 222, 232, 237, 244, 248–260, 262–264, 267, 286–290, 293, 297, 299, See also Cold; Freezing; Refrigeration; Shortening, cold; Temperature
  capacity, 257
  conventional, 122, 128, 139, 178, 249, 256
  delayed, 26, 50, 77, 115, 164, 242–243
  differential rate of, 142
  early, 26, 75, 122
  efficiency of, 250
  in a tunnel, 262
  rapid, 3, 5, 8, 20, 22, 27, 48, 51, 81, 103–106, 121, 127–129, 142, 160, 165, 174, 199–200, 206–207, 232, 242, 250, 253, 288–290
  slow, 20, 81, 106, 115, 162, 172, 288–290, 299
  space, 106, 177, 179, 251, 253, 267
  time, 258
Choice. See Carcass, Choice
Circulation of fluid, 251
Clip, jaw, 111. See also Electrode; Nose; Probe
Clostridium botulinum. See Microorganisms
Clostridium perfringens. See Microorganisms
Clots, supercontraction, 175
Coagulation, 32–34, 175
  heat, 51
Coagulum, 32
Cohnheim, fields of, 192
Cold, contracture, 49, See also Muscle, contraction; Shortening, cold
  effect, 3, 76
  -induced changes, 297–298
  processed, 248–249, See also Boning; Carcass, processing; Processing
  -reduced autolysis, 46–47, 51–60, 62, 64–66
  response, 21
  ring, 142, See also Heat, ring
  room, See Refrigeration
  shortening, See Shortening, cold
  toughening, 24–25, 30, 32, 39, 46–47, 50, 64–65. See also Shortening
Coliforms. See Microorganisms
Collagen, 30–32, 36, 62, 186, 188, 191, See also Connective tissue
  small molecular weight, 62
  soluble, 62
  thermolabile fraction, 62

Collagen net, configuration, 31
  crimped, 30
Collagenase, 62
Color, 27, 75, 127, 141–144, 146–147, 149–153, 160, 172–173, 201, 206–207, 214–215, 219–220, 246, 248, 281–283, 294. See also Discoloration; Lean scores, color; Pigments
Color of cured meat, 248
Commercial cut. See Primal cuts; Subprimal cuts
Commercial feasibility, 126
Commercial installation and application, 281
Commercial practices, 121, 124, 126–127, 135, 150–153, 160, 161, 178–180, 237, 266, 303. See also Industry; Packing plants; Technology, applied
Comminuted products, 138, 266
Compressor motor, input to, 257
Concentration gradient, 284
Concepts, 278
Condenser, water inlet conditions, 259
  water pump, power to, 256
  water supply temperature, 255
Conditioning, 2, 9, 14, 19–20, 25, 34–35, 38, 45–47, 53, 58–66, 80–81, 99, 101, 110, 121, 161–165, 198, 242–243, See also Aging; Cooler conditioning; High temperature conditioning
  accelerated, 111–114
  subzero, 101
Connectin, 30, 64, 300
  degradation, 57–58
Connective tissue, 1–2, 38, 48, 51, 62, 78, 136–137, 168, 175, 186–187, 191, 245, 255, 280, 285, See also Collagen; Elastin
  endomysial, 193
Consumer, 39, 127, 137, 151–153, 234, 267, 281–282
  complaints, 151–153
  markets, 260–261
  packs, 10
  panel, 177
  satisfaction, 266–267
  studies, 185
Consumption, 159, 179
Contamination, 173, 242–243, 245, See also Cross-contamination; Microbiology
  fecal and urine, 232, 234
Continuous chain. See Chain

Contractile, *See also* Muscle, contraction
  mechanism, 296
  response, 92
  tissue, 38
Contraction. *See* Muscle, contraction
Contraction state, 2, 31. *See also* Muscle
  contraction
Contracture bands. *See* Bands, contracture
Conveyors, 223–224, 253, 258
Cooked, after thawing, 6, 9
  before thawing, 9, 14, 20, 25, 81, 84, 102–103
  character, 31
  material, 4, 6, 24–25, 27, 28–31, 32–33, 35–36, 37, 64, 78, 81–82, 84, 104, 114, 161, 172, 175–177, 198
  prerigor, 28, 64, 161, 175–176, 211, 299
  water bath, 84
Cookery, 175–176
  dry heat, 138
  microwave, 175
  moist heat, 138
Cooking, losses, 172, 175–176, 248
  properties, 177
Cooler. *See* Chilling; Refrigeration
Cooler conditioning, 45–47, 49–54, 63–66. *See also* Aging; Conditioning
Cooling, *See also* Chilling; Cold; Refrigeration
  airflow, 251, 267
  conveyorized, 266
  efficiency, 286
  equipment, capacity, 259
    design, 256
  floor, ventilated, *See* Refrigeration
  period, 251, 267
  susceptibility, 286
  systems, conventional, 256–257
    counterflow conveyorized, 251, 254–255, 262, 266–267
Corporate leaders, 266
Corpse, resuscitation of, 73–74
Cosmetic improvement, 283, 291, 294–295. *See also* Color; Grading; Lean scores; Marbling; Quality
Cost. *See* Economics
CP, 15, 77, 140, 196–197, 243
C-proteins, 57
Creatine phosphate. *See* CP
Cross-bridges, 193–196, 201–202
Cross-contamination, 173. *See also* Contamination; Microbial
Cross-linking, 34, 296. *See also* Myosin; Protein

Cross-striations. *See* Striations
Cryostat, 12
Cryovac, 20
Cube rolls, 20–21
Curare, 86
Cured meat, 248, 266. *See also* Processing
Current, 108, 200, 230, 233, 280, 283, 292–293, 298
  density, 89
  distribution, 94, 166–167
  flow, 110
  measurement of, 89
  nontetanizing, 298
  tetanizing, 298
Cuts, shape, 172, 177, 244–245, 267. *See also* Boning; Primal cuts; Subprimal cuts
Cutting, 151, 179, 242, 244–245, 268
Cutting lines, 244–245
Cutting systems, 244–247
  American, 244–247
  British, 244–246
  French, 245–247
  Meat and Livestock Commission, 246

Daltons, 50–170K region, 57
  30K region, 55–56, 63
  23–28K region, 57
Dark-cutting, 141, 178, 207
  early identification, 178
Decapitation, 285
Degradation, 196. *See also* Protein
Dehiding. *See* Hide
Delta state, 296
Denaturation. *See* Degradation; Heat, denaturation; Protein, denaturation
Denmark, 123
Density, 255–256
Dephosphorylation of ATP, 77, 101
Depolarize, 200
Depolymerization, 38
Depreciation, 264
Desiccation, 173
Design, calculations, 258–259
  values, 257–259
Desmin, 28, 64
Destination point, 260
Diffusion, 245
Dinitrophenol, 16
Diphosphopyridine nucleotide tetrazolium reductase. *See* DPNH-TR

312 INDEX

Discoloration, 20, 146–147
Display. See Retail, display
Distribution, 147, 159–160, 173, 179, 237, 264–266
  costs, 265–266, 268
  of electrical stimulation equipment, 235
  sector, 266
DNP. See Dinitrophenol
Document, 280
Downstream processing. See Processing
DPNH-TR, 189
Dressing, aerial top, 3. See also Carcass
Drip loss, 8–9, 13, 153, 245, 248, 296. See also Carcass; Purge
Dry heat. See Cookery
Dry Ice-alcohol, 13
Dunedin, 2
Dynamic system, 284–285

Economics, 237, See also Energy
  of chilling, 251
  of grading, 144, 219
  of hot boning, 159–161, 178–180, 251, 261, 262
  of plant location, 260–266, See also Transportation
  of stimulation, 151–153, 219, 244, 280, 303
Economy, grassland, 2
Edible product, 160, 252–253
EDTA, 28, 37, 204
EEC. See European Economic Community
EGTA, 16
Elastin, 186, 191
Electric fence-charger. See Gallagher Energizer
Electric motors, 255
Electrical, arcing, 231
  contact, 108–109, 111–112, 126, See also Electrode; Probe
  discharge, 280
  pathway, 90, 108
  rubbing, 107, See also Electrode; Probe; Rub bar
  sparking, 107, 231
  stimulation, See Stimulation
  supply, 92, 95
Electricity, 261, 262, 264
Electrode, 107–108, 111, 114, 126, 221, 280–281, See also Cane hooks; Chain; Electrical, contact; Mechanical, claw; Nose; Probe; Rub bar
  contact, 94, 226, 231, 234
  pithing rod, 281
  polarization, 93
  spring-loaded pan, 280
  unsegmented, 109
Electron transport chain, 205
Electrophoresis. See Dalton; Gels; Protein, fragment
Emulsifying capacity, 163, 178, 245
Emulsion, 245–247
  coarse, 247
  fine, 247
  semifine, 247
  stability, 245
Endomysium, 48, 186–188, 191–192
End-products, 284
Energy, 201, 205–206, 237, 244, 250–268, 284–285, 298
  calculations, 259–260
  conservation, 237, 244, 250–253, 261, 263
  conserving site, 205
  costs, 159–161, 175, 178–179, 243–244, 250
  for cooling, 253–260, 263
  for transportation, 260–263
  input, 94, 160, 177, 251
  -intensive, 266
  of activation, 84–85
  of contraction, 186
  potential chemical, 196
  saving, 259–261, 264–268
  sources, 197, 214, 251
  transfer, 251
Engineering, 251
  genetic, 277
England. See Britain
Enthalpy of activation, 61
Enzyme, 28, 57, 197, 280, 292–293, 300, 302, See also Calcium-activated factor (CAF); Glycolytic, enzymes; Lysosomal enzymes; Proteolytic enzymes
  catabolic, 280
  endogenous, 66, 74
  optima, 300
  system, 280
Epimysium, 77, 186–187
Epithelium, 187
Equipment costs, 160
Escherichia coli. See Microorganisms
Esophagus, 110
Ethylene diamine tetraacetate. See EDTA

INDEX

Ethylene glycol tetraacetate. *See* EGTA
Europe, 289
European Economic Community, 20, 289
Evisceration, 232
Exports. *See* Trade
Exsanguination. *See* Bleeding; Sticking

Fabrication, 159–160, 177, 179, 237, 248, 253
F-actin, 193
Fan, 258, 267
Fan motor, input to, 256–257
Faradic response, 280
Fascia, 187
  lumbar-dorsal, 207
Fasciculi, 186–187. *See also* Muscle fiber, bundles
Fat, 179, 245, 252–253, 255, 261, 267, *See also* Adipose tissue; Finish; Marbling
  color, 172
  cover, 21–22, 27, 50, 266, 286, 289
  external, *See* Fat, subcutaneous
  globules, 245
  insulation, 21, 27, 50
  internal, 252
  intramuscular, 143, 214, 282–283, *See also* Marbling
  intramuscular distribution, 283
  staining, 160
  subcutaneous, 27, 142, 282
Fatness. *See* Finish
Feather bones, 232
Feed supplies, 289
Feeding history, 139
Feeding regimens, 130–132, 134
  forage, 22, 125, 130, 136–137, 140
  grain, 21–22, 130–131, 140, 143, 291
  long-fed, 131
  short-fed, 130–131
  time-on-feed, 132, 134, 249
Feet, 232
Fiber, *See* Muscle fiber
  diameter, *See* Muscle fiber, diameter
  stretched. *See* Muscle fiber, stretched
Filaments, *See also* Actin, filaments; Gap filaments; Myosin, filaments; Thick filaments; Thin filaments
  actin, 85
  gap, 29–39
  myosin, 32
  overlap of thick and thin, 66
Finish, 21, 164, 288–290, 300, 302. *See also* Fat

Firmness. *See* Lean scores
Fixation, 187
Fixatives, 12
Flavor, 45, 140–141, 151, 160, 163, 168, 185, 201, 213–215, 244–245, 281
Flavor enhancers, 213–214
Fluid retention. *See* Water holding capacity
Food poisoning, 177
Force score. *See* MIRINZ tenderometer; Shear force
Forequarter. *See* Quarter, front
Fracture, fiber, *See* Fragmentation; Muscle fiber
  internodal, 296
Fragment. *See* Electrophoresis; Protein, fragment
Fragmentation, 28, *See also* Muscle, fragmentation; Protein, fragment
  gel, 32
  myofibrillar, 57, 64–65
  myosin, 32
  values, 53
France, 123–124, 139
Frankfurters, 246–246, 278. *See also* Sausage products
Franklin, Benjamin, 74, 122–123
Free leg. *See* Leg, free
Free water. *See* Water, unbound
Freezing, 3, 6–9, 13, 24–25, 48, 75–77, 80–82, 91–92, 97, 99–104, 107, 110–115, 121, 128–129, 164–167, 173–174, 176–177, 199–200, 206, 237, 299, *See also* Conditioning, subzero
  blast, 3, 6–10, 13–14, 20, 81, 129
  chilling phase, 101, *See also* Chilling
  delayed, 110–113, 115
  extreme, 13
  prerigor, 13, 24, 122
  rapid, 3, 6–9, 24, 101, 115, 121, 165
  shell, 24
  slow, 3, 7, 101
  time, 101–103
  works. *See* Abattoir
French cutting system, 245–247
Frequency, 89–95, 280, 292, 298–299, 301–302, *See also* Current; Pulse
  cycle, 229
  optimum, 93
  pulse, 93–96, *See also* Pulse
  response, 89
Front quarter. *See* Quarter, front
Frozen, lamb, mutton, 2–3, 6–8, 75–76, 81, 84, 92, 113, 121, 129, 164, 199
  storage, 101–102

314 INDEX

Gallagher Energizer, 128–129
Galvanic response, 280
Gambrel, 24, 91, 108–109
Gap filaments, 29–39, 51, 57–58, 64–65, 300, See also Connectin; Titin
  degradation, 57–58
  heating of, 31–33, 57–58
  shielding of, 58, 66
  ultrastructure, 58
Gels, SDS polyacrylamide, 55, 57. See also Fragmentation, gel
Genetic engineering. See Engineering, genetic
Geometry, of carcass, 251–253, 286
  of product, 251
G-filaments. See Gap filaments
Glucose, 196–197, 249
Glycogen, 88–90, 162, 178, 197, 200, 243
Glycolysis, 14–19, 27, 75, 77, 82, 84–85, 89, 141, 144, 161–162, 165, 196–197, 199–201, 206–207, 213–215, 242, 250, 284–287, 290–295, 298–302, See also Metabolism
  aerobic, 196
  slow, 302
Glycolytic, enzymes, 85, 207
  fibers, 189
  intermediates, 292
  pathway, 75
  rate, 288, 293, 295
  response to stimulation, 89
  response to temperature, 84–85
Good. See Carcass, Good
Grade, 139, 143–145, 148, 152–153, 160, 178, 185, 214, 221, 244
  factors, 127, 291
  home, 144, 234
  U.S. Choice-U.S. Good grade line, 144–145, 266
Grading, 22, 130–131, 142, 144, 151–153, 166, 168, 214, 221, 281–283, 291, 294, 303, See also Carcass; Fat; Grade; Lean scores; Marbling
  federal, 142, 151, 178
  private, 144, 151, 234
  systems, 289
Grilling. See Cooked, material
Ground beef, 173–174, 176–177, 214, 248–249, See also Patties
  hot-boned, 173–174, 176–177, 248
  physical properties, 177
  sensory properties, 177
Ground line, 226
Ground meat, 138
Ground substance matrix, 62

H and H Meat Products, 126–127, 221
Ham, canned, 248. See also Cured meat; Processing
Hanging. See Suspension
Hanging chains. See Chain, hanging
Harsham-Deatherage, 74, 279–281. See also Patent
H-band, 192–194
Head, 232
Heat, capacity, 258
  conduction, 297
  cooking, 137
  denaturation, 34–36, 38
  dissipation, 22
  dissociation, 32
  rate, 255
  ring, 75, 127, 142–144, 148–149, 151, 152–153, 185, 207, 215, 221, 243, 281, 294, See also Cold, ring
  toughening, 26, 36
  transfer, 250–251, 253, 286
  transfer coefficient, 251, 255
  -trauma, 294
  waste, 262
Helix. See α-helix
Hematoxylin, 189. See also Iron hematoxylin
Hide, 2, 107, 224, 231, 252, See also Skinning
  pulling, 224
    downward, 94, 111
  removal, 135, 231
  scorching, 111
High energy bond, 196
High energy phosphate compounds, 196, 243
High temperature conditioning, 45–47, 50–62, 66, 136–137, 162, 177, 242–243, See also Aging; Conditioning; Cooler conditioning
  postrigor, 59, 61–62, 65–66
  prerigor, 59–60, 65–66
High temperature processing, 160. See also Boning
High voltage. See Voltage, high
Hindquarter. See Quarter, hind
Histochemistry, 13
Histology, 12–13, 27, 37–38
History, of electrical stimulation, research and development, by country, 123
  scientific reports, by country, 13, 124, 139, 178, 220–222, 243–244, 251, 253, 260, 264
  of hot boning, locations, 161
Hocks, 111

INDEX 315

Hog stunner. *See* Stunner
Hoggets, 24
Holding, 99–100, 251, 253, 255, *See also* Conditioning
 poststimulation, 101. *See also* Conditioning
Homogenates, 1, 12, 56, 82
Homogenizing, 292
Hot boning. *See* Boning
Hot cutting, 160. *See also* Boning
Hot processing, 160, 248–249, 266–268. *See also* Boning; Processing
Hot-boxes, 151
Hydrolysis, 197, 205
 of connectin, 57
 of troponin-T, 55
Hygiene, 19, 111, 114, 160–161, 179, 231, 233–234, 267–268. *See also* Sanitation; Sterilization
Hyperthermia, 294
H-zone, 194. *See also* Pseudo H-zone

I-band, 31–33, 35, 37, 51, 187–188, 192–195, 208–211
I-filament, 30–32, 37
IMP, 214
Imports. *See* Trade
Impulse. *See* Pulse; Stimulation
Industrial application, 219–235. *See also* Commercial installation and application
Industrial standard, 14
Industry, 2, 9, 12, 22, 25, 26, 39, 76, 96, 99, 110–115, 121–122, 124, 127, 143, 149, 151–153, 161, 164, 178–180, 185, 219–235, 242–244, 250, 253, 264–268, 279, 281, 303
 food, 178
 overall cost and energy requirements, 264
 segmentation, 266
Inedible product, 252. *See also* By-products
Inhibition, 197–302
Inhibitors, 16, 18, 28
 neuromuscular, 86
Injury, 294
Inosine monophosphate. *See* IMP
Inspection, federal, 235
Insulation, *See also* Carcass
 fat, *See* Fat
 mass, 21, 286
 muscle, 77, 286

Interaction sites. *See* Actin
Intercellular spaces, 208
Interest, 179
Interfilament connections, 297
Intermediate fibers. *See* Muscle fiber, intermediate
Internal ice. *See* Restraint, internal ice
Intracellular spaces, 204, 208, 215
Inventory, 160, 179
Iodoacetate, 82
Iron hematoxylin, 187
Ischemic heart muscle, 18

Joints, 244–245
Juiciness, 160, 163, 168, 175–177

Kill floor, 262
Kymograph, 15

Labor, 159–160, 178–179, 245, 263, 266, 268
 costs, 160–161, 179, 245, 263–264
 -intensive, 266
Laboratory, 279–280, 288–289, 291
Lactate, 111, 197
Lactate ions, 77
Lactic acid, 197, 243, 285
Lactobacillus. *See* Microorganisms
Lamb. *See* Carcass, lamb
Langmuir probe, 89
Lateral linkages, 34. *See also* Protein, linkages
Lean area, 266
Lean color. *See* Lean scores, color
Lean firmness. *See* Lean scores, firmness
Lean maturity. *See* Lean scores, maturity
Lean muscle separation. *See* Lean scores, muscle separation
Lean scores, color, 141, 148–153, 185, 206–207, 214, 219–221, 243
 firmness, 142–143, 150–151, 214, 220, 242–243
 maturity, 141, 150–151, 214, 243, 266
 muscle separation, 142–143
 texture, 142–143, 151, 214, 220
Lean texture. *See* Lean scores, texture
Leanness, 27, 104, 288–289. *See also* Fat; Finish

316   INDEX

LeFiell-USDA Test Unit, 124, 126–127, 221. *See also* Stimulators, electrical
Leg, directly stimulated, 89–90, 108
  free, 89–90, 108, 111, 231
  shackled, 89–90, 107, 231
Legging operation, 111
Length-toughness relationship, 5
Liquid air. *See* Air, liquid
Liquid nitrogen. *See* Nitrogen, liquid
Livestock production, 159. *See also* Production
Load, building, 257, 259–260, 268
  calculation of, 256–260
  circulating fan, 257, 259
  cooling, 258, 267–268
  equipment, 257, 259
  heat, 257
  infiltration, 257, 259
  personnel, 257, 259
  product, 257, 259
  transmission, 257, 259
Loading factor, 253, 258
Location, plant, 124, 126, 221, 260–266. *See also* Packing Plants; Region
Longitudinal striations. *See* Striations
Low voltage. *See* Voltage, low
Lumbar-dorsal fascia. *See* Fascia, lumbar-dorsal
Lysosomal enzymes, 29, 53, 58–59, 62, 123, 150, 208, 211, 215, 287
  β-galactosidase, 62
  β-glucuronidase, 62, 208
  cathepsin-C, 208
  free activity, 53–54, 65–66, 208
  specific activity, 208
  total activity, 208
Lysosomal membrane, 65, 150, 208
Lysosomal vesicles, 208
Lysosome, 28, 59

Macromolecular changes, 48
Magnification, 186
Mandible, 78
Man-hours, 245, 263
Manufacturing of electrical stimulators, 219–220, 222–231, 234. *See also* Stimulators, electrical, manufacturers
Marbling, 21, 143–145, 149–151, 185, 214–215, 219–221, 243, 266, 281–283
  setting up, 143–144, 151, 219

Margin, 264
Markets. *See* Consumer; Trade
Marketing, 159–160, 178, 234, 268
Mass, 253, 286. *See also* Insulation
Material costs, 160–161, 179
Materials handling system, 266
Maturity, 78. *See also* Aging, physiological; Lean maturity; Lean scores
Meat, *See also* Carcass; Muscle
  curing, *See* Cured meat
  products, 219, 234, 237, *See also* Comminuted products; Restructured meat
  properties of, 186, 278, 296
Meat and Livestock Commission, 244, 246
  cutting system, 246
Meat Industry Research Institute of New Zealand. *See* MIRINZ
"Meat Industry Special Guide—Electrical Stimulation Equipment," 222
Meat plants. *See* Packing plants
Mechanical, claw, 108
  deboning, 267
  tenderization, 242–243. *See also* Blade tenderization
Medical treatments, 277
Medicine, human, 12
Membranes, 197, 200, 245, 284–285
  deterioration, destruction of, 285
  permeability, 284
Merchandiser. *See* Retailer
Mesophilic microorganisms. *See* Microorganisms
Metabolic, *See also* Glycolysis
  differences, 87
  processes, 196
  rate, 93
  states, 13
Metabolism, postmortem, 16, 291–295
Metabolites, 284–285, 292
Methods, Millipore filter, 17
Mg^{2+} ions, 201
Microanatomy, 78
Microbes. *See* Bacteria; Microorganisms
Microbial, *See also* Bacteria
  contamination, 163, 242, 245
  counts, 162–163, 173–174, 249–250
  growth, 163, 174, 178, 214, 242–243, 267, 287
  spoilage, 106, 151, 174, 248–249
  standards, 174
Microbiology, 20, 75, 173–174, 248–250
  of processed meat, 250

Microflora, 173. *See also* Bacteria; Microbial; Microbiology
Micrographs, 4, 12, 37, 192–194, 210–213
Microorganisms, 173–174, 177, 248–250, *See also* Bacteria
  Clostridium botulinum, 177
  Clostridium perfringens, 249
  coliforms, 174, 249
  Escherichia coli, 174, 249
  food poisoning, 177
  Lactobacillus, 248
  mesophilic, 174, 249–250
  Pseudomonas, 248
  psychrotrophic, 173–174, 249–250
  Staphylococcus aureus, 249
  Staphylococcus sp., 177
  Streptococcus, 249
Microscopy, fixing for, *See* Fixation; Fixatives
  light, 186–187, 189, 191–192, 298
  phase contrast, 1
  scanning electron, 186
  sections for, 13
  staining for, *See* Staining
  transmission electron, 1, 186, 189, 192–194, 210–213
Middle East, 161
Millipore filter method. *See* Methods, Millipore filter
MIRINZ, 25, 82, 91, 111–114
MIRINZ tenderometer, 20, 36, 77–78
Mitochondria, 16–19, 191, 202, 204–206, 215, 243
  anoxic, 202
  heart, 18
  liver, 18
  swelling of, 89
Mitochondrial, preparations, 203
  respiration, 205
  synthesis, 205
  uncouplers, 202
M-line, 28, 34, 38, 192–194
Modus operandi, 278
Moist heat. *See* Cookery
Moisture, *See also* Water holding capacity
  content, 163
  surface, 267
Molecular weight, 30, 62, 248. *See also* Daltons
Morphology, 29, 186
Most probable numbers (MPN), 174, 249
M-proteins, 57
Muscle, *See also* Carcass; Muscle cell membrane; Muscle fiber; Muscle fibrils; Myofibril; Myofibrillar; Myofilament; Myosin
  adductor, 22–23
  as a food, 73
  autolysis, *See* Autolysis
  back, 94
  beef, 1, 4, 6, 10–12, 14–18, 20, 22, 25, 26–27, 32, 34, 36–37, 56–58, 60, 65, 74, 77–79, 81–85, 89, 92–97, 104–105, 107, 112–113, 115, 121, 123–125, 127, 129–152, 163–168, 172–179, 188, 190–191, 194, 202–203, 208–215, 260–261, 296, 299
  biceps femoris, 8, 17, 23, 87, 100
  brachiocephalicus, 96
  contraction, 1–6, 11, 13, 16, 18–19, 22, 26, 31–32, 34–35, 47, 49, 73, 76, 128, 161, 186, 188, 191, 193, 196–199, 201–202, 206, 211, 215, 279–280, 284–285, 287, 295, 297–299
    irreversible, 90
    massive, 135
    nodes, 297
    -relaxation cycle, 196, *See also* Muscle, relaxation
    second phase, 297–298
    sliding filament theory, 1, 29
    super, 298–299
    supraphysiological, 285, 292
    violent, 149, 232
  contracture bands, 48, *See also* Bands, contracture
  cutaneous trunci, 87
  deer, 115
  disruption, 215, 294–295, 297, 299–302
  energy supply, 111, *See also* Energy; Glucose; Glycogen; Glycolysis
  equilibrium length, 77
  excised, 2, 4, 5, 14, 25–26, 31, 48–49, 77, 81, 161–163, 167, 242–243, 293, 295, *See also* Boning
  extensor, 22
  extensor carpi radialis, 245
  fragmentation, 63, 65, *See also* Fragmentation; Muscle fiber; Protein
  frog, 13, 73, 296
  gastrocnemius, 86, 90, 92–93
  gluteus medius, 8, 23, 100
  glycolysis, *See* Glycolysis
  goat, 115, 123, 127, 129–131, 140–141
  groups, 244
  heart, ischemic, 18
  horse, 75

318 INDEX

Muscle (cont.)
  in vivo, 284–285, See also Animal, live
  infraspinatus, 8, 23
  longissimus, 6, 8, 13–14, 22–23, 53, 56–57, 60, 77, 81, 87–89, 91, 95, 97–105, 107, 138, 142, 144, 149–150, 163, 168, 172, 189, 194, 208–213, 286, 293
  masseter, 87–89
  neck, 77
  pig, 190–191
  poultry, 12–13, 15, 26, 74, 93, 122
  prerigor, 30–31, 34, 37, 59–62, 65, 77–79, 148, 162, 176, 185, 197–199, 202, 204–205, 214, 245, 248, 286–287, 291–292, 295, See also Boning; Processing, hot; Rigor, mortis
  proteins, 53, 55, 57–58, 65–66, 187, 192, 207–208, See also Protein
  psoas, 2, 15, 23–24, 48, 192–193
  rabbit, 10, 12, 15–16, 18, 25, 28, 36, 93, 123, 125, 191–193, 202, 204
  rat, 28, 89–90, 92–93, 98
  rectus abdominus, 26, 96
  rectus femoris, 23
  red, 16, 18, 202–204, 211, 215, See also Muscle fiber
  relaxation, 192, 196, 199, 201–202, 211, 215, 298, See also Muscle, contraction
  response to stimulation, 86, See also Muscle, contraction
  restraint, 22, 26, 49, 77, 161, 211, 293
  round, 136–138, 146–147, 172, 214
  sartorius, 13
  seaming technique, 245
  semimembranosus, 8, 23, 56, 89, 96, 163, 175, 245
  semitendinosus, 6–8, 23, 32, 60, 138, 163, 175, 243
  separation, 142–143, See also Lean scores
  sheep, lamb, 6–10, 12–14, 20, 22, 25, 75, 78, 80, 86–93, 95, 97–103, 112–113, 115, 121, 123–125, 127–128, 130–131, 140–141, 164–165, 202, 204, 211, 215, 278
  skeletal, 186–187, 189, 194–195, 205
  sternocephalicus, 75
  sternomandibularis, 5, 10–11, 14–19, 26–27, 32, 34, 37, 77–79, 82–85, 87–89, 92–95, 175
  stretching, 4, 6, 22, 30, 36–37, 50–51, 57, 63, 77, 297
  structural damage, 149, 215, 288
    submicroscopic, 300
  structural integrity, 186
  structural units, 186, 193
  structure, 29–39, 48–49, 65, 73, 175, 186–196, 198, 208–213, 215, 285, 292, 295–303
  supraspinatus, 8
  swine, 12–13, 25, 75, 123–125, 127, 129, 138, 140–141, 143, 147, 162–163, 178, 189, 190, 202, 294
  systems, 267, See also Muscle, groups
  tearing, 4, 199, 211, 215, 231, 298–299
  temperature, 53, 56, 58–59, 84, 148, 162, 199, 211, 253, 285–295, 298, 302, See also Carcass, temperature; Conditioning; Temperature
  tension, 11, 22, 49, 86, 188, 296, 298, See also Tension
  triceps, 7
  triceps brachii, 8, 23, 96
  twitch response, See Nerve; Nervous system; Tetanus; Twitch
  type, 115, See also Muscle, red; Muscle, white; Muscle fiber, type
  vastus lateralis, 23, 96
  whale, 8, 13
  white, 16, 18, 202–205. See also Muscle fiber
Muscle cell membrane, 186, 191, 201. See also Sarcolemma
Muscle fiber, 4–5, 16–19, 31–39, 58, 77, 87–89, 134, 137, 186–197, 280
  actively shortened, 48, 198
  bundles, 186
  contracted, 195, 198
    crimpled, 5, 31, 89
  contractile, 285
  density, 32
  diameter, 189
  direction of, 78, 245
  disruption, 211, 296, 299–302
  fast, 191
  fracture, 49, 285, 287, 290, 296–300
  glycolytic, 189–190
  intermediate, 17, 189–191
  length, 296
  oxidative, 198–190
  passively shortened, 48, 198
  red, 5, 17, 189–191, 202–203, See also Muscle, red
  reticular, 186, 191
  rupture, 288, 294
  slow, 191
  smooth, 192
  stretched, 29, 32, 35–38, 195, 208, 287
  structure, 29–39, 48–49, 188–197
  supercontracture, 89, 285, 294, 296–297

tension, 296, *See also* Muscle, tension; Tension
type, 87, 189, *See also* Muscle, red; Muscle, type; Muscle, white; Muscle fiber, intermediate; Muscle fiber, red; Muscle fiber, white
white, 5, 17, 87, 189–191, 202–203
Muscle fibrils, 201
Muscular contraction. *See* Muscle, contraction
Myofibril, 1–2, 12, 16, 26, 28–29, 31–32, 35–36, 39, 51, 57, 187–188, 192–193, 208–211, 215
  disintegration, 65
  disruption, 48, 215
  superstretching, 208
Myofibrillar component. *See* Myofibril
Myofibrillar fraction, 57. *See also* Protein, fragment
Myofibrillar fragmentation. *See* Fragmentation, myofibrillar
Myofibrillar proteins. *See* Protein, myofibrillar
Myofibrillar strength, 32
Myofilament, 191–192, 208
  macromolecular changes, 48
  overlap, 48
Myoglobin, 282–284. *See also* Pigments
Myosin, 28, 32–34, 57–58, 65, 189, 194–196, 198, 201, 206, 245
  binding strength, 245
  cross-bridges, 196, 201, 202
  cross-linking, 34
  crude, 245
  degradation, 56–57
  disassembly of, 38
  filaments, 32, 58, 66, 195, *See also* Thick filaments
  fragments, *See* Fragmentation, myosin
  head, 32, 195–196, 206
  light chain-1, 57
  light chain-2, 57
  molecule, 32, 195, 202
  tail, 32, 37
Myosin-actin, *See also* Thick filaments; Thin filaments
  bonds, 30
  interaction, 31, 63
  overlap, 66

NaCl, 245
National electric supply authority, 114
Neck pinning, 179

Nerve, 87, 92, 98
  -orchestrated system, 297
  sciatic, 98
Nervous system, 86–87, 90, 231, 285
  recruitment, 86
  stimulation, 11, 86, 201, 284
  transmission, 297
  twitch response, 11, 87–89
Netherlands, 290
Neuromuscular blocker. *See* Curare; Inhibitors, neuromuscular
Neutral protease. *See* Calcium-activated factor (CAF); Proteases
New Zealand, 2–3, 9, 13, 24, 64, 74–76, 81, 84, 99, 101–111, 114–115, 121, 123–124, 127–128, 139, 164–165, 198, 279
  Meat Producers' Board, 3
  Specification for Conditioned and Aged Beef, 20
  Specification for Conditioning and Ageing, 19
Nitrogen, atmosphere, *See* Atmosphere, nitrogen
  liquid, 13
Nitroso, 248
N_2-line, 38
North America, 121–122
Norway, 123
Nose, clamp, 230–231
  clip, 111, 280
  hook, 229
  pincer, 231
Nuclei, 191

Occupational Safety and Health Administration (OSHA), 233
Offal, 252
Operating costs, 261, 264, 268
Organelles, 17–19, 284
Organoleptic. *See* Palatability; Sensory panel
Orifices, 281. *See also* Nose; Probe
OSHA. *See* Occupational Safety and Health Administration
Oxidation, 196
Oxidative fibers, 189
Oxidative phosphorylation, 16, 196
Oxygen, 16, 76, 196, 202, 284, *See also* Aerobic; Anaerobic
  -dependent, 302
  deprivation, 76–77
Oxymyoblobin. *See* Pigments

Packaging, 106, 173, 267
  consumer pack, 10
  Cryovac pack, 20
  loose wrap, 167
  oxygen-permeable film, 172
  shrink-wrap, 39
  systems, 173
  vacuum, 147, 149, 159–164, 166–168, 172, 177–179, 249, 252, 263, 266–267
Packer, 121, 127, 145, 151–153, 221–222, 234. See also Processor
Packing plants, 268, 281, See also Abattoir
  beef, 266
  capacity, 220, 227, 234–235
  construction, 266, 268
  design/layout, 104, 106, 110, 160, 177, 262
  federal inspection of, 235
  location, 260–266, See also Location, plant; Region
  operation, 110, 261, 266, See also Operating costs
  renovation, 266, 268
  throughput, 99, 104
Palatability, 21, 124, 127, 137, 144, 149, 151, 168, 173, 176–177, 206, 219, 235, 282–284, 291, 293. See also Flavor; Sensory panel; Tenderness
Pale, soft, exudative. See PSE
Palmitoyl CoA, 18
Patent, 25, 123–124, 164, 185, 279–281
Pathology, 12
Patties, 267
  from hot- vs cold-boned meat, 248
  shrinkage, 248
  yield, 248
Peak power demand, 257, 260, 268
Pelt. See Hide
Pelting. See Hide, removal
Per rostrum ad rectum, 281
Perimysium, 186–187
pH, 9, 13–15, 20, 26–28, 29, 36–37, 50, 53, 56, 58, 64–66, 77, 82–101, 103, 114, 138, 148–150, 162, 165, 172, 176, 197–204, 206–207, 211, 213–215, 220, 231, 241–243, 278, 280, 282, 287, 290, 292–295, 300–303
  ΔpH, 82–96, 101, 103
  "dwell," 13
  neutral, 28, 302
  physiological, 13, 17, 28
  rigor, 28
  /temperature profile, 292

Phosphate, 16
  high energy compounds, 196, 243
  inorganic, 15–16, 18, 205
Phosphofructokinase, 197
Phosphorus, acid-labile, 15, 197
Phosphorylase, 197
Phosphorylation, 196
Photoelectric cells, 114, 225–226, 228
Photomicrograph, 189–199, 211
Physiology, 73, 278, 281, 296, 298–299
  nervous response, 11
  tension, 90
Pigments, 248. See also Myoglobin
Pistola cut, 245
Plants. See Location, plant; Packing plants; Research, and development, in-plant tests, locations
Polarity, 91–93
  alternating, 92–93, 105, 115
Polyphosphate, 248
Pork. See Carcass, pork
Portion-controlled products. See Product, portion-controlled
Posture, See also Suspension
  altered, 19–20, 24–25
  crouching, 24
  squat, 122
  standing, 20, 24
Potassium, 284
Potential chemical energy. See Energy
Potential reducing sugar content, 207
Prandtl number, 255–256
Prechill processing, 160. See also Boning
Prefabrication. See Fabrication
Prerigor excision, 160. See also Boning; Muscle, excised
Pressure difference, 251
Pressure-heat treatment, 37–38
Primal cuts, 20, 78, 82, 104, 115, 138, 159, 163, 172–173, 176, 179, 244, 252
  bone-in, 266
  uniform boneless, 266
  vacuum packaging of, 159, 177
Primal joints, 25
Prime. See Carcass, Prime
Probe, 224–225, 227, 229, 232, 234, 281, See also Electrode
  Langmuir, 89
  rectal, 90, 231, 281
Processed meat, 178, 250. See also Cured meat; Processing; Sausage products
Processing, See also Boning; Carcass, processing
  capacity, 257
  centralized, 159

conditions, 180
costs, 159, 237, 250, 264–266
downstream, 111
lines, 106–111, 267
methods, 159–160, 162, 165, 173, 179–180, 237, 268
of beef, 237–238, 244–245, *See also* Processors, electrically stimulated beef
of hams, 163
of pork, 178, 238
prior to rigor mortis, 160, *See also* Boning
properties, 127, 147–148, 160–161, 178, 245–248
rapid, 122
sector, 266
specifications, 111–115
time, 106
Processor, 96, 147, 163, 174, 177
Processors, electrically stimulated beef, 74, 123, 126–127, 185, 221–222, 234–235
Product, *See also* Inventory; Meat, products
flow, 164, 172
integrity, 173–174, *See also* Hygiene; Quality
portion-controlled, 234
turnover, 160
Production, areas, 159–160, 179, 260–261
of beef, 260–261
Propylene glycol, 178
Proteases, 28, 37, 57, 301–302, *See also* Calcium-activated factor (CAF); Lysosomal enzymes; Proteolysis
exogenous, 30, 57
extract, 37
indigenous, *See* Proteases, exogenous
muscle, 57–58
neutral, 28–29, 301
systems, 58
Protein, 252, *See also* Collagen; Connectin; C-proteins; M-proteins; Muscle; Titin
catabolism, 28, 302
coagulation, 32–34, 51, 175
connecting, 194
contractile, 192, 284–285
cytoskeletal, 63–64
denaturation, 32–35, 38, 207, 302
efficiency ratio, 147
filaments, 29, 175
fragment, 55, 57
linkages, 32–34
mitochondrial, 18

myofibrillar, 51, 55, 57, 62, 65, 189, 208, 278
quality, 147
regulatory, 284
resynthesis, 302
salt soluble, 163, 245
-salt interaction, 278
solvents, 30
structural, 28, 300
synthesis, 284
turnover, 302
Proteolysis, 1, 28, 32, 36–38, 51, 53, 55–60, 62–63, 208, 284, 286, 292, 295, 300–303
acid, 56, 58–59
carboxyl, 57
neutral, 56, 58
prerigor, 301–303
Proteolytic enzymes, 250, 287, 300–302, *See also* Calcium-activated factor (CAF); Lysosomal enzymes
activity of, 58–59
rate, 287
PSE, 27, 75, 278, 294
in beef, 75
Pseudo H-zone, 194. *See also* H-zone
Pseudomonas. See Microorganisms
Psychrotrophic microorganisms. *See* Microorganisms
Pulse, 106, 111, 115, 126, 150, 196–197, 232–233
duration, 92, 221, 229
interval, 221
peak, 93, 105
shape, 91–98
sine-wave, 92, 106, 115
spike, 92
square-wave, 92, 115
timers, 230
train, 92
bidirectional, 92–93, 196
cathodic, 93
unidirectional, 92–93, 98, 105–106, 109, 115
voltage, 92
Purge, 147–149, 160, 172. *See also* Drip loss
Purveyor, 127
Putrefaction, 249. *See also* Microbial; Microorganisms
Pyrogenesis, biological, 278
Pyruvate, 196

Quality, 1, 10, 20, 22, 26, 64, 75–76, 110–111, 113–115, 127, 144,

Quality (*cont.*)
    152–153, 160–161, 178, 180, 201, 219, 235, 242–244, 251, 278, 280, 282, 286–287, 290, 292, 294–296, 299–300, 302–303, *See also* Product, integrity
    assurance programs, 113–115, 268
    control, *See* Quality, assurance programs
    eating, 144, 282–284, 293, 295
    -enhancing, 284
    grade, *See* Grade
    -indicating characteristics, 144, 148–151, 186, 219, 221, 282
    indicators, *See* Quality-indicating characteristics
    of protein, 147
    of sausage products, 245
Quantity, 1
Quarter, front, 90, 244–247
    5-rib, 245
    hind, 244–247, 251
    8-rib, 245
    pistola cut, 245

Rail, 223
    earthed, *See* Rail, grounded
    grounded, 89, 107–109, 111, 226
    live, 280
Rapid processing, 160. *See also* Boning
Raquette, 245
Rectal probe. *See* Probe
Red fibers. *See* Muscle, red; Muscle fiber, red
Redox potential, 250
Reflectance, 207
Refrigeration, 2–3, 19–20, 39, 75, 84, 178–179, 250–260, 262, 281, *See also* Chilling; Cold; Freezing
    cold room, 46–47, 59, 135, 151–153, 160, 164, 173, 177, 232, 251, 253, 256–257, 262, 267, 288
    cooler design, 251
    cooling floor, ventilated, 3, 19
    holding rooms, 19, 251, 253, 256, 267
    space, 253
    systems, 178, 251–260
    units, 251
Region, deficit, 260–261
    geographical, 288
    surplus, 260–261
Relaxation. *See* Muscle, contraction; Muscle, relaxation

Rendering, 267
Reports, scientific, electrical stimulation, by country, 13, 124, 139, 178, 220–222, 243–244, 251, 253, 260, 264
Research, *See also* Reports, scientific
    and development, 280, 285, 295, 303
    in-plant tests, locations, 13, 123, 124, 126, 221
    basic, 13, 277–278, 285, 287, 294–295, 299
Resistance, 108, 111, 280, *See also* Electrical, contact
    critical threshold, 111
    internal, 251
Resistive heating, 89
Respiration, 284–285
Respiratory substrate, 205
Restaurateur, 127
Restraint, *See also* Muscle, restraint; Posture; Suspension
    from external fat, 24
    internal ice, 9, 13
    skeletal, 7, 9, 13, 19, 22, 24, 77, 244
Restructured meat, 266
Retail, appearance, 147
    caselife, 75, 146–147, 160, 164, 173, 177–178, 214, 242, 250, 281
    of sausage products, 245
    of vacuum packaged products, 266
    cuts, 127, 138, 185, 244–245, 252
    display, 146, 281
    self-service, 185
    outlets, 159–160
Retailer, 127, 147, 151, 174
Reticular fibers, 186, 191. *See also* Muscle fiber
Ribbing, 143–144, 150, 206, 221–222, 253
Ribeye, 136, 142–145, 185, 207, 214, 220
Rigor, alkaline, 200
    bonds, 48, 51
    effects of electrical stimulation, 82–84
    mortis, 1–5, 7, 10, 14, 19–20, 22, 25–32, 37–39, 45–49, 59–62, 73, 75–78, 81–84, 129, 144, 162, 164–165, 175–176, 186, 197, 199–200, 202, 206, 215, 219, 242, 248, 278, 286, 288, 292, 294, 296, 299–301
    prerigor, 9, 12, 24, 37, 50, 59–62, 65, 75, 77, 164–165, 175–176, 197, 199, 215, 245, 248, 279, 286–287, 292, 295, *See also* Carcass; Muscle

rapid, 75, 82–84, 91–92, 98–99, 104–105, 300
thaw, 9, 15, 19, 76–77
Roasting. See Cooked, material
Robotic handling, 266
Roller brand, 234
Rub bar, 107–109, 222–224, 227–228, 232, 234, 280

Safety, circuits, 114
  control door, 225
  enclosure, 224, 227
  features, 226–229
  product, See Hygiene
  requirements for electrical stimulation, 114–115, 233–234
  threshold, See Voltage
  worker, 136, 150, 233–234, 281
Sales, 151–152, 266
Salt, 245
Salt soluble proteins. See Protein
Sanitation, See also Hygiene; Sterilization
  cabinet, 224
  program, 267
Sarcolemma, 186, 191–192, 201. See also Muscle cell membrane
Sarcomere, 2, 5, 22, 24, 26–27, 29–32, 34–35, 37, 49–54, 58, 64, 66, 88–89, 188–189, 192–193, 195–198, 208, 211, 296–298
Sarcoplasm, 188, 191–192, 201
Sarcoplasmic reticular membranes, 203–206
Sarcoplasmic reticular vesicles, 202
Sarcoplasmic reticulum, 15–19, 191, 201–206, 215, 243
Sausage manufacture, 266
Sausage products, 147, 178, 245–248, 250. See also Emulsion; Frankfurters; Szynkow; Zwyezajna
Saw, band, 244
Sciatic nerve. See Nerve
Scribing, 179
Seam between muscles, 245
Sections. See Microscopy, sections for
Sensory panel, 6–7, 53–55, 65, 78, 128–130, 133–137, 139–140, 164, 168, 174–176, 201, 207, 243, 250, 282–284
Sensory properties, 161, 173, 177, 201. See also Palatability
Serine protease, 28

Shackle, 89, 107, 231
Shackled leg. See Leg, shackled
Shank, 24
Shear force, 4–9, 14, 20–21, 23–27, 32, 35, 37, 52–54, 77–82, 97, 100, 102, 104–105, 107, 128–137, 139, 168, 175, 198, 205, 207
  of raw meat, 32, 51
Sheep. See Carcass, lamb; Muscle, sheep, lamb
Shelf-life. See Retail, caselife
Shipping, 160. See also Carcass, shipping; Distribution; Transportation
Shortening, 6, 9–10, 12–39, 46–53, 58, 64–66, 75–79, 129, 161, 165, 175, 188, 198–207, 215, 286, 289, 293, 295–299, 301, See also Thaw toughening
  Ca^{2+}-induced, 243
  cold, 2, 4–22, 27, 29, 32, 36–38, 46–52, 58, 64–66, 75, 76, 81, 83–84, 91, 100–101, 103–106, 114, 122, 127, 129, 134, 138, 150, 161, 163–165, 176, 185, 191, 198–203, 205–207, 211–213, 215, 241–243, 250, 278–279, 288–291, 293–295, 297, 299–302
  -elimination process, 299
  heat, 5, 20, 27, 36, 47, 77, 299
  -induced toughening, 289–290, 299, 301
  percentage, See Shortening, region
  pressure-heat treatment, 37–38
  region, 30, 34, 48, 198–199, 296–297
  rigor, 9–10
  second stage, 298
  thaw, 5, 8–14, 16, 19, 24, 26, 75–77, 81, 91, 100–101, 114, 165, 296–297
  /toughening structural relationship, 297
Shoulder, 245
Shrinkage. See Carcass, shrink
Shrouding, 179
Sine wave. See Pulse, sine-wave; Waveform
Skeletal attachments, 188, 293
Skeletal muscle. See Muscle, skeletal
Skeletal restraint. See Muscle, restraint; Restraint, skeletal
Skeleton, 6, 19–20, 77, 244. See also Restraint, skeletal
Skids, 108–109
Skin, 187

Skinning, 280
  cold, 130
  hot, 130
Slaughter, 268, 280, 294, 302, See also
    Abattoir; Carcass, processing;
    Packing plants
  capacity, 266
  -dressing sequence, 136, 149–151, 219,
    231–233
  lines, 267
  procedures, See Carcass, processing
  -processing system, 266
  statistics, 262
Sliding filament theory. See Muscle, contraction, sliding filament theory
Soluble components, 248–249
South America, 121, 161
Space, 160–161
Specific heat, 255–256. See also Heat
Spike. See Pulse; Waveform
Spine, broken, 94, 111. See also
    Vertebrae
Splitting, 135, 232. See also Carcass,
    split
Square wave. See Pulse, square-wave;
    Waveform
Squat posture. See Posture, squat
Staining, 187–190, 192, 194
Stains, glycolytic, 190
  oxidative, 190
Standard. See Carcass, Standard
Staphylococcus aureus. See
    Microorganisms
Staphylococcus sp. See Microorganisms
Startup procedure, 114
Sterilization, 111, 222, 224, 226, 228,
    231, 234. See also Hygiene
Sternum, 78
Sticking, 111
Stimulation, automatic, 126–127, 151–
    153, 222–224, 226–228
  direct, 200, See also Leg
  duration of, 91, 93, 95, 98, 100, 109,
    113, 122, 126, 149–151, 208,
    229–230, 232–233
  effective, 92, 114, 200, 281, 291
  electrical, See History, of electrical
    stimulation; Stimulators,
    electrical
  footplates, 114
  key switches, 114
  manual, 126–127, 151, 152, 224–231
  method of application, 208, 222–231
  parameters, 84–99, 106–107, 113, 126,
    177, 232–233, 289, 295, See also
    Current; Frequency; Polarity;
    Pulse; Voltage; Waveform

  period, See Stimulation, duration of
  research, summary of, 125, 169
  sites, 106–107, 112, 126, 135–136, 139,
    149–151, 231–233
  systems, 107–112, 114, 222–235
  time of application, 97–99, 108, 113,
    115, 135, 150–151, 167, 200, 295
  tunnel, 114, 224
  voltage. See Voltage
Stimulators, electrical, See also Stimulation; Voltage
  high-voltage, 222
  low-voltage, 124, 126, 127, 221–226,
    229–230, 231, 235
  manufacturers, 124, 126, 127, 221–
    226, 227, 228, 229, 234–235. See
    also Manufacturing of electrical
    stimulators
Storage, 160–162, 172, 173, 249, 262
Streptococcus. See Microorganisms
Stress, 75–77, 82, 90, 178, 201, 294, See
    also Animal
  antemortem, 76
  cold, 243
  exercise, 90
  nervous, 75
Striations, 189, 298
  cross, 187, 192
  longitudinal, 187
Stunner, hog, 122, 124, 221, 229. See also
    Best and Donovan Hog Stunner
  modified for electrical stimulation, 229
Stunning, 111, 150, 231, 285, 294–295,
    See also Stunner
  attachment, 229
  electrical, 110
Subprimal cuts, 138, 147, 159, 162–163
  bone-in, 266
  uniform boneless, 266
  vacuum packaged, 147, 159, 162–163,
    166–167
Substrates, 284–285
  structural protein, 300
  used by bacteria, 243, 245, 249
Supercooled solution, 178
Supermarkets, 185. See also Retail
Superphosphate, 3
Supplies, 179
Surgical procedures, 277
Suspension, 22–26, 49–50, 107, 110–111,
    188, 253, See also Posture
  by aitchbone, 24
  "crouching-posture," 24
  horizontal, 22, 24
  normal, 25–26
  obturator foramen, 50, 53, See also
    Aitchbone

INDEX 325

pelvic, 22–24, 26
Tenderstretch. *See* Tenderstretch posture
Sweden, 96–97, 123–124, 139, 289–290
Synthesis, 196
Szynkow, 247

Tallow, 2
TAMU, 121–131, 134–136, 138–139, 142, 146, 150–151, 222
Taste panel. *See* Sensory panel
TCA cycle, 196
Techniques, 279
Technological advances, 280–281, 303
Technology, 303, *See also* Commercial practices
  applied, 126–127, 163, 279, 281, 285
  electronic, 277
  existing, 278
Teeth, 187
Temperature, 2, 9–10, 15–18, 20, 22, 25–29, 31–37, 45–66, 76, 80, 82, 84–85, 122, 142, 162, 170–171, 173, 176–177, 179, 198–199, 202–213, 215, 241–244, 248–249, 251, 255–257, 262, 267, 286, 291, 295, 300, 303, *See also* Carcass; Conditioning; Muscle
  air off the coil, 256
  air onto conveyor, 258
  air to coil, 256
  ambient, 13, 37, 286–287
  carcass, 76, 82, 151, 255, 280, 285, 288
  coefficient, 20, 242
  conditioning, 45–66, 122, 168
  dependence, 84
  elevated, 47, 51–62, 64, 66, 122, 129, 177, 208–213, 242, 250
  evaporating, 256
  physiological, 26, 28, 58, 106, 173, 284, 294
  potential for energy transfer, 251
  prerigor, 46, 51–53, 58
  profile, 173, 292
  refrigerant condensing, 256
  sensitivity, 289
  -shortening relationship, 47
  storage, 101–102, *See also* Frozen, storage; Storage
  water inlet, 256
Tenderness, 1–9, 12, 19–39, 45–66, 74–82, 84, 91–92, 97, 99–108, 110–115, 121–140, 148–153, 160–168, 175–177, 185, 189, 196, 201, 205–213, 215, 219, 242–243, 278–283, 286, 288–291, 293–294, 296–303
  objective measurement of, *See* Fragmentation, values; Shear force
  subjective measurement of. *See* Sensory panel
Tenderstretch posture, 22, 121–122, 136
Tendon, 187. *See also* Skeletal attachments
Tensile properties, 27, 29–30, 39, 58
Tension, 86, 90, 92–95, 98
  isometric, 296
  peak, 86
  peak tetanic, 94
  response, 90, 93
  sustained, 93
  traces, 86
Terminal cisternae, 201
Terminals, 229
Testing, in-plant, 124
Tetanus, 298–299. *See also* Twitch
Texas A&M Tenderstretch. *See* Tenderstretch posture
Texas A&M University. *See* TAMU
Texture, 27, 142, 151, 178, 214, 220. *See also* Lean scores, texture
Thaw, 299
Thaw toughening, 14, 24–25, 128. *See also* Rigor, thaw; Shortening, thaw
Thermal. *See* Heat
Thermal conductivity, 255–256
Thermal shrinkage of collagen, 31
Thermal stability of collagen, 62
Thermodynamics, 251, 262–263
Thick filaments, 48, 65, 189, 196, 201, 296–298. *See also* Myosin, filaments
Thin filaments, 187, 189, 193–194, 196, 201, 296–298
Thiol reducers, 30
Thoracic region, 82
Time-on-feed. *See* Feeding regimens
Timer, cycle and pulsation, 230
Time-temperature relationships, 286
Tissue thickness, 251, 253
  effect on cooling rate, 251
Titin, 30, 57, 64
Toughness. *See* Tenderness
Trabecula, 187
Trade, 2–3, 9, 19–20, 76, 121–122, 143, 282
  block beef, 137
  frozen meat, 2, 9, 84, 121
  international, 2, 282
  names, 234, *See also* Stimulators, electrical, high-voltage
  North American, 19, 21

Transformer core saturation, 93
Transition from muscle to meat, 292
Transportation, 147, 159–160, 179, 260–263, *See also* Location, plant; Packing plants; Region
costs, 260–261
Tricarboxylic acid cycle. *See* TCA cycle
Trimming, 151, 176, 249
Tripolyphosphate, 245
Tropomyosin, 57, 245
Troponin-C, 57, 201, 206
Troponin-I, 57, 201, 206
Troponin-T, 57, 60, 63, 65
degradation, 55–56, 60, 63
Troponin-tropomyosin complex, 201, 206
Truckers, 179
T-tubule system, 201
Turnover mechanism, 284
Twitch, 92–93, 298–299
fast twitch muscles, 87–89
slow twitch muscles, 87–89
/tetanus concept, 299

Ultrastructure, 89, 186, 211
of gap filaments, 58
Unbound water. *See* Water, unbound
Unidirectional pulse train. *See* Pulse
United States, 3, 121–124, 126, 128, 139–140, 143–144, 161, 164–165, 177–178, 219–220, 222, 234–235, 237, 260–262, 281–282, 288–290
United States Department of Agriculture, Agricultural Research Service, 221
Food Safety and Inspection Service, 233
Meat and Poultry Inspection, 232–233, 235
Equipment Group, 233, 235
Facilities Group, 233, 235
Meat Grading Branch, 143
United States Department of Commerce, 178

Vacuum barrier bag, 267
Vacuum packaging. *See* Packaging
Velocity, 251
of air, 251, 258, 286
Vertebrae, 6, 232

Viscosity, absolute, 256
kinematic, 256
Visual characteristics, 283–284, 291, 295. *See also* Color; Cosmetic improvement; Grade; Heat-ring; Lean scores; Marbling
Voltage, 84–101, 103–104, 122, 126, 165, 200, 208, 221–235, 280, 285–292, *See also* Stimulation
gradient, 89
high, 84, 87, 89, 92, 94–101, 103–115, 122, 128, 148–153, 221–229, 231–235, 280, 284, 290–291, 299
intermediate, 128
low, 84, 86, 89–90, 92–96, 98–99, 101, 103–111, 115, 148–153, 229–231, 233–235, 280–281, 284, 289–291, 298–299
peak, 92
safety threshold, 84, 96
stepwise increases of, 97

Wages, 179
Waste products, 285
Water, unbound, 168, 207
Water bath. *See* Cooked
Water holding capacity, 148, 168, 170–172, 176, 178, 278
Water spray, 111
high pressure jets, 107
Watery pork. *See* PSE
Waveform, 90–93, 231
frequency, 91
Weekend cattle, 143
Weight loss, 147, 168, 172, 287
evaporative, 289
Whale muscle. *See* Muscle, whale
White fibers. *See* Muscle, white; Muscle fiber, white
Wholesale cuts, 127, 173. *See also* Primal cuts; Subprimal cuts
Wholesaler, 151, 173
Wool, 2, 107
scorching, 197
World War II, 3

Yield, 179, 197, 246–248, 253, 264
cooking, 175, 248
of hams, 163, 248
of sausage, 246–247
point, 27, 30, 32, 37–38, 78

Z-disc, 189, 198–199
Z-line, 28, 30–32, 34, 36–38, 48, 63, 188–189, 192–195, 208–211, 296–297

Zone, fracture, 297
  stretch, 297
  supercontracted, 294, 296–297
Zwyezajna, 247

MIX
Papier aus verantwortungsvollen Quellen
Paper from responsible sources
FSC® C105338

If you have any concerns about our products,
you can contact us on
ProductSafety@springernature.com

In case Publisher is established outside the EU,
the EU authorized representative is:
**Springer Nature Customer Service Center GmbH
Europaplatz 3, 69115 Heidelberg, Germany**

Printed by Libri Plureos GmbH
in Hamburg, Germany